Springer Series in Plasma Science and Technology

Plasma Science and Technology covers all fundamental and applied aspects of what is referred to as the "fourth state of matter." Bringing together contributions from physics, the space sciences, engineering and the applied sciences, the topics covered range from the fundamental properties of plasma to its broad spectrum of applications in industry, energy technologies and healthcare.

Contributions to the book series on all aspects of plasma research and technology development are welcome. Particular emphasis in applications will be on high-temperature plasma phenomena, which are relevant to energy generation, and on low-temperature plasmas, which are used as a tool for industrial applications. This cross-disciplinary approach offers graduate-level readers as well as researchers and professionals in academia and industry vital new ideas and techniques for plasma applications.

Francesco Romanelli

Physics of Nuclear Energy

Foundations Towards Fusion Energy

Francesco Romanelli
Department of Industrial Engineering
University of Rome Tor Vergata
Rome, Italy

ISSN 2511-2007 ISSN 2511-2015 (electronic)
Springer Series in Plasma Science and Technology
ISBN 978-981-97-9611-3 ISBN 978-981-97-9609-0 (eBook)
https://doi.org/10.1007/978-981-97-9609-0

This Springer imprint is published by the registered company Springer Nature Singapore Pte Ltd.
The registered company address is: 152 Beach Road, #21-01/04 Gateway East, Singapore 189721, Singapore

To my wife Paola and my son Giovanni

Preface

This book collects my lecture notes of the course of *Physics of Nuclear Energy* at the Industrial Engineering Department of the University of Rome *Tor Vergata* and has been written for master students in engineering. Concepts such as *probability distribution function* and *cross section* are introduced at the very beginning as they do not belong to the typical curricula of engineers. In the same spirit, the basic concepts of electromagnetism and special relativity are reviewed. Although excellent books exist on plasma physics applied to nuclear fusion, I have tried in this course to give the students the general perspective of the challenges of fusion energy that go well beyond those of plasma physics alone.

The first 10 chapters give the student an introduction to quantum physics and nuclear phenomena. Using the liquid drop model, the energetics of the nuclear decays and of the main nuclear reactions (including fission and fusion) are discussed. Many of the concepts introduced here are used in the second part devoted to fusion physics and engineering. The presentation on fusion is focused on tokamak confinement and includes the classical collisional effects (resistivity, slowing down and runaway electrons), the power balance of a fusion power plant, particle motion in magnetic fields, plasma equilibrium and plasma edge physics. The main engineering challenges associated with the electromagnetic forces and stresses and the heat load on plasma facing components are underlined. The final chapter is devoted to the nuclear aspects of fusion.

At the end of the course, the student is expected to be able to provide a definition of the engineering parameters of a fusion power plant based on magnetic confinement.

I would like to acknowledge all my postdocs and Ph.D. students who have helped during all these years in making the course more clear by raising intelligent questions. I would like to acknowledge the following people for the final reading of the manuscript: Paola Batistoni, Giacomo Dose, Michele Lungaroni, Simone Noce, Aldo Pizzuto, Gianmario Polli, Marco Ripani, Giovanni Romanelli.

Rome, Italy
April 2024

Francesco Romanelli

Glossary

Activity The number of radioactive decays per unit time. It is measured either in becquerel (1Bq = 1 decay per second) or in curie (1Ci = 3.7×10^{10}Bq); 1Ci approximately corresponds to the activity of a gram of ^{226}Ra.

Alpha decay Radioactive decay in which an alpha particle is emitted by a nucleus.

Atomic mass Mass of an individual isotope of a given element.

Atomic mass unit Defined as 1/12 of the mass of the ^{12}C isotope: $1AMU = 1.660539 \times 10^{-27} kg$.

Atomic weight Average mass of an element weighted on the natural abundance of each stable isotope.

Avogadro number The number of molecules in a mole: $N_A = 6.02214076 \times 10^{23}$ particles /mole.

Beta decay Radioactive decay in which an electron (positron) is emitted together with an antineutrino (neutrino) leaving the nucleus with one more (less) proton and the same number of nucleons.

Barn Nuclear cross section unit; $1 barn = 10^{-28} m^2$.

Binding energy per nucleon The total binding energy of a nucleus divided by the number of nucleons.

Black body A system in which matter and radiation are in thermodynamic equilibrium.

Bohr radius Radius of the innermost orbit of an hydrogen atom; $a_o = \epsilon_o h^2/(\pi m_e e^2) \approx 5.29 \times 10^{-11} m$.

Classical electron radius The classical electron radius is defined as $r_{cl} = e^2/(4\pi \epsilon_o mc^2) \approx 2.8 \times 10^{-15} m$ and is the radius at which the electrostatic energy of the electrons equals its rest energy.

Compton effect In the scattering of a photon by a charged particle the photon energy is changed. The effect is larger for photon energies closer to the rest energy of the charged particle. The effect is a manifestation of the existence of quanta. Classical electromagnetism predicts no change in the scattered radiation frequency.

Cross section The effective area that is shown by a particle for a specific interaction with another particle. It measures the probability that a certain interaction takes place.

Decay constant The exponential rate of decay of a radionuclide.

Dose The energy absorbed per unit mass due to the exposure to a radioactive source. It is measured in gray $1Gy = 1J/1kg$. The dose weighted on the biological effects on biological tissues is measured in sievert.

Endoergic A process that is energetically unfavourable.

Exoergic A process that is energetically favourable.

Fine structure constant Dimensionless constant defined as $\alpha \equiv e^2/(2\epsilon_o hc) \approx 1/137$.

Gamow factor The correction that accounts for the finite probability of two nuclei to tunnel through their mutually repulsive Coulomb potential.

Half-life The time after which the number of radiunuclides is reduced by a factor two. It is related to the decay constant λ by the relation $\tau_{1/2} = \ln(2)/\lambda$.

$\hbar$ Reduced Planck constant $\hbar = h/(2\pi)$.

Impact parameter In collision theory the impact parameter is the distance measured in the centre of mass system between the scattering centre and the initial (unperturbed) trajectory of the incoming particle.

Inner Fuel Cycle The part of the tritium processing system of a fusion power plant that does not include the breeding system.

Isotopes Nuclei of a certain element with the same atomic number Z but different atomic mass A.

Lethargy The logarithm of the energy after a collision divided by the initial energy.

Mass defect Difference between the mass of a nucleus and the sum of the masses of the protons and neutrons. The associated energy is the binding energy of the system.

Moderator An element used to reduce the energy of neutrons in a fission system. Typical moderators are water or graphite.

Neutron/proton drip lines Lines beyond which the loss of one neutron/proton becomes energetically favourable. The neutron and proton drip lines bound respectively from below and above the region of nuclei in the radionuclide chart.

Nucleon A generic term indicating both the neutron and the proton.

Outer Fuel Cycle The part of the tritium processing system of a fusion power plant that does include the breeding system.

Phase velocity The velocity at which the surfaces corresponding to the same phase of a wave propagate in a medium.

Plasma A superposition of gas of negative electrons and positive ions. In neutral plasmas negative and positive charges balance in arbitrarily small volumes.

Radionuclide The nucleus of an element undergoing a nuclear decay.

Radioactive chain The series of alpha and beta decays undergone by a parent nucleus until a stable nuclide is reached.

Range Distance at which a massive charged particle comes at rest in the propagation through a medium.

Reduced mass In a system composed by two bodies of mass m_1 and m_2 the reduced mass is defined as $m_r = m_1 m_2/(m_1 + m_2)$.

Retarded time The time at which a change in the status of an electromagnetic source (charge density or current) is reflected in the electromagnetic field at a distance r from the source.

Secular equilibrium The long term equilibrium solution achieved in a radioactive chain in which the parent nucleus has a half-life much longer than the child nucleus.

Stopping power The average value of the energy lost per unit length.

Straggling Fluctuation in the range due to the fluctuation in the energy loss.

Thermal velocity For a Maxwellian distribution of particles with mass m at temperature T the thermal velocity is defined as $v_t = (2T/m)^{1/2}$.

Contents

Acronyms

appm	Atomic parts per million
bcc	Body centred cubic
com	Centre of mass
DBTT	Ductile-to-brittle transition temperature
DC	Direct Current
dpa	Displacement per atom
e.m.f.	Electro motive force
fcc	Face centred cubic
FLiBe	Fluorine-Lithium-Beryllium
FLiNaBe	Fluorine-Lithium-Sodium-Beryllium
FPY	Full Power Year
hcp	Hexagonal close packed
HTHE	High Temperature Helium Embrittlement
i.e.	Id est
IFC	Inner Fuel Cycle
IFMIF	International Fusion Material Irradiation Facility
ITER	International Thermonuclear Experimental Reactor
JET	Joint European Torus
MHD	Magnetohydrodynamic
NIF	National Ignition Facility
ODS	Oxide Dispersion Strengthened
OFC	Outer Fuel Cycle
PFC	Plasma Facing Component
PKA	Primary knock-om atom
r.h.s.	Right hand side
RAFM	Reduced activation ferritic-martensitic steels
SOL	Scrape-off layer
TBR	Tritium Breeding Ratio

Chapter 1
Introduction to Nuclear Energy

Abstract *This chapter presents an historical overview of the scientific developments that led to our present knowledge of the atomic structure. It also provides an introduction to the basic concept of the motion of charged particles in electric and magnetic fields.*

This book is all about the *principle of equivalence between mass and energy* synthesised in the equation [1]

$$E = mc^2$$

This equation has become so popular that can be found everywhere. It will be discussed in detail in Chap. (4) but reference to this principle will be made also in this chapter.

The relevance of the principle for the energy production is that if in a reaction the mass of the reactants is larger than the mass of the products such a difference can be exploited to produce energy. Even a tiny difference in mass can lead to a huge energy release because the velocity of light is a large number.

This chapter deals mostly with the evidences that accumulated up to the first half of the 20th century about the difference in mass between different elements that eventually led to the use of nuclear energy.

1.1 The Atomic Model

The concept of *atom* was introduced by the Greek philosophers Democritos and Leucippos. However, it was only in the 18th century that the evolution of chemistry posed the basis for a scientific description of matter in terms of atoms. The formulation of some general laws of chemistry between the end of the 18th century and the beginning of the 19th century led to the proposal of the atomic model.

Antoine-Laurent Lavoisier (1743–1794) was the first to propose in 1789 the *principle of conservation of mass* [2].

F. Romanelli, *Physics of Nuclear Energy*, Springer Series in Plasma Science and Technology, https://doi.org/10.1007/978-981-97-9609-0_1

In a chemical reaction the sum of the masses of the reactants is equal to the sum of the masses of the products

The law of conservation of mass seems to contradict the every day experience. Burning wood leads to ashes that have much lower weight! It is the merit of Lavoisier having recognised that it is necessary to include in the balance *all* the products of the reactions (so also the gases involved in the combustion of wood).

The conservation of mass suggests that in a chemical reaction the involved elements do not disappear but simply lose their individual identity to merge in the compound substance.[1]

A few years later Joseph Louis Proust and John Dalton proposed the *law of definite proportion* and the *law of multiple proportions* [3]

If two elements combine with each other to form more than one compound, the weights of one element that combine with a fixed weight of the other are in a ratio of small whole numbers

For example, 1 g of carbon and 1.33 g of oxygen produce 2.33 g of CO whereas 1g of carbon and 2.66 g of oxygen produce 3.66 g of CO_2, with the ratio of oxygen in the two compounds being exactly 2.66/1.33 = 2. For the same amount of carbon, the amount of oxygen required to produce CO_2 is exactly double the amount required for CO. You cannot produce a different quality of CO or CO_2 by mixing different proportions of carbon and oxygen. A compound substance is always characterised by a well defined proportion of the elements it is made of. If we now imagine to look at smaller and smaller volumes of the compound we are tempted to conclude that at the elementary level there is one particle of carbon and one (two) particles of oxygen that combine to give an elementary particle of CO (CO_2).

Therefore, these observations led Dalton to formulate the atomic hypothesis [3]:

- Matter is made of elementary particles called atoms;
- All the atoms of the same elements are identical and have the same mass;
- It is not possible to change an atom of an element into an atom of a different element;
- Atoms of an element can combine only with an integer number of atoms of another element;
- In a chemical reactions atoms cannot be created or destroyed but are simply transferred in the new compounds.

The Dalton hypothesis is illustrated in Fig. 1.1. As we will see later in this chapter, nuclear physics has falsified some of the Dalton hypotheses. Indeed, atoms are not

[1] We know today that mass and energy are equivalent (see Chap. 4). Thus, this law is not exactly true. Indeed it neglects the change in the binding energy of the products with respect to the reactants. This change however only provides a correction of order 10^{-9} that was not measurable by Lavoisier.

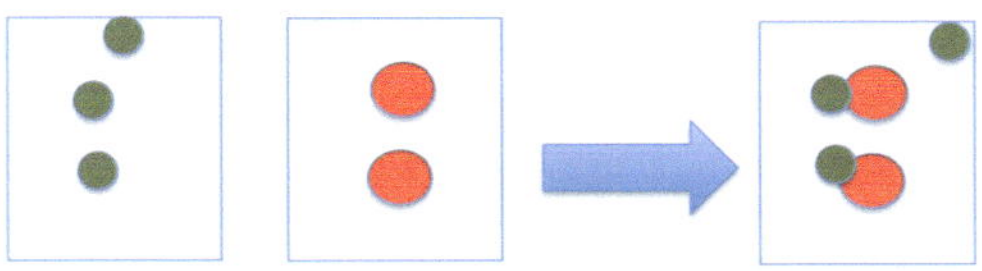

Fig. 1.1 Illustration of the Dalton hypothesis. When two elements combine the law of multiple proportion tells that the ratio of the weights of the two elements is the same independently of the size of the sample. This suggest that an elementary particle exists that characterises the element: the atom

elementary (but rather made of other elementary particles), atoms of the same elements appears with slightly different masses (different isotopes), it is possible to change an atom of an element in an atom of another element through nuclear reactions. Nevertheless the Dalton model has paved the way to a deeper understanding of the structure of matter that has culminated in the second half of the 19th century in the first formulation of the periodic table.

The Dalton atomic theory was not immediately accepted. The main reason for skepticism was due to the difficulty in identifying a well defined atomic weight for the various elements due to the erroneous assumptions made by Dalton on the stoichiometric composition of the compounds. The clarification came from the explanation of the Gay-Lussac *law of combination volumes* [4] stating that

When gases combine chemically, they do so in numerically simple volume ratios. If gases combine to form gases, the volumes of the products are also in simple numerical ratios to the volume of the original gases

If from 1 litre of element A plus 1 litre of element B 1 litre of the compound AB would be obtained, it would be possible to conclude that different gases at the same temperature and pressure have the same number of atoms for unit volume (Berzelius, 1813 [5]). However, this is not true e.g. in the case of the formation of HCl from H and Cl (1l of H plus 1l of Cl yields 2l of HCl). The solution to this problem was found by Avogadro (1811) by introducing the concept of molecules [6]

Different gases at the same temperature and pressure have the same number of molecules per unit volume

As both H and Cl in gaseous form appear as bi-atomic molecules, the Avogadro principle now explains the observation (Fig. 1.2).

The Avogadro principle suggests a simple and accurate method to measure the relative molecular weights of two substances by comparing the weight of two equal volumes of gases at the same temperature and pressure. Since the number of particles is the same, the ratio of the two weights is equal to the ratio of individual molecules. The absolute weight is more difficult to determine and requires to know the number of molecules in the volume, i.e. the Avogadro constant that was determined by Josef Loschmidt [7]. The Avogadro principle was accepted only in 1860 at the Karlsruhe Congress on chemistry when Stanislao Cannizzaro presented his refined table of atomic weights measured using the Avogadro principle [8].

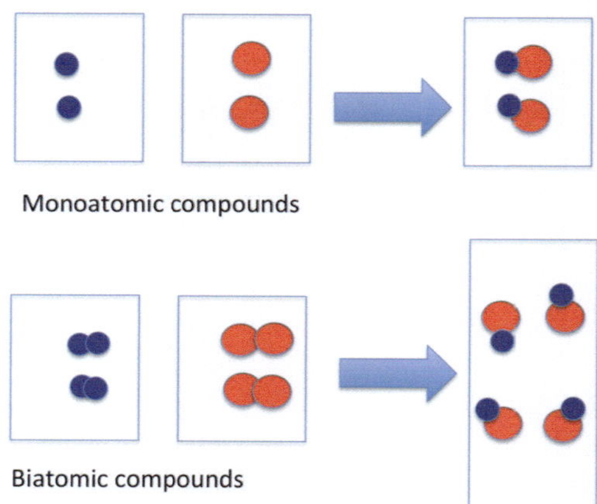

Fig. 1.2 Illustration of the Avogadro principle for *HCl*. From a volume of hydrogen and a volume of chlorine it is obtained a double volume of hydrochloric acid. This result (perfectly in line with the Gay-Lussac principle) contradicts the hypothesis that the number of atoms in the same volume (at the same temperature and pressure) is the same for all the compounds. The solution is that both hydrogen and chlorine are diatomic molecules. To react they have to dissociate thus doubling the volume

1.2 The Periodic Table of Elements

Until the end of the 18th century the distinction between what we call today a *chemical element* and a generic substance was not clear. About 10 elements were known since prehistory ($C, S, Fe, Cu, Ag, Sn, Au, Hg, Pb$ and perhaps *Zn*). Lavoisier [2] was the first to define an element as a substance that could not be further broken down through chemical reactions. Nevertheless his table of elements reported 23 true chemical elements together with a few compounds and even light and heat!

Most of the chemical elements were discovered in the 200 years between 1669 and 1871. Understandably, in the 19th century scientists started asking if such a large number of elements had any rationale behind them. It was known these elements could be grouped according to their chemical properties (e.g. halogens or alkali metals). In addition, the atomic weight of most elements had been measured within an accuracy that was substantially improved after the acceptance of the Avogadro principle.

A few unsuccessful attempts were made to organise the elements according to their atomic weight (Meyer 1864, Newlands, 1864). In 1869 a Russian chemist, Dmitrij Mendeleev, tried to organise the known elements in *groups* according to their chemical properties, ordering the elements in each group according to their atomic weight [9] (Fig. 1.3). Elements were therefore arranged in a two-dimensional array with one direction corresponding to *groups* and the other to *periods*. He noted that some of the places in this 2-D array were empty. The missing elements were discovered a few years later, providing a strong support to the Mendeleev approach.

			Ti=50	Zr=90
			V=51	Nb=94
			Cr=52	Mo=96
			Mn=55	Rh=104.4
			Fe=56	Ru=104.4
			Ni=Co=59	Pd=106.6
H=1			Cu=63.4	Ag=108
	Be=9.4	Mg=24	Zn=65.2	Cd=112
	B=11	Al=27.4	?=68	Ur=116
	C=12	Si=28	?=70	Sn=118
	N=14	P=31	As=75	Sb=122
	O=16	S=32	Se=79.4	Te=128?
	F=19	Cl=35.5	Br=80	I=127
Li=7	Na=23	K=39	Rb=85.4	Cs=133
		Ca=40	Sr=87.6	Ba=137
		?=45	Ce=92	
		?Er=56	La=94	
		?Yt=60	Di=95	
		?In=75.6	Th=118?	

Fig. 1.3 Mendeleev table from Ref. [9]. Mendeleev arranged the elements known in 1869 in groups (according to their chemical properties) and ordered the elements in each group according to their atomic weight. Each group could be then placed adjacent to other groups in such a way that elements on each vertical column (period) appeared in the right atomic mass order. Some of the positions were empty and were filled a few years later thanks to the discovery of new elements. However the Mendeleev table had some problems. The position of neighbouring elements in the same period had to be interchanged to keep them in the right group. Note that the modern Mendeleev tables are rotated by 90° with respect to the original one

However, Mendeleev himself noted some issue as he had e.g. to invert the position of tellurium and iodine in order to bring them in the proper group (the same problem arose with cobalt and nickel or with argon and potassium—argon was not yet known in 1869). It was clear that the atomic weight was only an approximate ordering parameter.

The solution to the problem of what was the true ordering parameter had to wait a few decades with the introduction of the concept of atomic number.

1.3 The Discovery of Sub-atomic Particles and Radioactivity

Up to Mendeleev the classification of elements had been a matter mostly for chemists. Towards the end of the 19th century also the physicists started to give their contribution. During the 19th century, physics had made a major progress in understanding the common nature of seemingly uncorrelated phenomena such as electricity and magnetism. This progress culminated in the formulation in 1861 of the Maxwell equations of electromagnetism [10].

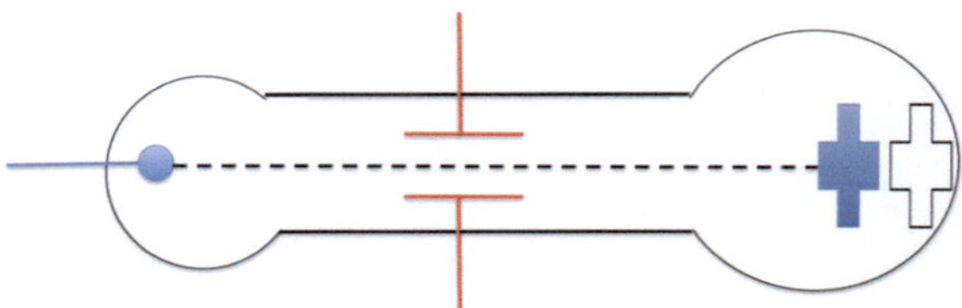

Fig. 1.4 Cathode ray tube. The progress in vacuum technology made it possible to evacuate the tube down to pressure of $10^{-6} atm$. In these conditions Krook observed fluorescent light emitted from the end of the tube. This was due to a new kind of radiation since upon interposing an object, the shadow of the object was reproduced on the tube

The technologies developed to support the experimental effort was crucial to the discovery of elementary particles, with the first being the electron discovered by Thomson [11] using a cathode ray tube.

The cathode ray tube (see Fig. 1.4) was a development of the gas discharge tube. Michael Faraday in 1938 noted that in a glass tube in which air had been partially evacuated a glow discharge could be established between two oppositely charged electrodes [12]. Further reducing the gas pressure down to values below $0.1 Pa$, in 1870 W. Crook observed a new phenomenon. The diffuse glow disappeared while the end part of the glass tube started to emit fluorescent light. The origin of the light was a new kind of ray that were called cathode rays because, emitted from the negatively charged cathode, were stopped by interposing an object (as shown in Fig. 1.4 for the cross-shaped anode). J.J. Thomson investigated the nature of this radiation and discovered that could be deflected by electric and magnetic fields (so it was negatively charged). Upon assuming that the charge was an elementary charge, from the ratio e/m Thomson concluded that the mass of the cathode ray particles was much less than the mass of the hydrogen atom. The *electron* had been discovered!

Thomson had to assume something about the elementary charge. The existence of an elementary charge had been already pointed out in 1874 by Stoney [13] on the basis of the electrolysis (he was indeed the person who proposed the term *electron*). The basis for the estimate is the Faraday law of electrolysis that states that the mass of a substance subject to electrochemical reactions is proportional to the electrical charge that has flown through the circuit. Thus, measuring the amount of mass eroded by the anode during electrolysis and knowing the atomic weight of the eroded substance it is possible to determine the number of atoms that took part in the process. Upon comparing this number to the electric charge exchanged during the process, Stoney concluded that the elementary charge was about $10^{-20} C$ (not too far from the exact value of $1.6 \times 10^{-19} C$).

Since the evaluation of particle orbits in electric and magnetic fields will come a few times in this book, it is worthwhile discussing the J.J. Thomson experiment in quantitative terms.

The geometry is shown in Fig. 1.5. In the absence of electric and magnetic fields, the cathode rays travel along the x axis at constant velocity v_x. They arrive at the

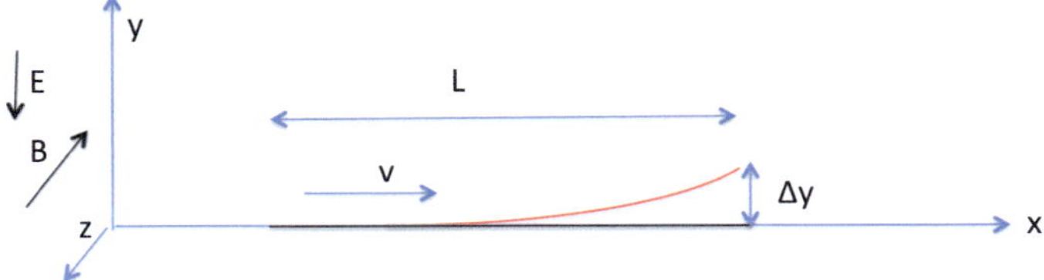

Fig. 1.5 Cathode ray tube geometry. Electrons are emitted from the cathode and travel to the anode. If an electric field along y is imposed the electrons are deflected (red curve). The deflection can be avoided (black trajectory) by simultaneously producing a magnetic field along z in such a way that the Lorentz force equilibrates the force due to the electric field

screen location (placed at distance L from the cathode) after a time $T = L/v_x$. If we now superimpose an electric field along the y axis, the cathode rays are subject to a constant acceleration qE/m and are deflected by a distance

$$\Delta y = \frac{qE}{m}\frac{T^2}{2}.$$

However the time T is unknown until the velocity v_x is determined. This can be accomplished by superimposing a magnetic field B_z along the z axis that will produce a Lorentz force $qv \times B$. The total force acting in the y direction on the particles of the cathode rays will be

$$F_y = qE_y - qv_xB_z$$

choosing B_z appropriately, the total force can vanish and the rays will arrive undeflected on the screen. This clearly occurs for

$$v_x = \frac{E_y}{B_z}$$

which allows to determine the ray velocity and, in turn, the flight time T and, from the measurement of the deflection, the ratio q/m.

At the end of the 19th century another discovery was going to change the views on the nature of matter: radioactivity. In 1896 A. H. Bequerel was investigating the phenomenon of phosphorescence (the delayed emission of light after a material has been exposed to light) using uranium-potassium sulphate ($K_2UO_2(SO_4)_2$). He accidentally discovered that the uranium salt emitted radiation even when not irradiated by light [14]. The radiation was capable of penetrating through matter (similarly to the X-rays discovered one year before by Röntgen [15]). The term *radioactivity* was introduced by Pierre and Marie Curie who discovered in 1898 two new elements (polonium and radium) that are more active than uranium [16]. The radioactive emission was investigated by Ernest Rutherford who in 1899 was able to distinguish between *alpha* (positively charged) and *beta* (negatively charged) radiation [17].

Finally, one year later (1900) P.U. Villard discovered a third kind of radiation, neutral and strongly penetrating [18] that was named *gamma* radiation. The evidence accumulated led Rutherford [19] to propose that radioactivity was due to the disintegration of atoms [19]. One of the pillars of the Dalton atomic hypothesis was abandoned! Many years later (1919) Rutherford was also able to produce the first induced transmutation by bombarding nitrogen with alpha radiation [20]. The reaction produced a heavy oxygen isotope plus a hydrogen nucleus and is considered the discovery of the proton.

The mechanisms that produce these three kinds of radiation will be investigated in more detail in the following chapters. What Rutherford was able to understand was that alpha radiation was made by helium nuclei and beta radiation by electrons. Experiments showed that alpha radiation was easily shielded (a sheet of paper was enough). To shield beta radiation required a thin metal shield whereas gamma radiation was the most penetrating among the three and required a thick shield of heavy metal like lead.

1.4 The Internal Structure of the Atom

After the discovery of the electron, a particle much lighter than hydrogen that could be emitted by atoms in radioactive decay processes, it was natural to ask how its existence was related with the atomic hypothesis. Atoms could not be elementary particles as supposed by Dalton. Thus, Thomson [21] proposed a model for the atomic structure that resembled that of a plum pudding [21]. He assumed that atoms were made by a diffuse cloud of positive charge in which electrons were inserted as plumes in a pudding. Starting in 1908 Rutherford carried a series of experiments that demonstrated that the plum-pudding model was wrong and in 1911 proposed a different model of the atom [22].

Hans Geiger and Ernest Marsden, under the direction of Rutherford, bombarded (using alpha particles generated in a radioactive decay) a thin gold foil and, using a detector invented by Geiger (the Geiger detector was indeed a spin off of this experiment), measured the number of alpha particles deflected in the various directions. If the Thomson model were true, the alpha particles should have passed through the gold foil almost un-deflected. The diffuse positive cloud would have acted as a homogeneous medium with no deflection expected. The light electrons would have not deflected the alpha particles either, in the same way as a flying cannon ball is not deflected by the collision with the air molecules.

What Geiger and Marsden measured was different [23]. Although most of the particles passed through the gold foil un-deflected, a small fraction of them was substantially deflected with respect to the original direction. They even detected particles scattered backwards!

The only way to understand the result was to assume that the positive charge, rather than being spread throughout the atom volume, was concentrated at its centre in a massive nucleus. We will go back to a quantitative analysis of the Rutherford problem in Chap. 10.

On the basis of the results Rutherford proposed a model of the atom similar to the solar system with the positively charged nucleus playing the role of the Sun and the electrons playing the role of the planets. Unfortunately gravitation and electromagnetism are not exactly equivalent—electrons accelerated by the nucleus electric charge were predicted to fall on the nucleus in fractions of a second—and the Rutherford model was not immediately accepted (see Chap. 5). It was only in 1913 that, thanks to the work of Bohr explaining the atomic spectra [24], the Rutherford model was accepted (and it is indeed known as the Bohr-Rutherford model).

Rutherford not only proposed a qualitative picture of the atomic structure. From the measurements of the gold foil experiment he was also able to give an estimate of the charge of the gold nucleus. Since atoms are electrically neutral we know that the charge of the nucleus (in elementary units) must be equal to the number of electrons around the nucleus. His estimate was about 100 elementary charges, not too far from the true value (79).

A few months after the Rutherford proposal of atomic structure, van den Broek proposed [25] that the number of elementary charges of the nucleus was equal to the position in the periodic table and called this quantity the atomic number Z. But how to measure the atomic number? The solution came from the measurements of the atomic spectra.

Since spectroscopy is going to play a crucial role in the development of quantum mechanics, let us make a step back and consider its evolution during the 19th century.

1.5 Spectroscopy and the Measurement of the Atomic Number

The first physicist to use the term *spectrum* was Isaac Newton [26]. He noted that light passing through a glass prism is decomposed into different colours. At the beginning of the 19th century physicists understood that radiation extended beyond visible light. Herschel [27] discovered the infrared radiation [27] and Johan Ritter the ultraviolet radiation [28]. In 1865 the Maxwell's theory of electromagnetism led to the identification of light as electromagnetic radiation, one of the major achievements of 19th century science.

However, the decomposition of the spectrum using the dispersion of light (i.e. the property that different spectral components travel at different velocity and therefore follow different paths) did not allow determining the frequency and/or the wavelength of radiation.

A substantial advance in spectroscopy occurred when prisms were replaced with gratings made of parallel narrow slits that reflect light and produce constructive interference at well defined angles. In 1821 Joseph von Fraunhofer [29] decomposed the spectrum by diffraction rather than by dispersion. The quantitative analysis of this phenomenon will be given in Chap. 4. What it is important to understand is that by changing the spacing of the grating (specifically by decreasing the spacing) it was possible to separate electromagnetic waves at shorter wavelength. At some point a limit was met on the spacing that could be realised with the available technologies and physicists turned to naturally occurring gratings such as those offered by crystals. In 1911 L. Bragg and W.H. Bragg devised a method to determine the wavelength of X-rays using a crystal lattice [30].

The use of diffraction gratings allowed not only to extend the measurable electromagnetic spectrum but also to increase the resolution of the spectral measurements in the already known part of the spectrum (i.e. to measure the difference of the intensity of rays at very close frequencies). The result was that the spectrum of light emitted by stars no longer appeared continuous. Spectra were interrupted at very well defined frequencies (wavelengths) by black lines. Physicists soon realised that spectral lines were the fingerprint of different elements and around 1860 started to classify all the known elements on the basis of the lines that they emitted (or absorbed) when excited. All this effort led to the first determination of the chemical composition of stars.

In 1913 the Bohr theory of the atomic structure [24] provided a way to predict the wavelength of the spectral lines emitted (or absorbed) by atoms. Specifically, it predicted that the wavelength emitted by the electrons closer to the nucleus was proportional to $1/Z^2$, with Z the nuclear charge. For high-Z atoms the wavelength falls in the X-ray region. The Bragg diffraction method had been proposed two years earlier and a young Oxford physicist, H. G. J. Moseley, set up an experiment to measure the radiation emitted by the elements from aluminum to gold under electron bombardment. The result was not only the confirmation of the $1/Z^2$ law but also the determination of the exact number (15) of lanthanides with the separation of neodymium and praseodymium and the prediction of new elements with Z = 43(technetium), 61(promethium), 72(hafnium) and 75(rhenium) [31].

The van den Broek hypothesis had been confirmed and the atomic number Z finally measured!

1.6 The Mass Spectrometer. Isotopes and Nuclear Binding Energy

After the discovery of radioactivity it was soon realised that substances having different atomic masses had the same chemical properties. This led Frederick Soddy (1913) to call them *isotopes* since they occupy the same cell of the periodic table [32].

The existence of isotopes has the consequence that the atomic weight of an element corresponds to the average mass of its isotopes. Thus, the atomic weight depends on the natural abundance of the isotopes. Taking as an example chlorine, two stable isotopes exist: ^{35}Cl (with atomic mass $5,8067 \times 10^{-26}$ kg) is the more abundant (75.76%) with the remaining 24.24% being ^{37}Cl (with atomic mass

6.1383×10^{-26} kg). Thus, the *Cl atomic weight* (to be distinguished from the *atomic mass* of the individual isotopes) of the element is given by

$$75.76\%\ 5,8067 \times 10^{-26}\,\text{kg} + 24.24\%\ 6.1383 \times 10^{-26}\,\text{kg} = 5.8871 \times 10^{-26}\,\text{kg}$$

The measurement of the mass of the isotopes is associated with the name of Francis Aston who started in 1909 to work on the measurement of the atomic mass at the Cavendish laboratory using a simple apparatus first developed by Thomson [33]. Aston improved the apparatus all over his life, with the latest version able to measure the atomic mass with a precision of 3×10^{-4} to be compared with a 10% accuracy of the original Thomson spectrometer. The scheme of a mass spectrometer is shown in Fig. 1.6.[2]

Ions produced by the source S pass through a region where both an electric field E and a magnetic field B' are present. As in the case of the cathode ray tube this region acts as a velocity selector: only the particles with velocity $v = E/B'$ are not deflected and can pass through the two slits s and s'. Beyond the slit s' only a magnetic field B is present. Under the effect of the Lorentz force particles follow a circular trajectory of radius $R = v/(qB/m)$ and hit a photographic plate where they leave a trace. From the measurement of their position, knowing E, B' and B, it is possible to determine the

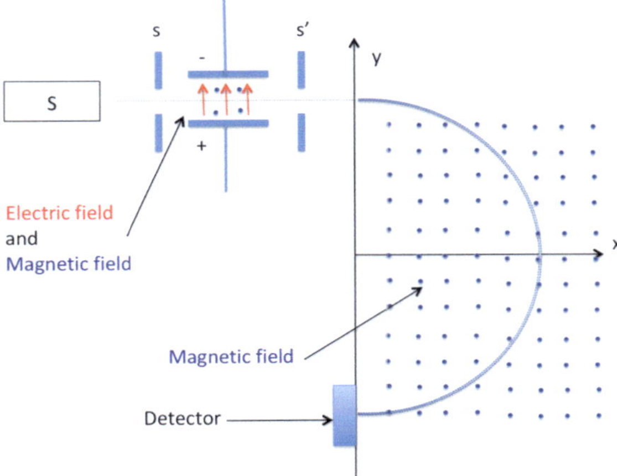

Fig. 1.6 The mass spectrometer. The spectrometer shown in the figure is not a scheme of the Aston mass spectrometer but rather the version developed late by Bainbridge. The ions are emitted from an hot source and pass through a velocity selector where the simultaneous use of an electric and magnetic field allows the passage through the slit only of particles with a well defined velocity. After the slit the ions move under the effect of a Lorentz force making a circular trajectory to end on a photographic plate. From the position of the arrival point it is possible to deduce the quantity q/m

[2] This is not actually the Aston mass spectrometer but rather the model built by Bainbridge, the principle being basically the same.

ratio q/m. Since q must be an integer multiple of the elementary charge, it is possible to determine the mass m. Clearly the measurement cannot distinguish particles with the same q/m.

Aston used his apparatus to make a systematic scan of all the elements. Already using the Thomson spectrometer to analyse a gas containing neon he had measured (1912) two lines corresponding to mass 20 and mass 22. Mass 22 could be due to the presence of CO_2. By purifying the gas from its CO_2 content he measured the same lines, ruling out the CO_2 hypothesis. It was known from density measurements that neon had an atomic weight 20.2. Thus Aston started working on the idea that this value was not due to a single species but rather to the combination of two species of atomic weight 20 and 22, respectively, and the former species being nine times more abundant than the latter.

After World War I Aston built the first version of his mass spectrometer and was able to demonstrate that within the experimental accuracy all the isotopes he was able to separate were approximately integer multiples of the *atomic mass unit AMU* (at that time defined as 1/16 of the ^{16}O mass—the definition as 1/12 of the ^{12}C came only in 1962), the only exception being the hydrogen (protium) itself that was about 8×10^{-3} higher. He was eventually able to measure masses with 0.1% accuracy that was instrumental for the discovery of nuclear energy.

To complete the picture of the atomic structure it was necessary to understand what made the mass of the nucleus besides the Z protons. Rutherford made the hypothesis that the rest of the mass was made by particles (called *neutrons*) with a mass close to that of the proton and zero electric charge that he (erroneously) assumed made by electron-proton pairs.

Neutrons were observed first only in 1931–1932 when Bothe-Becker [34] and Joliot-Curie [35] discovered a new kind of neutral radiation by bombarding light elements (lithium, boron and beryllium) with alpha particles. They erroneously identified this radiation with gamma rays but in 1932 J. Chadwick pointed out that this could not be the case and identified the new particle with the neutron, the particle proposed by Rutherford to explain the composition of the nucleus [36]. As protons and neutrons have similar masses we will often refer to them for simplicity as *nucleons*.

The number of protons and neutrons in a nucleus is the *mass number* A. In spite of its name the mass number is only an *approximate* measure of the mass of an isotope. First, the proton and the neutron have different masses and we need the exact composition in order to have the correct value of the mass of the constituents. But there is a more important effect. Since the nucleons form a bound system, in order to separate the nucleus energy must be spent. The situation is similar to a group of heavy balls at the bottom of a (gravitational) potential well. At the bottom of the well they form a bound system. If we want to extract a ball we have to provide energy to overcome the attractive potential. This energy is the *binding energy* of the ball. Thus, the energy of the bound state must be lower than the energy of the system with a nucleon being extracted. The binding energy *reduces* the total energy of the system. We have anticipated at the beginning of this chapter that mass and energy are equivalent (we will discuss in detail this equivalence in Chap. 4). Thus, the mass of the nucleus will be reduced with respect to the mass of the constituents by an

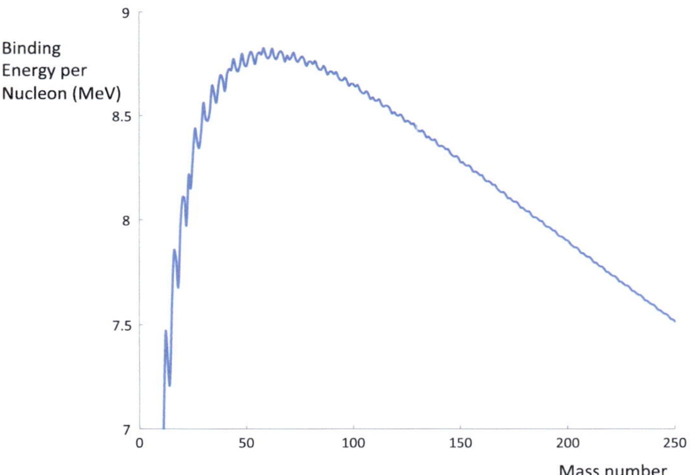

Fig. 1.7 Binding energy per nucleon. Aston spent his life to improve the mass spectrometer and was eventually able to measure the atomic mass with sufficient precision to observe the difference due to the binding energy. The binding energy divided by the mass number (binding energy per nucleon) turned out to be a non monotonic curve. Very heavy nuclei have a binding energy per nucleon smaller than the nuclei around $A \approx 100$. Similarly, light nuclei are less bound than intermediate A nuclei. Thus, splitting a heavy nucleus (fission) or fusing two light nuclei (fusion) leads to a more tightly bound system and an excess of energy that can be used to produce electricity

amount equivalent to the binding energy and this was what Aston determined by further development of his apparatus.

With the second and third version of his mass spectrometer Aston was able to measure with sufficient precision the binding energy. The difference turned out to be negative (*mass defect*) as expected. The Aston measurements were so precise that he was able to measure the difference in the binding energy per nucleon of the various isotopes. If we divide the binding energy by the number of nucleons we have the *binding energy per nucleon* (or *packing fraction*). This quantity measures how much energy on average must be used to extract a nucleon from the nucleus of a certain isotope. The early observation by Aston that the atomic mass is approximately the multiple of the AMU except for protium indicated that the binding energy per nucleon is approximately a constant. But Aston went further.

The curve determined by Aston is shown in Fig. 1.7: the binding energy per nucleon is *almost* constant and equal to 8 MeV. However, the binding energy per nucleon tends to be lower for very heavy nuclei and for very light nuclei with a maximum for nuclei placed in the region of iron ($A \approx 50 - 60$). To extract a nucleon from an iron nucleus require more energy than in the case of uranium or in the case of the hydrogen isotopes. Going from a less bound to a more bound state some energy will be released. Thus energy can be produced either by *fusion* of light elements or by *fission* of heavy elements.

Table 1.1 Mass of elementary particles

	Mass (kg)	Mass (AMU)
Electron	9.1094×10^{-31}	5.5×10^{-4}
Proton	1.6726×10^{-27}	1.00728
Neutron	1.6749×10^{-27}	1.00866

1.7 The Atomic Structure-Summary

At this point we can summarise all the above findings in the nuclear/atomic model that can be found in all the physics/chemistry textbooks. This was the result of a research effort that has taken about 150 years from Lavoisier to the discovery of the neutron.

The atom is made by a nucleus of charge $+Ze$ and Z electrons of charge $-Ze$. The mass number A of the nucleus is the number of protons and neutrons in the nucleus.

We will indicate a certain element with the symbol ^{A}X with X the element (which is equivalent to give the atomic number Z).

The mass is measured in Atomic Mass Units (AMU) defined as 1/12 the mass of ^{12}C.

$$1AMU = 1.660539066 \times 10^{-27}\,\mathrm{kg}$$

The quantity of matter is measured in *moles*. One mole is the amount of molecules corresponding to $12g$ of ^{12}C and correspond to N_A molecules with

$$N_A = 6.02214076 \times 10^{23} \quad particles/mole$$

the *Avogadro number*.

The masses of the elementary particles that we have encountered so far are summarised in the following Table 1.1

The isotopes are nuclei of a certain element that have different number of neutrons (same Z but different A). The *natural composition* of an element is the fraction of the various isotopes that can be found in nature. The natural composition includes only the isotopes that are stable or have a radioactive decay that occurs on a very long time scale. The atomic weight of an element is the mass of the various isotopes weighted with the relative abundance.

For example, Carbon has only two stable isotopes ^{12}C (98.93%) and ^{13}C (1.07%). The mass of ^{12}C is 12 AMU (by definition of AMU) and that of ^{13}C is 13.003 AMU. Thus the mass of natural carbon is

$$12AMU \times 98.93\% + 13.003AMU \times 1.07\% = 12.011AMU$$

1.8 Elementary Particles

The discovery of electrons, protons and neutrons demonstrated that the atom is not an elementary quantity but is rather made of smaller constituents. The analysis of cosmic rays led to the discovery of other particles. A number of different particles where found as soon as accelerator were made available to perform high-energy collisions.

All the elementary particles are classified according to the following properties:

- the mass;
- the electric charge (in units of the proton charge)
- the intrinsic angular moment (*spin*) usually expressed in units of $h/2\pi$ with h being the Planck constant that we will discuss later in this book.

The present understanding of the structure of matter recognises *leptons* (literally "light particles") and *quarks* as (truly) elementary particles. They are divided into three families each composed by two leptons and two quarks.

All the particles in Table 1.2 have spin equal to $1/2$. For each particle there is a corresponding *anti-particle* characterized by the same values of mass and spin but opposite electric charge.

Quarks are not observed as isolated particles but only through a combination of a quarks and an anti-quark (*mesons*) or of three quarks (*baryons*). Outside the nucleus, mesons are short lived particles. Baryons can be on the contrary very stable. Specifically, the baryons of the first family are the proton and the neutron. The proton is considered stable. Its lifetime is larger that 10^{20} times the age of the universe. The neutron can be stable inside the nucleus. As an isolated particle it decays in about 15 minutes. Protons are made of two up quarks and one down quark, neutrons are made of two down quarks and one up quark.

Almost all the matter in the universe is made of the constituents of the first family. The particles made of the other two families can be found only in high-energy collisions.

Table 1.2 Elementary particles. Mass in unit of the proton mass

	Family 1	Family 2	Family 3
Quark (q = 2/3)	Up (u) – m = 0.0047	Charm (c)) – m = 1.6	Top (t) – m = 189
Quark (q = –1/3)	Down (d) – m = 0.0074	Strange (s) – m = 0.16	Bottom (b) – m = 5.2
Lepton (q = –1)	Electron (e) – m = 0.00054	Muon (μ) – m = 0.11	Tau (τ) – m = 1.9
Lepton (q = 0)	Electron neutrino (ν_e) – m $\leq 10^{-8}$	Muon neutrino (ν_μ) – m $\leq$0.0003	Tau neutrino (ν_τ)) – m $\leq$0.033

The number of leptons of each family is a conserved quantity. A value +1 is assigned to the electron and the electron neutrino leptonic number L_e. The corresponding anti particles have leptonic number $L_e = -1$ whereas the up and down quarks have $L_e = 0$. Similarly for the leptonic number L_μ of the second family and the leptonic number L_τ of the third family.

Also the number of baryons of each family is conserved. For example, for the first family a values $B = 1$ is assigned to the proton and the neutron and a value $B = -1$ to their anti particles ($B = 0$ for the leptons) and in any transformation the number must be the same before and after the reaction.

For illustration we consider the case of the beta decay of a neutron into a proton, an electron and an anti-neutrino. The baryon number is +1 before and after the decay. Similarly the lepton number is zero before, after the decay the electron and the antineutrino have lepton number +1 and −1 respectively yielding again a total lepton number equal to zero.

The interaction between particles can occur through four kinds of forces: gravitational, weak, electromagnetic and strong force. All the particles interact via gravitation but this interaction is extremely weak and it shows itself only with bodies of very large mass. Therefore it will be totally neglected in this book. The electromagnetic force involves particles that have a non zero electric charge. It is much stronger than gravitation (and also of the weak force). The other two forces show their effect only at distances of the order of the atomic nucleus. The weak force is responsible for phenomena like the beta decay that we will discuss later in this book. The strong force is responsible for keeping together the constituents of the atomic nucleus and does not act on leptons.

The interaction due to the four forces takes place via the exchange of particles. The electromagnetic interaction is due to the exchange of *photons*. The weak interaction involves the W and Z vector bosons. The strong interaction is associated with the exchange of *gluons* and the gravitational interaction via the exchange of *gravitons*.

A large effort has been devoted to the derivation of general laws that make the four forces the manifestation of a single universal interaction. In the 1960 this effort has successfully unified the weak and electromagnetic interaction into a single electroweak force. Theories have been proposed to unify the electroweak and strong interaction but a demonstration is still lacking. The unification of all the four forces is even more challenging [37].

1.9 Suggestions for Further Readings

A simple introduction to atomic physics phenomena can be found in the monographs of Max Born [4]. A specific discussion about the present understanding of elementary particle can be found in the book of Brian Greene [37].

1.10 Exercises

Problem 1.1 Hydrogen has two stable isotopes: protium ($A = 1$) and deuterium ($A = 2$). The hydrogen atomic weight is 1.00794 AMU. Determine the amount of deuterium in a litre of water.

Problem 1.2 Given a cathode ray tube $1m$ long and an electric field of 10 kV/m the deflection of an electron along y can be compensated by adding a magnetic field along z of intensity $1G(= 10^{-4}T)$. Determine the deflection in the absence of the magnetic field. Determine the electron velocity along x.

Problem 1.3 Using reasonable values for E and B determine the position of the lines of the mass spectrum of Fig. 1.6.

References

1. A. Einstein, Ist die Trägheit eines Körpers von seinem Energieinhalt abhängig? Annalen der Physik **18**(13), 639–641 (1905)
2. A.-L. Lavoisier, Traité élémentaire de chimie (1789)
3. J. Dalton, *A New System of Chemical Philosophy* (1808)
4. J.L. Gay-Lussac, Mémoire sur la combinaison des substances gazeuses, les unes avec les autres. Mémoires de la Société d'Arcueil **2**, 207–234 (1809)
5. J.J. Berzelius, Ann. Philos. **2**, 443–454 (1813); **3**, 51–2, 93–106, 244–255, 353–364 (1814)
6. A. Avogadro, Journal de Physique, de Chimie et d'Histoire naturelle t. **73**, 58–76 (1811)
7. J. Loschmidt, Zur Grüsse der Luftmoleküle. Sitzungsberichte der Kaiserlichen Akademie der Wissenschaften Wien. **52**(2), 395–413 (1865). English translation: J. Loschmidt with William Porterfield and Walter Kruse, trans.: On the size of the air molecules. J. Chem. Educ. **72**(10), 870–875 (1995)
8. S. Cannizzaro, Lettera del Prof. Stanislao Cannizzaro al Prof. S. de Luca; Sunto di un corso di filosofia chimica fatto nella Reale Universitá di Genova dal Professore S. Cannizzaro. Il Nuovo Cimento **7**, 321–366 (1858)
9. D. Mendeleev, Ueber die Beziehungen der Eigenschaften zu den Atomgewichten der Elemente [On the relations of properties of the elements to their atomic weights]. Zeitschrift für Chemie **12**, 405–406 (1869)
10. J.C. Maxwell, *A Treatise on Electricity and Magnetism*, vol. 1 (Clarendon Press, Oxford, 1873)
11. J.J. Thomson, Cathode rays. Philos. Mag. **44**, 293–316 (1897)
12. M. Faraday, VIII. Experimental researches in electricity.–Thirteenth series. Philos. Trans. R. Soc. Lond. **128** 125–168 (1838)
13. G.J. Stoney, Philos. Mag. (5), II, 381–390 (1881)
14. H. Becquerel, *On the Radiation Emitted in Phosphorescence*. In H. Becquerel Compte rendus de l'Academie des Science, Paris, vol. 122, pp. 420–421 (1896)
15. W. Röntgen, Ueber eine neue Art von Strahlen. Vorläufige Mitteilung. Aus den Sitzungsberichten der Würzburger Physik.-medic. Gesellschaft Würzburg, pp. 137–147 (1895)
16. P. Curie, M.S. Curie, Sur une nouvelle substance fortement radio-active, contenue dans la pechblende. Comptes rendus de l'Academie des Sciences, Paris, (26 December), **127**, 1215–1217 (1898)
17. E. Rutherford, Uranium radiation and the electrical conduction produced by it. Philos. Mag. **47**, 109–163 (1899)

18. P.U. Villard, Sur la réflexion et la réfraction des rayons cathodiques et des rayons déviables du radium. Compte rendus de l'Academie des Science, Paris **130**, 1010–1012 (1900)
19. E. Rutherford, F. Soddy, The cause and nature of radioactivity I. II. Philos. Mag. **4**, 370–396; 569–585 (1902)
20. E. Rutherford, Collision of α particles with light atoms, IV: An anomalous effect in nitrogen. Philos. Mag. **37**, 581–587 (1919)
21. J.J. Thomson, On the structure of the atom: an investigation of the stability and periods of oscillation of a number of corpuscles arranged at equal intervals around the circumference of a circle; with application of the results to the theory of atomic structure. Philos. Mag. **6, 7**(39), 237–265 (1904)
22. E. Rutherford, The scattering of α and β particles and the structure of the atom. Philos. Mag. **21**, 669–688 (1911)
23. H. Geiger, E. Marsden, On a diffuse reflection of the α-particles. Proc. R. Soc. Lond. A **82**(557), 495–500 (1909)
24. N. Bohr, On the constitution of atoms and molecules, part I. Philos. Mag. **26**, 1–25 (1913)
25. A. van den Broek, The number of possible elements and mendeléff's "Cubic" periodic system. Nature **87**, 78 (1911)
26. I. Newton, Opticks: or, a treatise of the reflexions, refractions, inflexions and colours of light (1704)
27. W. Herschel, Experiments on the solar, and on the terrestrial rays that occasion heat; with a comparative view of the laws which occasion them, are subject, in order to determine whether they are the same, or different. Part I and Part II. Philos. Trans. R. Soc. **90**, 293–326; 437–538 (1800)
28. J. Frercksa, H. Weberb, G. Wiesenfeldt, Reception and discovery: the nature of Johann Wilhelm Ritter's invisible rays. Stud. Hist. Philos. Sci. Part A **40**(2), 143–156 (2009)
29. J. Fraunhofer, Kurtzer Bericht von the Resultaten neuerer Versucheuber die Gesetze des Lichtes, und die Theorie derselbem. Gilbert's Ann. Phys. **74**, 337–378 (1823). J. Fraunhofer, Uber dieBrechbarkeit des Electrishen Lichtes. K. Acad. D. Wiss. Zu Munchen, 61–62 (1824)
30. W.H. Bragg, W.L. Bragg, The reflection of X-rays by crystals. Proc. R. Soc. Lond. A **88**(605), 428–438 (1913)
31. H.G.J. Moseley, The high-frequency spectra of the elements. Philosophical Magazine. 6th series. **26**, 1024–1034 (1913)
32. F. Soddy, The atomic weight of "Thorium" lead. Nature **98**(2468), 469 (1917)
33. G. Squires, Francis Aston and the mass spectrograph. J. Chem. Soc. Dalton Trans. 3893–3899 (1998)
34. W. Bothe, H. Becker, Künstliche Erregung von Kern-γ-Strahlen [Artificial excitation of nuclear γ rays]. Z. Phys. **66**, 289 (1930)
35. I. Curie, F. Joliot, Émission de protons á grande vitesse par les substances hydrogénées sous l'influence des rayons γ trés pénétrants. C. r. hebd. séances Acad. sci. Paris **194**, 273 (1932)
36. J. Chadwick, Possible existence of a neutron. Nature **129**, 312 (1932)
37. B. Greene, *The Elegant Universe: Superstrings, Hidden Dimensions, and the Quest for the Ultimate Theory* (Random House Inc., Vintage Series, 2000)

Chapter 2
The Statistical Description of Many Body Systems

Abstract *This chapter presents an introduction to statistical physics. The Maxwell-Boltzmann distribution is derived first and its properties discussed. The Fermi-Dirac and Bose-Einstein distributions are also derived.*

2.1 The Distribution Function

The systems considered in this book are composed by a very large number of particles. This makes it impossible a description based on the individual particle motion. Therefore we will use a statistical approach by defining the distribution function $f(\mathbf{r}, \mathbf{v})$ as follows.

The probability dP of finding a particle with position between $\mathbf{r}$ and $\mathbf{r} + d\mathbf{r}$ and velocity between $\mathbf{v}$ and $\mathbf{v} + d\mathbf{v}$ is

$$dP = f(\mathbf{r}, \mathbf{v})d\mathbf{r}d\mathbf{v} \tag{2.1}$$

If we sum over all the possible values of position and velocity, the total probability must be 100%. Therefore

$$\int f(\mathbf{r}, \mathbf{v})d\mathbf{r}d\mathbf{v} = 1 \tag{2.2}$$

with the integral extended over all velocities and positions. Let us consider first a homogeneous system (f independent of $\mathbf{r}$) and focus on the velocity dependence of the distribution function. The velocity distribution function was determined first by Maxwell using a heuristic derivation and later by Boltzmann using a statistical approach.

In order to apply probability theory it is convenient to model the continuous energy space as a set of k discrete boxes. Each velocity has the same *a priori* probability to be achieved (Fig. 2.1).

F. Romanelli, *Physics of Nuclear Energy*, Springer Series in Plasma Science and Technology, https://doi.org/10.1007/978-981-97-9609-0_2

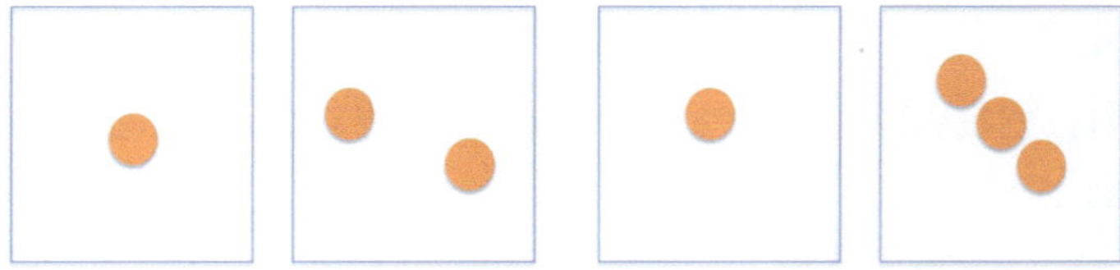

Fig. 2.1 The velocity distribution function can be evaluated by dividing the velocity space in k discrete intervals and distributing a number N of particles. The distribution that corresponds to the largest number of combinations is the most probable distribution. The figure illustrate the procedure in the case $k = 4$ and $N = 7$

Let us consider a system of N particles and total internal energy U. We want to distribute the particles in the k boxes with the only constraints that the total number of particles and the internal energy are fixed. At the end of this operation the first box will contain N_1 particles, the second box N_2 particles, etc. with the sum $N_1 + N_2 + ... + N_k = N$. Then we count the number of combinations in which such distribution can be realised. The distribution that can be realised with the largest number of combinations will be the most probable distribution. Each box can have a number of sub boxes corresponding to states with the same energy but different parameters (e.g. in a one dimensional case at the same energy correspond two different directions of the velocity). The number of different states corresponding to the same energy will be called the *degeneracy* of the state and indicated with g_i.

Let us assume at first that particles are *distinguishable* (i.e. a label can be given to each of them). We start by filling the first box with N_1 particles. The first draw has N possibilities, the second $N - 1$ and so on until the last of the N_1 draws that has $N - N_1 + 1$ possibilities. The total number of possibilities will be the product $N \times (N - 1) \times ... \times (N - N_1 + 1)$. However, since the order in which the particles are chosen is irrelevant, we have to divide this number by the number of permutations of N_1 particles i.e. by $N_1!$. In addition, we need to multiply by the number of states corresponding to the distribution of the N_1 particles in the g_1 sub boxes. If there are no constraints in distributing the particles in the sub boxes (which is correct in the classical case discussed here but not for the quantum statistics discussed later) this number can be estimated taking into account that there are g_1 possible choices for each of the N_1 particles with a total number of possibilities given by $g_1^{N_1}$. Thus the total number of possibilities W_1 to distribute N_1 particles in the first box is

$$W_1 = \frac{N!}{(N - N_1)!N_1!} g_1^{N_1} \tag{2.3}$$

The procedure is continued with the draw of N_2 particles for box 2. The total number of combinations, accounting for all the boxes, is given by the product of all W_i

$$W \equiv \prod_{i=1}^{k} W_i = \frac{N!}{(N-N_1)!N_1!} g_1^{N_1} \times$$
$$\times \frac{(N-N_1)!}{(N-N_1-N_2)!N_2!} g_2^{N_2} \times \ldots \times \frac{(N-N_1-\ldots-N_{k-1})!}{(N-N_1-\ldots-N_k)!N_k!} g_k^{N_k} =$$
$$= \frac{N!}{\prod_{i=1}^{k} N_i!} \prod_{i=1}^{k} g_i^{N_i} \tag{2.4}$$

Since the number of particles by assumption is very large $N!$ can be approximated by the Stirling formula

$$\ln N! \approx (N+1/2)\ln N - N + (1/2)\ln(2\pi N) \approx N \ln N - N \tag{2.5}$$

to obtain

$$\ln W \approx N \ln N - N - \sum_{i=1}^{k} (-N_i \ln g_i + N_i \ln N_i - N_i) \tag{2.6}$$

Finally, the most probable distribution can be determined by maximising Eq. (2.6) with respect to N_i with the constraints

$$N = \sum_{i=1}^{k} N_i \tag{2.7}$$

$$U = \sum_{i=1}^{k} E_i N_i \tag{2.8}$$

The minimisation is obtained using the Lagrange multipliers method by solving the equation

$$\frac{d}{dN_i}(\ln W - \mu \sum_{i=1}^{k} N_i - \beta \sum_{i=1}^{k} E_i N_i) = 0 \tag{2.9}$$

with the two Lagrange multipliers μ and β determined by Eqs. (2.7) and (2.8). Upon substituting Eq. (2.6) into Eq. (2.9) the expression for N_i is obtained.

$$N_i = g_i e^{-\mu-\beta E_i} \tag{2.10}$$

Let us now illustrate this procedure in the case of a one dimensional system along x. In this case each box has an extension Δv_x fixed.

For $\Delta v_x \to 0$ the summation can be replaced by an integral and, as for the same energy the velocity can have two directions, $g_i = 2$. Then Eqs. (2.7) and (2.8) yield

$$N = \int_0^\infty \frac{dv_x}{\Delta v_x} 2e^{-\mu-\beta\frac{mv_x^2}{2}} \tag{2.11}$$

$$U = \int_0^\infty \frac{dv_x}{\Delta v_x} 2\frac{mv_x^2}{2} e^{-\mu-\beta\frac{mv_x^2}{2}} \tag{2.12}$$

From the equations above it is possible to solve for $e^{-\mu}/\Delta v_x$ in terms of N. Furthermore, the quantity β can be identified as the inverse of the temperature ($\beta = 1/T$).[1] The average energy is given by the ratio of Eqs. (2.11) and (2.12), yielding

$$< E >= \frac{U}{N} = \frac{T}{2} \tag{2.13}$$

In the case of a $3D$ configuration the distribution function must be the product of the distribution functions of the velocity in each dimension

$$f(v_x, v_y, v_z) = f(v_x)f(v_y)f(v_z) \tag{2.14}$$

Thus, for a system of particles of mass m at the thermodynamic equilibrium corresponding to a temperature T, the distribution in velocity space is given by a *Maxwell-Boltzmann distribution*

$$f(v) = \frac{n_o}{(\pi^{1/2}v_t)^3} e^{-\frac{v^2}{v_t^2}} \tag{2.15}$$

where $v_t = (2T/m)^{1/2}$ is the *thermal velocity* and $n_o = N/V$ the density (number of particles per unit volume).

2.2 Thermal Velocity and Average Velocity

The thermal velocity v_t in the definition of the distribution function is a measure of the incoherent thermal motion of particles. The average velocity for the distribution function (2.1) is zero. This can be easily shown by multiplying for each of the components of the velocity (v_x, v_y, v_z) and integrating over all the velocity space. Taking for example the x component we have

[1] In this book we will measure the temperature in energy units therefore the Boltzmann constant will not appear.

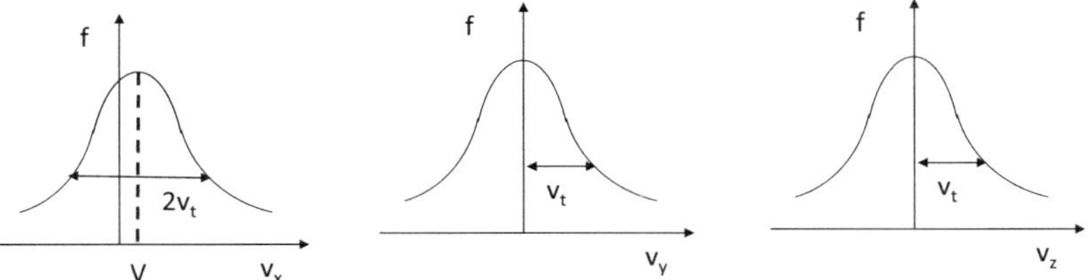

Fig. 2.2 An example of distribution function with non zero average velocity is the shifted Maxwellian

$$< v_x >= \int dv f(v) v_x = \frac{n_o}{\pi^{1/2} v_t} \int dv_x v_x e^{-\frac{v_x^2}{v_t^2}} = 0 \tag{2.16}$$

Thus, there is no average motion associated with the distribution function (2.1). What is non-zero is the mean-square velocity

$$< v_x^2 >= \int dv f(v) v_x^2 = \frac{n_o}{\pi^{1/2} v_t} \int dv_x v_x^2 e^{-\frac{v_x^2}{v_t^2}} = \frac{T}{m} \tag{2.17}$$

As an example of distribution function with non zero average velocity we consider the case of a shifted Maxwellian illustrated in Fig. 2.2. In this case the average velocity along y and z is zero but that along x is V.

Since the distribution function is symmetric in v_x, v_y and v_z we have

$$< v^2 >= 3 < v_x^2 >= 3\frac{T}{m} \tag{2.18}$$

This result can be also expressed in terms of the average kinetic energy

$$< E >= \frac{m < v^2 >}{2} = \frac{3}{2} T \tag{2.19}$$

which states that at the thermodynamic equilibrium there is a contribution equal to $T/2$ for each degree of freedom.

2.3 The Kinetic Theory of Gases

The statistical approach together with the microscopic description is a powerful tool to understand the basic properties of a system. It can be used for example to recover the equation of state of ideal gases. Making reference to Fig. 2.3 we can evaluate the

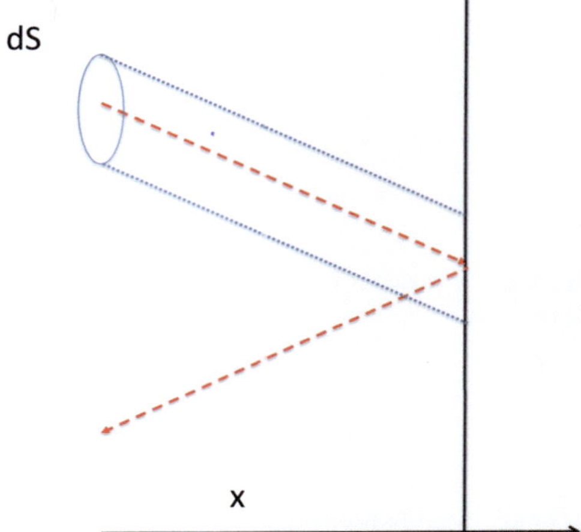

Fig. 2.3 The microscopic description of the gas dynamics and the statistical description are able to reproduce the macroscopic equation of state of an ideal gas

pressure exerted on the wall of a cubic container by the particles that hit the wall and are reflected.

The number of particles impinging on the element dS of the wall in the time interval dt are those in the cylinder shown in the figure with base surface dS and height $v_x dt$. If the density is n (homogeneous in space) the number of particles will be

$$dN = n dS v_x dt \tag{2.20}$$

When a particle is reflected at the wall it transfers an amount of momentum $2mv_x$. The transfer of momentum per unit time and unit surface is the pressure p exerted on the wall. From Eq. (2.20) we have

$$p = \frac{2mv_x n dS v_x dt}{dS dt} = 2\, mnv_x^2 \tag{2.21}$$

We now have to integrate over all the possible values of the velocity v_x corresponding to the particles moving towards the wall (so with positive v_x), weighting the effect on the distribution function

$$p = \frac{2\, mn}{\pi^{1/2} v_t} \int_0^{+\infty} dv_x v_x^2 e^{-\frac{v_x^2}{v_t^2}} = \frac{NT}{V} \tag{2.22}$$

or

$$pV = NT \tag{2.23}$$

which is the equation of state of an ideal gas.

2.4 The Equipartition Theorem

So far we have considered a system of free particles. What happens to a system under the influence of a force? Assume for simplicity a one dimensional harmonic oscillator force. The energy of a particle will be

$$E = \frac{mv_x^2}{2} + \frac{kx^2}{2} \tag{2.24}$$

We can follow the same steps in the derivation of the most probable distribution except that the energy is now given by Eq. (2.24) and the distribution is now a function of the two variables v_x and x. The average energy U/N is the sum of two contributions respectively linked to the kinetic and potential energies. Each of them provide a contribution $T/2$ to the average energy. In general, each quadratic term in the expression of energy contribute for a factor $T/2$ to the average energy. In the case of a free particle (three degrees of freedom) the average energy is $3T/2$ whereas in the case of a monidimensional harmonic oscillator the average energy is T.

This is a specific application of the *equipartition theorem* in statistical mechanics.

2.5 Bose-Einstein and Fermi-Dirac Statistics

In the derivation of the Maxwell-Boltzmann distribution function we have made use of the assumption that particle are *distinguishable* i.e. a label can be put on each of them. In the case of quantum mechanical systems this is no longer true. Furthermore two further assumptions can be made that define two different kind of particles:

- A single state cannot be occupied simultaneously by more than one particle (*Pauli exclusion principle*). Particles of this nature obey the *Fermi-Dirac* statistics and are called *fermions*. Fermions are particles with half-integer values of the spin.
- There is no limit to the number of particles that can occupy the same state. These particles obey the *Bose-Einstein* statistics and are called *bosons*. They are characterised by integer values of the spin.

In the case of the Fermi-Dirac statistics, we need to evaluate the number of combinations of N_i particles in g_i sub boxes (with $N_i < g_i$ due to the exclusion principle) with the constraint of maximum one particle per sub box.

We start with the first particle that can be accommodated in one of the g_i sub boxes available. For the second particle we have $g_i - 1$ possible choices, and so on. The total number is $g_i!/(g_i - N_i)!$. Taking into account that the particles are indistinguishable we have to divide by a further factor $N_i!$ yielding

$$W_i = \frac{g_i!}{N_i!(g_i - N_i)!} \tag{2.25}$$

The total number of microscopic configurations will be the product of all the number of microscopic configurations $W = \prod_{i=1}^{K} W_i$. The derivation of the most probable distribution can be made as for the case of Maxwell-Boltzmann using the same constraints given by Eqs. (2.7) and (2.8) yielding

$$N_i = \frac{g_i}{e^{\beta(E_i - E_F)} + 1} \tag{2.26}$$

with $E_F = -\mu T$. Equation (2.26) corresponds to the *Fermi-Dirac* statistics and is illustrated in Fig. 2.4.

In the case of the Bose-Einstein statistics the number of microscopic states can be calculated making reference to Fig. 2.5. The box corresponds to the $i-$th energy group. We divide the box in g_i sub-boxes by inserting $g_i - 1$ lines. Then we take N_i particles and start distributing them into each sub box. In this case there is no limit to the number of particles in each sub-box.

A microscopic distribution corresponds to a series of N_i dots (the particles) and $g_i - 1$ lines. For example, if we put first all the dots and then the lines we will get the configurations in which all the particles are in the first sub box and the other sub boxes are empty. The number of ways in which we can realise a certain series of dot and lines is the number of possible permutations of $N_i + g_i - 1$ objects. However, changing particles among themselves and lines among themselves lead to the same state, thus, we will have to divide for the number of permutations of N_i dots and $g_i - 1$ lines. The final number of microscopic configurations will be

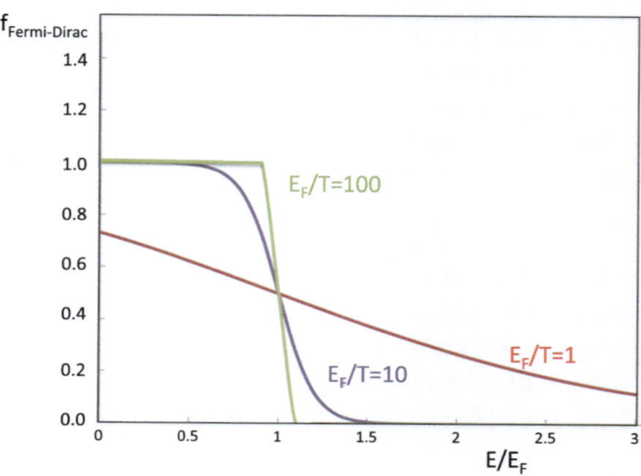

Fig. 2.4 Fermi-Dirac distribution function. The distribution function is shown for three values of E_F/T

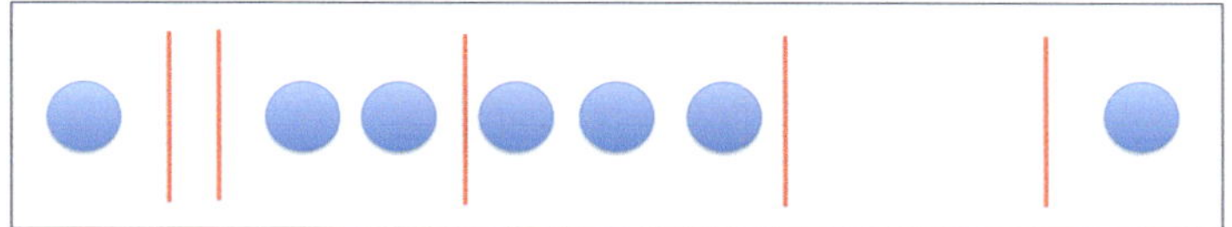

Fig. 2.5 In the case of the Bose-Einstein statistics there is no limit to the number of particles that can occupy a state. The number of combinations is given by the number of permutations of $N_i + g_i - 1$ objects with the N_i particles and the $g_i - 1$ lines being indistinguishable

$$W_i = \frac{(N_i + g_i - 1)!}{N_i!(g_i - 1)!} \tag{2.27}$$

The total number of microscopic configurations will be the product of all the number of microscopic configurations $W = \prod_{i=1}^{K} W_i$. The most probable distribution is obtained by maximising W again under the constraint that the total number of particles and the total energy is fixed. Proceeding as in the case of the Maxwell-Boltzmann statistics we obtain

$$N_i = \frac{g_i}{e^{\mu + \beta E_i} - 1} \tag{2.28}$$

After finding the solution for N_i, the two Lagrange multipliers μ and β can be derived from Eqs. (2.7) and (2.8). Equation (2.28) is the *Bose-Einstein* distribution that we will find again in Chap. 5 when we will consider the black body problem.

Note the difference in the sign of the second term in the denominator with respect to the Fermi-Dirac distribution.

2.6 Exercises

Problem 2.1 Evaluate the average velocity in the three directions and the average of the kinetic energy for a distribution function given by a shifted Maxwellian $f(\mathbf{v}) = n_o/(v_t \pi^{1/2})^3 \exp(-(\mathbf{v} - V\mathbf{x})^2/v_t^2$.

Problem 2.2 Evaluate the thermal velocity of ^4He ions at a temperature $T = 10\,\text{keV}$.

Problem 2.3 Evaluate the average kinetic energy (in eV) of neutrons at $T = 300\,^\circ\text{C}$ and their thermal velocity.

Problem 2.4 Evaluate the most probable velocity and energy of a Maxwell-Boltzmann distribution at given T.

Chapter 3
The Cross Section

Abstract *This chapter introduces the concept of cross section to measure the probability that a certain interaction can take place. An empirical and a microscopic definitions are provided. The dynamics of elastic collisions is reviewed and the expressions of the energy exchange is derived. Applications are presented to some basic nuclear phenomena.*

In this book we will examine systems in which different kinds of interaction can take place between particles. Examples of interactions are:

- The Coulomb interaction between charged particles;
- The fusion of two charged nuclei;
- The neutron-induced fission of a heavy nucleus;
- The ionization of an atom following a collision with an electron or a proton;
- The sputtering/deposition processes.

All these phenomena involve quite different physical processes. However there is a unified way to describe all of them by using the concept of *cross section*. The cross section measures how likely is the interaction between two particles.

Note that the same pair of particles can interact via different forces. In this case we will have to consider separately the probability associated with each kind of interaction.

3.1 The Center of Mass System

The evaluation of the collision dynamics between particle 1 and particle 2 is made easier by working in the centre of mass (com) frame with the centre of mass velocity defined by

$$M\mathbf{V}_{com} = m_1\mathbf{v}_1 + m_2\mathbf{v}_2 \tag{3.1}$$

with $M = m_1 + m_2$. If no external forces act on the system, the total momentum is conserved, $M\mathbf{V}_{com} = const.$. This approximation can be considered valid even if

F. Romanelli, *Physics of Nuclear Energy*, Springer Series in Plasma Science and Technology, https://doi.org/10.1007/978-981-97-9609-0_3

there are external forces provided during the collision they are small compared with the forces responsible for the collision.

In the following we will indicate with the suffix (com) quantities evaluated in the centre of mass system. Quantities without a suffix are evaluated in the laboratory system.

The velocities in the center of mass frame are

$$\mathbf{v}_1^{(com)} \equiv \mathbf{v}_1 - \mathbf{V}_{com} = \frac{m_r}{m_1}\mathbf{u} \quad \mathbf{v}_2^{(com)} \equiv \mathbf{v}_2 - \mathbf{V}_{com} = -\frac{m_r}{m_2}\mathbf{u} \tag{3.2}$$

where the relative velocity $\mathbf{u} \equiv \mathbf{v}_1 - \mathbf{v}_2$ and the reduced mass m_r have been introduced

$$m_r = \frac{m_1 m_2}{m_1 + m_2} \tag{3.3}$$

Note that the relative velocity is the same in all the systems. From Eq. (3.2) it can be observed that the momenta of the two particles in the com system is equal and opposite

$$\mathbf{p}_1^{(com)} \equiv m_1 \mathbf{v}_1^{(com)} = -\mathbf{p}_2^{(com)} \equiv m_2 \mathbf{v}_2^{(com)} \tag{3.4}$$

Thus, in the com frame particle moves in opposite directions and the total momentum is zero ($\mathbf{p}_1^{(com)} + \mathbf{p}_2^{(com)} = 0$).

The kinetic energy in the com system can be obtained from Eq. (3.2)

$$E_{1,2}^{(com)} \equiv \frac{m_{1,2}(v_{1,2}^{(com)})^2}{2} = \frac{m_r}{m_{1,2}} \frac{m_r u^2}{2} \tag{3.5}$$

Therefore the total energy in the com system is

$$E_{tot}^{(com)} \equiv E_1^{(com)} + E_2^{(com)} = \frac{m_r u^2}{2} \tag{3.6}$$

We can express the particle velocity in the laboratory frame in term of $\mathbf{V}_{com}$ and $\mathbf{u}$

$$\mathbf{v}_1 = \mathbf{V}_{com} + \frac{m_r}{m_1}\mathbf{u} \tag{3.7}$$

$$\mathbf{v}_2 = \mathbf{V}_{com} - \frac{m_r}{m_2}\mathbf{u} \tag{3.8}$$

Using Eqs. (3.7) and (3.8) it can be easily shown that the total energy in the com system is simply the difference between the total energy evaluated in the laboratory frame and the kinetic energy of the centre of mass

$$E_{tot}^{(com)} = E_1 + E_2 - \frac{MV_{com}^2}{2} \tag{3.9}$$

In this book we will consider forces that can be derived from a central potential $V(|\mathbf{r}_1 - \mathbf{r}_2|)$, i.e. forces that depend only on the relative distance between the particles (such as e.g. the Coulomb force). Upon defining the relative position $\mathbf{r} = \mathbf{r}_1 - \mathbf{r}_2$, and taking into account the equations of motion of the two particles

$$\frac{d\mathbf{r}_1}{dt} = \mathbf{v}_1 \tag{3.10}$$

$$\frac{d\mathbf{v}_1}{dt} = -\frac{1}{m_1} V\prime(r) \frac{\mathbf{r}_1 - \mathbf{r}_2}{|\mathbf{r}_1 - \mathbf{r}_2|} \tag{3.11}$$

$$\frac{d\mathbf{r}_2}{dt} = \mathbf{v}_2 \tag{3.12}$$

$$\frac{d\mathbf{v}_2}{dt} = -\frac{1}{m_2} V\prime(r) \frac{\mathbf{r}_2 - \mathbf{r}_1}{|\mathbf{r}_1 - \mathbf{r}_2|} \tag{3.13}$$

a closed system of equations for $\mathbf{r}$ and $\mathbf{u}$ can be obtained by subtracting Eq. (3.12) from Eqs. (3.10) and (3.13) from Eq. (3.11),

$$\frac{d\mathbf{r}}{dt} = \mathbf{u} \tag{3.14}$$

$$m_r \frac{d\mathbf{u}}{dt} = -\nabla V(r) \tag{3.15}$$

and the equations of motion for the two particles then reduce to the equation of motion of a single particle of mass m_r in a potential $V(r)$. Since the com velocity is a constant of motion, Eqs. (3.14) and (3.15) are sufficient to determine the evolution of the dynamical variables.

3.2 Definition of Collision. Impact Parameter

We will often use the term *collision*. The prototype of collision is the interaction between two hard spheres (e.g. two billiard balls of radius R_b). The two spheres moves independently of each other until the distance between their centres becomes equal to $2R_b$. When they touch they feel a sudden repulsive force that changes abruptly their velocity, thereafter again moving independently of each other.

Making reference to Fig. (3.1) the *impact parameter* b is defined as the distance between the particle unperturbed trajectory in the com system and the origin of the central potential that we will place at $\mathbf{r} = 0$.

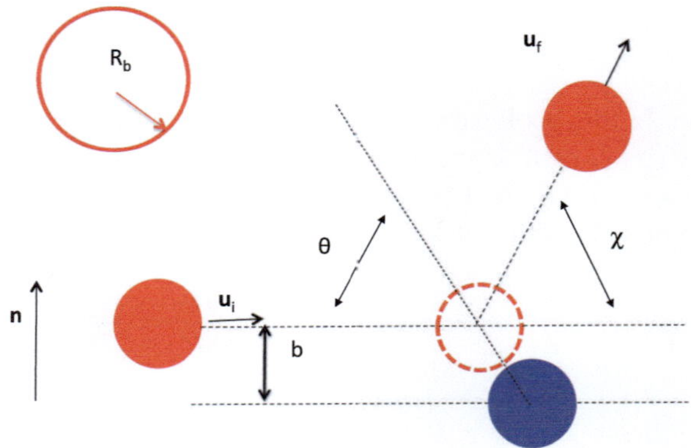

Fig. 3.1 Collision geometry in the case of hard spheres of radius R_b. The horizontal axis is taken along the initial direction of the relative velocity $\mathbf{u}_i$. The deflection angle depends on the impact parameter b

Head-on collisions correspond to $b = 0$ whereas if $b \geq 2R_b$ no interaction takes place and the incoming particle passes un-deflected. The deflection angle χ can be related to the impact parameter through

$$\chi = \pi - 2\theta \quad \sin\theta = \frac{b}{2R_b} \tag{3.16}$$

During the collision there is a change in particle momentum Δp due to the effect of the force F acting for a time Δt. For hard spheres the force is infinite and the duration of the collision is zero. In most of the cases however particles do not experience a sudden force but rather interact through a long-range force (such as the Coulomb force). To estimate the duration of a collisione we can assume that the maximum change in momentum occurs at the minimum distance $\approx b$. The time spent by the particle at this distance is approximately $\Delta t \approx b/v$. Provided the collision time as defined in this way is much shorter than the time interval between two collisions, the simple picture of collisions as the interaction between two hard spheres can be applied.

3.3 The Cross Section Empirical Definition

Let us consider a beam of particles A impinging on a target of width Δx made of particles B with density n_{target} (Fig. 3.2)

The beam flux Φ (particles per unit area and unit time arriving on the target) can be related to the beam density n_{beam} and the beam velocity v_{beam} by considering the number dN of particles that arrive on the surface dS during the time dt. The number dN is given by the beam particle density times the volume of the cylinder described by the beam footprint dS moving at velocity v_{beam} in a time dt: $dN = n_{beam}(v_{beam}dt)dS$. Thus, upon dividing by dS and dt the beam flux turns out to be $\Phi \equiv n_{beam}v_{beam}$.

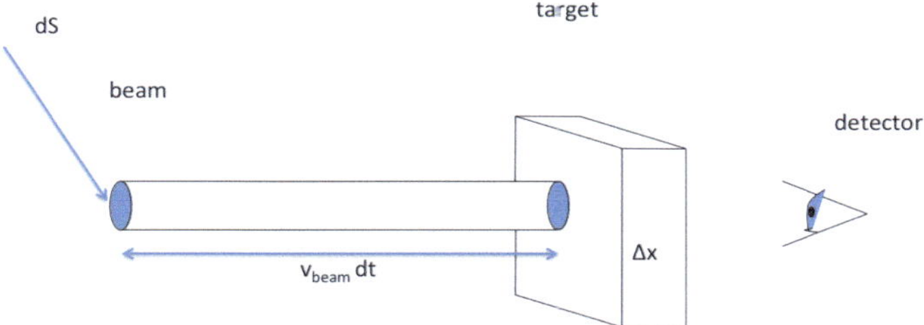

Fig. 3.2 A beam impinges on a target of width Δx. A detector is placed along the direction of the beam on the other side of the target. The interaction of the beam with the target reduces the particles along the original beam direction (through scattering or absorption). The relative beam attenuation will be proportional to the density of target particles, the target widths and a quantity with the dimension of a square length that defines the cross section

A detector is placed on the other side of the target along the direction of the incoming beam. The attenuation of the beam in passing through the target will be proportional to the density and the width of the target

$$\frac{\Delta\Phi}{\Phi} = -n_{target}\sigma\Delta x \tag{3.17}$$

The proportionality constant σ is the *microscopic cross section*[1] and has the dimension of an area. If we consider the target made of infinitesimal layers of width dx, we can take the continuum limit to obtain

$$\frac{d\Phi}{dx} = -n_{target}\sigma\Phi \tag{3.18}$$

which yields an exponential decay of the flux inside the target with a characteristic attenuation length given by $(n_{target}\sigma)^{-1}$.

The microscopic cross section provides a measure of how likely is a certain process. The larger is σ, the larger will be the attenuation of the beam. A convenient unit for the cross sections we will find in this book is the *barn* with $1b = 10^{-28}m^2$.

To understand better the physical meaning of σ it is useful to consider the microscopic description of the process.

[1] Nuclear physics texts make use of the so called *macroscopic cross section* Σ defined as the product $(n_{target}\sigma)$. Note that Σ is the inverse of the attenuation length and has not the dimension of an area!.

3.4 The Cross Section Microscopic Definition

Now let us consider the target as made of spheres of radius R_b (beam particles will be assumed to be point like). The area of each target particle seen by particle A is $\pi R_b^2 \equiv \sigma_o$. The target width dx is assumed to be infinitesimal (Fig. 3.3).

A beam particle that collides with a target particle is deflected and is therefore lost by the beam. Thus, to evaluate the attenuation of the beam we have to evaluate the fraction of the infinitesimal beam footprint area dS that is occupied by the target particles. This fraction is given by

$$(n_{target} dSdx)\frac{\sigma_o}{dS} \tag{3.19}$$

The first term is the number of target particles in a volume of area dS and width dx whereas the second term is the fractional area of each target particle. In a time dt the number of beam particles arriving on the area dS of the target is $\Phi dSdt$. Thus the change in the number of beam particles due to their interaction with the target is given by

$$dN_{beam} = -\Phi dSdt(n_{target} dSdx)\frac{\sigma_o}{dS} \tag{3.20}$$

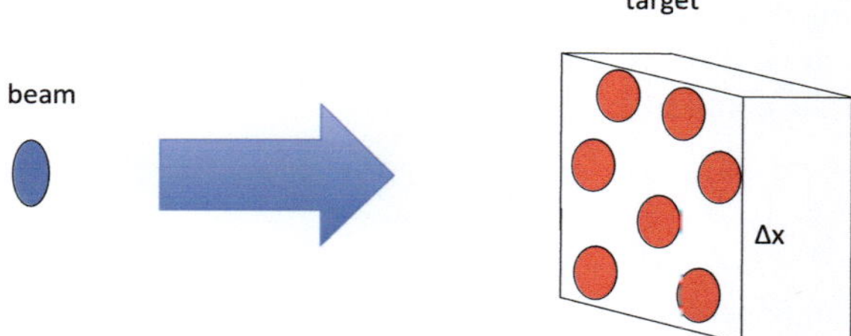

Fig. 3.3 Cross section—microscopic definition. A beam of particles A impinges on a target made of particles B. Each particle B is seen by the beam as a disk of finite extension σ_o. The number of collisions of a particle A with the target particles is given by the number of target particles in a cylindrical element of area dS and height $v_{beam} dt$ times the fractional area of each particle. The relative change in the beam flux has exactly the same expression obtained from the empirical definition provided we identify σ_o with the cross section

Fig. 3.4 Frequency of events. The number of collisions per unit time can be evaluated as made for the microscopic definition of cross section

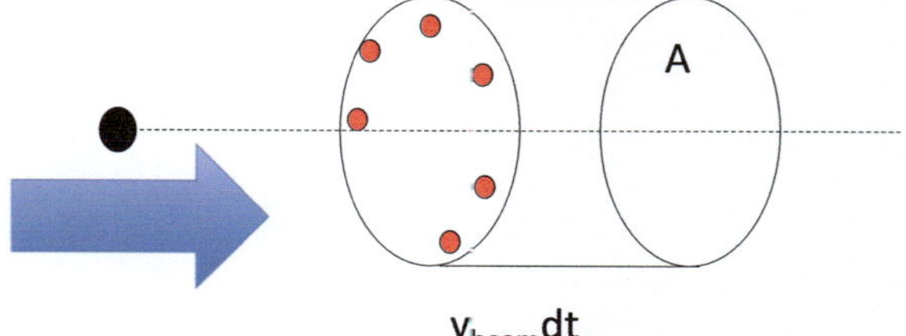

by definition this change is also $dSdtd\Phi$. Thus, we finally obtain

$$d\Phi = -\Phi n_{target}\sigma_o dx \tag{3.21}$$

This is the same equation for the attenuation that defined the cross section σ Eq. (3.18). Thus, σ can be identified with the effective area σ_o seen by particle A interacting with particle B. Note that the same value for σ would be obtained by interchanging the role of particle A and particle B.

3.5 Frequency of Events

Let us denote with v_A the velocity of the beam particles. For the sake of simplicity we will assume that the beam is monoenergetic—all the particles have the same velocity (Fig. 3.4).

A single particle of the beam has the probability given by Eq. (3.19) to interact with the target within a width dx. Since in a time dt the beam particles travel a distance $dx = v_A dt$, the frequency of events, i.e. the number of collisions per unit time between a particle A and the target can be evaluated as follows

$$\nu_{AB} = n_{target}\sigma v_A \tag{3.22}$$

Equation (3.22) is the number of collisions per unit time between a particle A and the target made of particles B. In Eq. (3.22) the velocity v_A is the velocity of the beam. Since the target is at rest this is also the relative velocity between particle A and particles B. In the rest of this book we will consider systems in which species A and species B are both moving. For this general case, it is indeed the relative velocity that matters.

The notion of the frequency of events can be generalized to the case in which the two species are both moving e.g. under the effect of their thermal motion. In this case the beam velocity v_A in Eq. (3.22) is replaced by the relative velocity $u \equiv |\mathbf{v}_A - \mathbf{v}_B|$. Furthermore, we anticipate that the cross section is in general a function of the total kinetic energy in the com frame $E_{com} = m_r u^2/2$. If the species are not mono energetic but rather both follow a Maxwell Boltzmann distribution, the frequency of events must be averaged over the distribution function of both species, yielding

$$\begin{aligned} < \nu_{AB} >= n_B \int\int \frac{d\mathbf{v}_A}{(\pi^{1/2}v_{tA})^3} e^{-\frac{v_A^2}{v_{tA}^2}} \frac{d\mathbf{v}_B}{(\pi^{1/2}v_{tB})^3} e^{-\frac{v_B^2}{v_{tB}^2}} \sigma(m_r u^2/2)u = \\ = n_B < \sigma u > \end{aligned} \tag{3.23}$$

The quantity $< \sigma u >$ is called *Maxwellian reactivity*. A quantity that may be of interest is the number of events for unit volume and unit time. This information is obtained by multiplying the density of species A by the number of AB reactions per unit time Eq. (3.23). For example the amount of power generated per unit volume and unit time by the fusion reactions between deuterium and tritium is given by

$$P_{fus} = n_D n_T < \sigma_{DT} u > 17.6\, MeV \tag{3.24}$$

where 17.6 MeV is the energy released in a single DT fusion reaction. If the temperature of the two reacting species is the same ($T_A = T_B = T$) the double integral in the expression of the reactivity reduces to a single integral via the change of variable $(\mathbf{v}_A, \mathbf{v}_B) \rightarrow (\mathbf{V}_{com}, \mathbf{u})$

$$< \sigma u > = \int \frac{d\mathbf{V}_{com} e^{-\frac{V_{com}^2}{v_{tM}^2}}}{(\pi^{1/2} v_{tM})^3} \int \frac{d\mathbf{u} e^{-\frac{u^2}{v_{tr}^2}}}{(\pi^{1/2} v_{tr})^3} \sigma(m_r u^2/2) u =$$
$$= \int \frac{d\mathbf{u} e^{-\frac{u^2}{v_{tr}^2}}}{(\pi^{1/2} v_{tr})^3} \sigma(m_r u^2/2) u \tag{3.25}$$

where $v_{tM} = (2T/M)^{1/2}$ and $v_{tr} = (2T/m_r)^{1/2}$. Figure 3.5 shows the DT fusion cross section as a function of the com energy and the associated Maxwellian reactivity as a function of temperature.

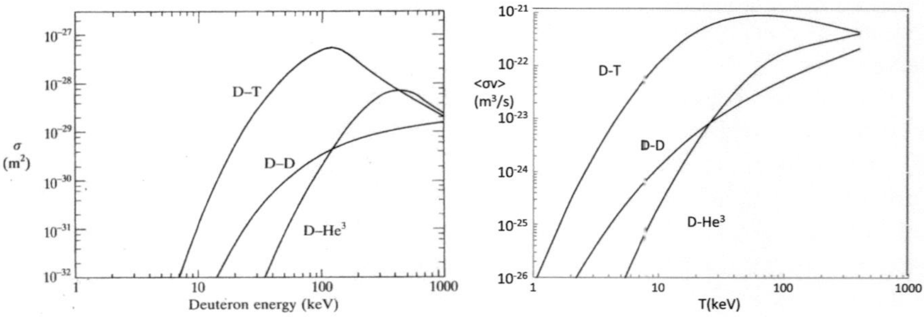

Fig. 3.5 Fusion cross section and Maxwellian reactivity. The figure on the left shows the cross section of three fusion reactions as a function of the energy in the centre of mass system. The figure on the right is the reactivity obtained by averaging the product σv over the Maxwellian distribution function of the reactants, with v the relative velocity

3.6 Mean Free Path

The mean free path is the average distance between two consecutive collisions. Let us assume that between collisions no macroscopic forces act on the particles and the trajectories are straight lines. The mean free path is given by the ratio between the average relative velocity and the collision frequency

$$\lambda_A = \frac{v_{tA}}{\nu_{AB}} \approx \frac{1}{n_B \sigma} \tag{3.26}$$

3.7 Impact Parameter and Deflection Angle

So far we have given an empirical definition of the cross section and we have seen that it can be justified on the basis of microscopic model of the interacting particles. Is it possible to determine the cross section from first principles? This is in principle possible if the interaction force is known as for example in the case of Coulomb interaction that will be discussed in detail in Chap. 10.

To illustrate how the cross section can be determined from the interaction force in the simplest possible case we consider the interaction of point particles with rigid spheres of radius R. The relation between the deflection angle and the impact parameter is given by Eq. (3.16) and is shown in Fig. 3.6.

In the case of hard sphere collisions the cross section, according to the microscopic model of Sect. 3.3 is simply πR_b^2. Since R_b corresponds to the value of the impact parameter above which no deflection is observed, we may be tempted to define the cross section for a generic interaction as πb_M^2 with b_M being the impact parameter above which there is no significant deflection. Looking at Fig. 3.6 it is clear that in the case of Coulomb interaction between charged particles there is no sharp transition in the deflection angle as in the case of hard spheres. The deflection angle decreases monotonically as a function of b/R with $R = q_1 q_2/(4\pi\epsilon_o E)$ (with E being the total energy in the centre of mass frame—see below). Such monotonic decrease is a reflection of the long range behavior of the Coulomb interaction and requires a specific treatment (see Chap. 10). For the time being we may assume that an *effective* cross section can indeed be defined using for b_M the value at which the deflection angle becomes a small fraction of the maximum deflection angle (e.g. taking b_M as the value corresponding to $\chi = \pi/3$).

3.8 Conservation Laws in Collisions Between Particles

In collisions we have to determine the six unknown components $(\mathbf{v}_{1f}, \mathbf{v}_{2f})$ of the particle velocities after the collision knowing the values $(\mathbf{v}_{1i}, \mathbf{v}_{2i})$ before the collision.

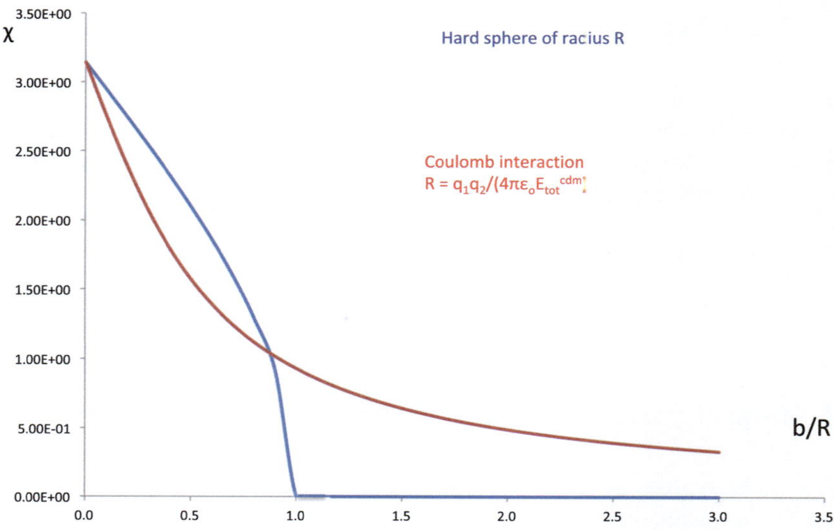

Fig. 3.6 Deflection angle versus impact parameter. The deflection angle is shown as a function of the impact parameter for two representative cases, the collision of a point particle with a hard sphere of radius R (blue) and the collision between two charged particles of charge q_1 and q_2 (red)

At the beginning of this chapter we have introduced the com frame and shown that the equations of motion can be reduced to the solution of a dynamical problem for a particle of mass equal to the reduced mass moving in a central potential, with the reduced equations written in terms of the relative position and the relative velocity. It is possible to deduce further constraints arising from the invariance of the dynamical equations that reduce the number of unknowns of the problem.

The first simplification comes from the conservation of the total momentum. Since $\mathbf{V}_{com} = const.$, it can be evaluated from $(\mathbf{v}_{1i}, \mathbf{v}_{2i})$ and the final velocities will depend only on the change of the relative velocity $\mathbf{u}$.

In the case of a central potential also the total angular momentum $\mathbf{L} = \mathbf{r} \times \mathbf{p} = m_r \mathbf{r} \times \mathbf{u}$ is conserved

$$\frac{d\mathbf{L}}{dt} = m_r \frac{d\mathbf{r}}{dt} \times \mathbf{u} + m_r \mathbf{r} \times \frac{d\mathbf{u}}{dt} = m_r \mathbf{u} \times \mathbf{u} + m_r \mathbf{r} \times \left(\frac{-\nabla V(r)}{m_r} \right) = 0 \tag{3.27}$$

Therefore the relative velocity remains, throughout the collision process, on the plane normal to $\mathbf{L}$. Since $\mathbf{L}$ is constant, it can be determined by the value of $\mathbf{r}$ and $\mathbf{u}$ at $t \to -\infty$. Thus, the plane is uniquely determined by the line corresponding to the unperturbed trajectory ($d\mathbf{r}^{unperturbed}/dt = \mathbf{u}_i$) and the point corresponding to the scattering center ($r = 0$). On this plane the x axis will be taken along $\mathbf{u}_i$. The unit vector on this plane in the direction orthogonal to x will be denoted with $\mathbf{n}$ (see Fig. 3.1).

Finally, since $V(r)$ must vanish when $r \to \infty$, from the conservation of energy

$$\frac{m_r u^2}{2} + V(r) = cost \tag{3.28}$$

it follows that $|u|$ has the same value before and after the collision ($u_f = u_i$) and the relative velocity after the collision can be written as

$$\mathbf{u}_f = \cos\chi \mathbf{u}_i + u_i \sin\chi \mathbf{n} \tag{3.29}$$

Thus, the three unknown components of the relative velocity $\mathbf{u}_f$ after the collision are reduced to a single quantity, the deflection angle in the com frame χ between the relative velocity after the collision and the relative velocity before the collision. The angle χ depends on the details of the interaction potential (see Chap. 10 for the case of the Coulomb interaction).

The velocities in the laboratory frame are determined from Eqs. (3.29), (3.7) and (3.8)

$$\mathbf{v}_{1f} = \mathbf{V}_{com} + \frac{m_r}{m_1}(\cos\chi \mathbf{u}_i + u_i \sin\chi \mathbf{n}) \tag{3.30}$$

$$\mathbf{v}_{2f} = \mathbf{V}_{com} - \frac{m_r}{m_2}(\cos\chi \mathbf{u}_i + u_i \sin\chi \mathbf{n}) \tag{3.31}$$

From Eq. (3.30) it is possible to determine the energy and momentum variation in a collision as a function of the deflection angle

$$\Delta E_1 \equiv E_{1f} - E_{1i} = -m_r \mathbf{u}_i \cdot \mathbf{V}_{com}(1 - \cos\chi) + m_r u_i \sin\chi \mathbf{n} \cdot \mathbf{V}_{com} \tag{3.32}$$

$$\Delta \mathbf{p}_1 \equiv (\mathbf{p}_{1f} - \mathbf{p}_{1i}) = -m_r \mathbf{u}_i (1 - \cos\chi) + m_r u_i \sin\chi \mathbf{n} \tag{3.33}$$

3.9 Applications: Slowing Down of Neutrons and Two Body Decays

Let us consider the scattering of a particle by a target at rest. For $\mathbf{v}_{2i} = 0$ we have $\mathbf{u}_i = \mathbf{v}_{1i}$ and

$$\mathbf{V}_{com} = \frac{m_r}{m_2}\mathbf{v}_{1i} \tag{3.34}$$

$$\mathbf{u}_f = \cos\chi \mathbf{v}_{1i} + v_{1i} \sin\chi \mathbf{n} \tag{3.35}$$

$$\mathbf{v}_{1f} = \mathbf{v}_{1i}\left(\frac{m_r}{m_2} + \frac{m_r}{m_1}\cos\chi\right) + \frac{m_r}{m_1} v_{1i} \sin\chi \mathbf{n} \tag{3.36}$$

Upon squaring both members of Eq. (3.36), a relation between the final and initial energy of particle 1 in the laboratory frame is obtained

$$\frac{E_{1f}}{E_{1i}} = \frac{|v_{1f}|^2}{|v_{1i}|^2} = \frac{m_1^2 + m_2^2 + 2m_1 m_2 \cos\chi}{(m_1 + m_2)^2} \tag{3.37}$$

Taking the two limiting case of back scattering ($\chi = 180\,°$) and particle undeflected ($\chi = 0$) the following admissible range for the ratio between final and initial energy is obtained

$$\alpha \equiv \frac{(m_1 - m_2)^2}{(m_1 + m_2)^2} \leq \frac{E_{1f}}{E_{1i}} \leq 1 \tag{3.38}$$

Thus, the maximum energy transfer occurs for equal masses of the two particles (and scattering at 180 °).

These findings can be applied to the problem of slowing down of neutrons in a thermal nuclear reactor. This process is called moderation. Neutrons are generated with energies of around 2 MeV and must be slowed down because the cross section for fission of ^{235}U increases with decreasing energy (see Chap. 8). The slowing down of neutrons takes place mostly through elastic scattering with the moderator.

We focus here on the elastic scattering of neutrons in order to determine how many collisions are needed to slow the neutron from 2 MeV down to thermal energy typical of a reactor core (operating say at 300 °C). Knowing the cross section for elastic scattering of neutrons the collision frequency and the characteristic time needed to a neutron to slow down can be evaluated. For the time being we will consider the time between two elastic collisions as the characteristic time step of this process and determine how many steps are involved.

What we know from Eq. (3.38) is that if after the i_{th} step the neutron energy is E_i, following the $i_{th} + 1$ collision the neutron energy could be any value between E_i and αE_i. Thus, we need some information about the probability for the neutron to end at some energy within this range. The experimental evidence shows that the neutron scattering in the com frame is *isotropic*, i.e. that all the angles between $\chi = 0$ and $\chi = \pi$ are equally probable. From Eq. (3.38) it follows that this is equivalent to say that the probability dW of having the energy at step $i + 1$ in the range between E_{i+1} and $E_{i+1} + dE_{i+1}$ is independent of E_{i+1} provided $\alpha \leq E_{i+1}/E_i \leq 1$, or

$$dW = k dE_{i+1} \tag{3.39}$$

for $\alpha \leq E_{i+1}/E_i \leq 1$ and

$$dW = 0 \tag{3.40}$$

otherwise. The constant k can be determined from the condition that the total probability to find the neutron at energies such that $\alpha \leq E_{i+1}/E_i \leq 1$ is 100%

$$1 = \int dW = \int_{\alpha E_i}^{E_i} k dE_{i+1} = k(1-\alpha)E_i \tag{3.41}$$

yielding $k = (A+1)^2/(4AE_i)$ and A being the mass number of the moderator.

The average energy variation in a single collision can now be evaluated. It is convenient to use the *lethargy* $u_L \equiv \ln(E_i/E)$, rather than the energy itself, because the average lethargy variation ξ after a collision is independent of the initial energy

$$\xi \equiv \int_{\alpha E_i}^{E_i} \ln\frac{E_i}{E}\frac{dW}{dE}dE = \frac{(A+1)^2}{4A}\int_{\alpha E_i}^{E_i} \ln\frac{E_i}{E}\frac{dE}{E_i} = 1 + \frac{(A-1)^2}{2A}\ln(\frac{A-1}{A+1}) \tag{3.42}$$

The average lethargy variation is $\xi = 1$ for hydrogen ($A = 1$) and $\xi = 0.15777$ for carbon ($A = 12$). Each collision decreases the average lethargy by ξ, thus, after n collisions, we have

$$< \ln(E_n) >= \ln(E_o) - n\xi \tag{3.43}$$

This expression can be used to evaluate the number of collisions needed to slow down a neutron (see exercises at the end of this chapter).

A second application of the formalism developed in the previous paragraph is in the two-body decay. If a nucleus disintegrates in two fragments we can estimate the ratio between the kinetic energy of the two fragments from the conservation of momentum. In the system in which the nucleus is at rest ($\mathbf{V}_{com} = 0$) after the decay the total momentum is conserved

$$m_1\mathbf{v}_{1f} + m_2\mathbf{v}_{2f} = 0 \tag{3.44}$$

Therefore the ratio between the kinetic energy of the two fragments is given by

$$\frac{E_{1f}}{E_{2f}} = \frac{m_1 v_{1f}^2}{m_2 v_{2f}^2} = \frac{m_2}{m_1} \tag{3.45}$$

Therefore, the largest fraction of the decay energy is taken by the lightest fragment.

Note that the argument above applies only in the case of a two-body decay. In the case of a three-body decay the energy of the fragments is not defined and it spans a continuous spectrum of values.

3.10 Exercises

Problem 3.1 A gas of ionized nuclei at $T = 20$ keV is made of 50% deuterium and 50% tritium. The gas density is $10^{20} m^{-3}$. Determine the number of fusion events per seconds.

Problem 3.2 The particles of species A can be absorbed by particles of species B with a cross section $\sigma = 100b$. A beam of particles A impinges on a target of particles B. The beam has a density $n_A = 10^{20}m^{-3}$ and intensity $10^{10} particlecm^{-2}s^{-1}$. The target has a density $10^{23}m^{-3}$. Evaluate the beam attenuation after crossing a width of 10 cm, the attenuation length λ and the velocity v_A of the beam particles.

Problem 3.3 Species A and species B interact through elastic collisions with cross section $\sigma_e = 10b$ and reacts (forming species C) with cross section $\sigma_r = 0.1b$. Evaluate how many elastic collision are needed to get a reaction $a + b \to c$.

Problem 3.4 A nucleus of ^{238}U decays in a ^{234}Th nucleus emitting an α particle. The energy released in the decay is 4.3 MeV. Determine the energy of the α particle.

Problem 3.5 A ^{5}He nucleus decays in a neutron and an α particle releasing 17.5 MeV. Evaluate the energy of the two fragments.

Problem 3.6 How many collisions are needed to slow down a 1 MeV neutron down to thermal energy ($T = 300\,°C$) using hydrogen ($A = 1$) or carbon ($A = 12$) as moderator?

Chapter 4
Electromagnetism and Special Relativity

Abstract *This chapter presents a short overview of the main concept of electromagnetism that will be used later in the book. Specific attention is given to the problem of radiation emission by an accelerated charge. The concept of dispersion relation is introduced and illustrated for the case of radio waves reflection by the ionosphere. The last part of the chapter presents an introduction to special relativity.*

The nature of light has fascinated philosophers since antiquity. Light has been represented either as made of particles or as made of waves. Isaac Newton [1] proposed the corpuscular theory whereas Robert Hook [2], Rene Descartes [3] and Christian Huygens [4] favoured the wave theory. This debate seemed to be settled in the 19th century when Young [5] and Fresnel [6] performed a series of experiments that demonstrated the wavelike character of light and Maxwell [7] proposed a comprehensive theoretical framework for electromagnetism that predicted light to be a wave phenomenon. Nevertheless, in the early 20th century Albert Einstein [8] brought back the Newton idea of wave as a corpuscular phenomenon.

As we will see in the next few chapters, the same phenomenon can simultaneously present corpuscular and wavelike features. Even matter can behave as a wave! For the moment we limit our discussion to understand how to discriminate if a phenomenon has a corpuscular or a wavelike nature.

Let us consider two persons separated by a wall that prevents the two to be in direct eye contact (Fig. 4.1).

If A tries to interact with B sending a series of bullets (i.e. tries to interact via corpuscles), the bullets will travel in straight trajectories above the tip of the wall and no interaction will take place. However, if A sends a sound signal, this will go around the wall and reach the ears of B. The sound path bends around the obstacle because the sound is a typical wavelike phenomenon. Each point of space reached by the sound wave becomes the source of a new sound wave that propagates in all the directions (Huygens-Fresnel principle, see Fig. 4.2). The superposition of all the waves generated in this way produces the observed pattern. This phenomenon was called by Grimaldi [9] *diffraction* and it is typical of waves. Thus, to determine whether a phenomenon has a corpuscular or wavelike character the test to be performed is

F. Romanelli, *Physics of Nuclear Energy*, Springer Series in Plasma Science and Technology, https://doi.org/10.1007/978-981-97-9609-0_4

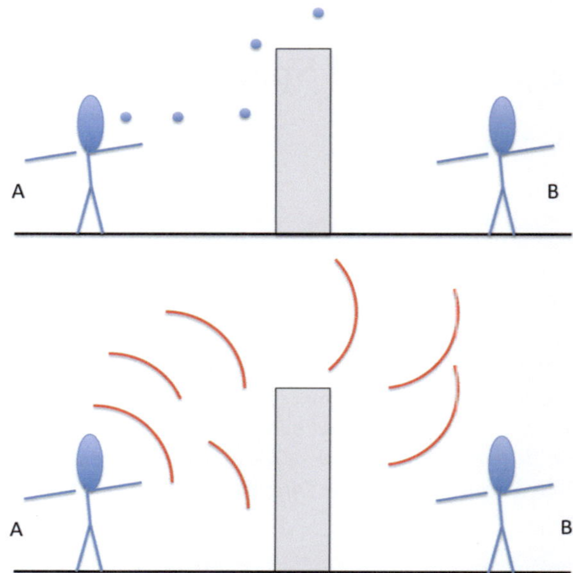

Fig. 4.1 Difference between wave and particle behavior. Two individuals A and B interact either via bullets (particles) or via sound waves. The particles sent by A along the line of sight AB are intercepted by the wall and do not reach B. The sound wave bends around the obstacle

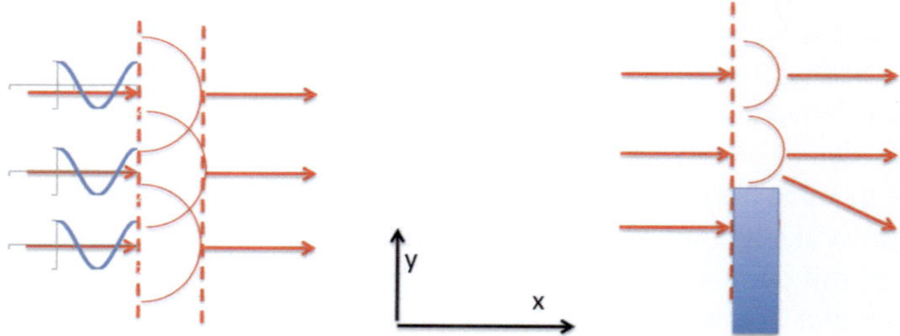

Fig. 4.2 Illustration of the Huygens-Fresnel principle. A plane wave propagates along x. The wavefront (the location of the points with the same phase) is a plane (dashed line). Each point of the wavefront becomes the source of a wave. The interference of the waves produces a new plane wave because the velocity component in the y direction of two neighboring waves are equal and opposite. If an obstacle is interposed the waves generated at the border of the obstacle have a velocity component along y that is not balanced by a neighboring wave. The new wavefront bends around the obstacle

how it propagates in the presence of an obstacle (a wall or a screen with slits and holes). This is what we will do with light later in this chapter. In order to interpret the diffraction phenomena it is convenient to recall first the main concepts about the propagation of electromagnetic waves.

In the following use will be made of Gauss and Stokes theorems.

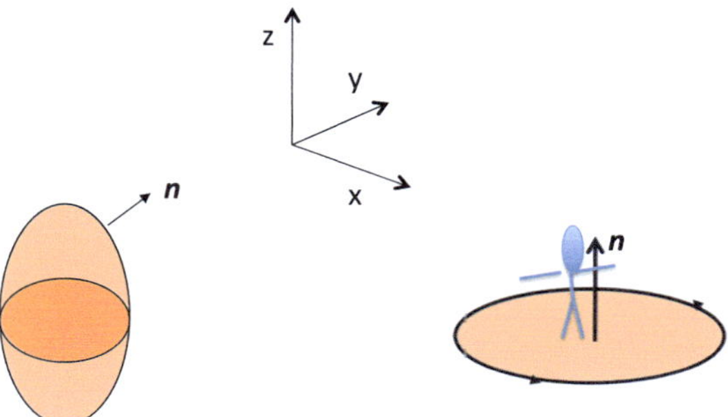

Fig. 4.3 Illustration of Gauss and Stokes theorems. Note that in the case of Gauss theorem the outgoing normal must be considered and that in the case of Stokes theorem the normal to the surface has the direction of the body if the closed contour is taken anti clock wise

Gauss theorem states that for a generic vector field $\mathbf{F}(\mathbf{r})$ the following identity holds

$$\int_V dV \nabla \cdot \mathbf{F} = -\int_S dS \mathbf{F} \cdot \mathbf{n} \tag{4.1}$$

where S is the surface delimiting the volume V and $\mathbf{n}$ is the *outgoing* normal unit vector on the surface S.

Stokes theorem states that for a generic vector field $\mathbf{G}(\mathbf{r})$ the following identity holds

$$\int_S dS \mathbf{G} \cdot \mathbf{n} = \oint_C \mathbf{dl} \cdot \mathbf{G} \tag{4.2}$$

where C is the closed contour delimiting the surface S and $\mathbf{n}$ is the normal unit vector on the surface S. The direction of $\mathbf{n}$ has to be taken in such a way that having your feet on the surface you follow the contour in the counter clock wise direction (Fig. 4.3)

4.1 Maxwell's Equations and Their General Solution

The electromagnetic phenomena are described by Maxwell's equations for the electric field $\mathbf{E}$ and the magnetic induction field $\mathbf{B}$ (related to the magnetic field $\mathbf{H}$ through the relation $\mathbf{B} = \mu_o \mathbf{H}$) [10]

$$\nabla \times \mathbf{E} = -\frac{\partial \mathbf{B}}{\partial t} \tag{4.3}$$

$$\nabla \times \mathbf{B} = \mu_o \mathbf{J} + \frac{1}{c^2} \frac{\partial \mathbf{E}}{\partial t} \tag{4.4}$$

$$\nabla \cdot \mathbf{E} = \frac{\rho}{\epsilon_o} \tag{4.5}$$

$$\nabla \cdot \mathbf{B} = 0 \tag{4.6}$$

where $\epsilon_o \equiv 8.8542 \times 10^{-12} F/m$, $\mu_o \equiv 4\pi \times 10^{-7}$H/m and $\epsilon_o \mu_o = 1/c^2$ with $c \equiv 2.99792 \times 10^8$m/s the velocity of light.

It is convenient to express Maxwell's equation in terms of the vector potential **A** and the scalar potential Φ. Equation (4.6) suggests to express **B** as

$$\mathbf{B} \equiv \nabla \times \mathbf{A} \tag{4.7}$$

Equation (4.7) identically satisfies Eq. (4.6). Upon substituting this definition in Faraday's law Eq. (4.3) we obtain

$$\nabla \times (\mathbf{E} + \frac{\partial \mathbf{A}}{\partial t}) = 0 \tag{4.8}$$

which can be satisfied identically provided

$$\mathbf{E} = -\nabla\Phi - \frac{\partial \mathbf{A}}{\partial t} \tag{4.9}$$

The equations for Φ and **A** are then obtained from Eqs. (4.4) and (4.5)

$$-\Delta\mathbf{A} + \frac{1}{c^2}\frac{\partial^2 \mathbf{A}}{\partial t^2} = \mu_o \mathbf{J} - \nabla\left[\nabla \cdot \mathbf{A} + \frac{1}{c^2}\frac{\partial \Phi}{\partial t}\right] \tag{4.10}$$

$$-\Delta\Phi = \frac{\rho}{\epsilon_o} + \frac{\partial \nabla \cdot \mathbf{A}}{\partial t} \tag{4.11}$$

where the identity $\nabla \times \nabla \times \mathbf{A} = \nabla(\nabla \cdot \mathbf{A}) - \Delta\mathbf{A}$ has been used. The definition of the potentials is not uniques since, upon taking $\mathbf{A}\prime = \mathbf{A} + \nabla\Psi$ and $\Phi\prime = \Phi - \partial\Psi/\partial t$, the same expression for **E** and **B** and the same equations for $\mathbf{A}\prime$ and $\Phi\prime$ are recovered. The choice of Ψ is referred to as the choice of the *gauge*. Thus, Maxwell's equations are invariant for the choice of the gauge. It is possible to take advantage of this invariance by choosing a proper gauge in order to simplify Eqs. (4.10) and (4.11). In the following the Lorentz gauge will be adopted

$$\nabla \cdot \mathbf{A} + \frac{1}{c^2}\frac{\partial \Phi}{\partial t} = 0 \tag{4.12}$$

that corresponds to the choice of a function Ψ satisfying the equation

$$-\Delta\Psi + \frac{1}{c^2}\frac{\partial^2 \Psi}{\partial t^2} = \nabla \cdot \mathbf{A}\prime + \frac{1}{c^2}\frac{\partial \Phi\prime}{\partial t} \tag{4.13}$$

The choice of the Lorentz gauge allows writing the equations for **A** and Φ in a symmetrical form

$$-\Delta\mathbf{A}+\frac{1}{c^2}\frac{\partial^2\mathbf{A}}{\partial t^2}=\mu_o\mathbf{J} \tag{4.14}$$

$$-\Delta\Phi+\frac{1}{c^2}\frac{\partial^2\Phi}{\partial t^2}=\frac{\rho}{\epsilon_o} \tag{4.15}$$

It is possible to show [10] that the general solution of Maxwell's equation can be written in terms of the sources as follows

$$\mathbf{A}(\mathbf{r},t)=\frac{\mu_o}{4\pi}\int d^3\mathbf{r}\prime\frac{\mathbf{J}(\mathbf{r}\prime,t\prime_r)}{|\mathbf{r}-\mathbf{r}\prime|} \tag{4.16}$$

$$\Phi(\mathbf{r},t)=\frac{1}{4\pi\epsilon_o}\int d^3\mathbf{r}\prime\frac{\rho(\mathbf{r}\prime,t\prime_r)}{|\mathbf{r}-\mathbf{r}\prime|} \tag{4.17}$$

with $t\prime_r=t-|\mathbf{r}-\mathbf{r}\prime|/c$ being the *retarded time*. The integrals extend over all the space but only the regions where the sources **J** and ρ are different from zero provide a finite contribution. Equations (4.16) and (4.17) relate the vector and scalar potential at point **r** and time t with the source elements evaluated at point $\mathbf{r}\prime$ and at a retarded time $t\prime_r$. The presence of a delay between the status of the source at time $t\prime_r$ and position $\mathbf{r}\prime$ and its effect on the electromagntic fields at time t and point **r** is a consequence of the finite velocity of propagation of electromagnetic signals.

Equations (4.3)–(4.6) have an integral invariant obtained by scalar multiplying Eq. (4.3) by **B** and Eq. (4.4) by **E** and summing

$$\frac{\partial}{\partial t}\left(\frac{B^2}{2\mu_o}+\frac{\epsilon_oE^2}{2}\right)=\frac{\mathbf{E}\cdot\nabla\times\mathbf{B}-\mathbf{B}\cdot\nabla\times\mathbf{E}}{\mu_o}-\mathbf{J}\cdot\mathbf{E}=-\nabla\cdot\frac{\mathbf{E}\times\mathbf{B}}{\mu_o}-\mathbf{J}\cdot\mathbf{E} \tag{4.18}$$

Equation (4.18) has a simple interpretation if the quantity $W=(B^2/\mu_o+\epsilon_oE^2)/2$ is identified with the energy density of the electromagnetic field and the Poynting vector $\mathbf{S}=(\mathbf{E}\times\mathbf{B})/\mu_o$ is introduced, yielding

$$\frac{\partial W}{\partial t}+\nabla\cdot\mathbf{S}=-\mathbf{J}\cdot\mathbf{E} \tag{4.19}$$

Upon integrating over the volume V and using Gauss theorem Eq. (4.19) states that the variation of the electromagnetic field energy inside V is equal to the flux of the Poynting vector across the surface delimiting the volume V minus the work done by the interaction of the electric field and the current inside V.

4.2 Electromagnetic Waves in Vacuum—Dispersion Relation

Vacuum is defined by the conditions $\mathbf{J} = 0$ and $\rho = 0$ and the propagation of waves can be analysed going back to Maxwell's equation for $\mathbf{E}$ and $\mathbf{B}$ Eqs. (4.3)–(4.6) and considering, for the sake of simplicity, solutions that depend only on the space coordinate x (i.e. $\partial/\partial y = \partial/\partial z = 0$) to obtain, after trivial algebra, the following set of equations

$$\left(\frac{\partial^2}{\partial t^2} - c^2\frac{\partial^2}{\partial x^2}\right) E_y = -c^2\mu_o\frac{\partial J_y}{\partial t} = 0 \tag{4.20}$$

$$\left(\frac{\partial^2}{\partial t^2} - c^2\frac{\partial^2}{\partial x^2}\right) E_z = -c^2\mu_o\frac{\partial J_z}{\partial t} = 0 \tag{4.21}$$

$$\frac{\partial B_z}{\partial t} = -\frac{\partial E_y}{\partial x} \tag{4.22}$$

$$\frac{\partial B_y}{\partial t} = \frac{\partial E_z}{\partial x} \tag{4.23}$$

$$B_x = 0 \quad E_x = 0 \tag{4.24}$$

The equations for the y and z component of the electric field are wave equations with x being the propagation directions. The equation

$$\left(\frac{\partial^2}{\partial t^2} - c^2\frac{\partial^2}{\partial x^2}\right) F = 0 \tag{4.25}$$

with c constant is a classical wave equation. Any arbitrary function of the form $F(x \pm ct)$ is solution of Eq. (4.25)as it can be verified by substitution. The solution corresponds to the propagation of a signal of shape $F(x)$ at $t = 0$ along x in the positive or negative direction in such a way that at time $t = t_1$ the shape of the solution is the same but centered around a point $x = x_1 = \pm ct_1$ (Fig. 4.4).

Electromagnetic waves propagating in vacuum are purely *transverse wave*, i.e. the electric and magnetic fields oscillate in the directions orthogonal to the propagation direction (x). Indeed Eq. (4.24) show that the electric and magnetic field components along x are identically zero.

The y and z components of the electric field are independent. Thus, the general solution is a superposition of a wave with the electric field in the y direction (and the magnetic field in the z direction) and of a wave with the electric field in the z direction (and the magnetic field in the y direction). The plane of oscillation of the electric field determine the *polarization* of the wave.

Since the equations are linear with constant coefficients solutions can be found of the form

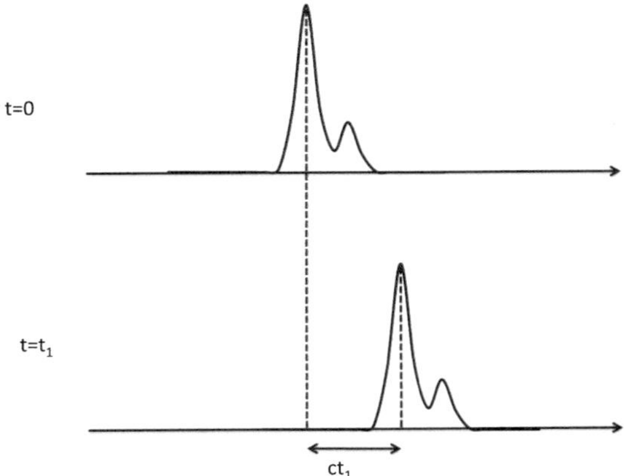

Fig. 4.4 Wave propagation. The solution of Eq. (4.25) is a packet that propagates without changing its shape at constant velocity c

$$F = Ae^{i(kx-\omega t)} + c.c. \tag{4.26}$$

Upon substituting into the equations the following condition is obtained

$$(-\omega^2 + k^2c^2)A = 0 \tag{4.27}$$

that has non-trivial solutions ($A \neq 0$) only if

$$\omega^2 = k^2c^2 \quad or \quad \omega = \pm kc \tag{4.28}$$

This is the *dispersion relation* of electromagnetic waves in vacuum.

In general, the linear propagation of a wave in a medium requires frequency and wave-vector be related through a dispersion relation of the form $\omega = f(k)$ stating that, in order to propagate, the wave-vector k must have a precise relation with the (angular) frequency ω. The name comes from the property that the propagation velocity of waves with different frequency (wave-vector) is not the same.

4.3 Standing Waves and Traveling Waves

Equation (4.25) is a second order partial differential equation that can be solved either as a time dependent problem or an eigenvalue problem. In the case of the time dependent problem the function $F(x)$ is given at time $t = 0$ in order to determine the solution for arbitrary values of t. In the case of the eigenvalue problem the function F or its derivative is imposed on a certain boundary and the eigenvalue ω is determined.

The waves described by Eq. (4.25) can be of either *traveling* or *standing*. A traveling wave describes the propagation of a signal in space and time. A standing wave describe the characteristic mode of oscillation of a system (as the chord of a guitar). A standing wave can be visualized as the superposition of two counter propagating waves of the same amplitude.

To describe a standing electromagnetic wave let us consider a one dimensional problem in which an electromagnetic wave oscillates between two perfectly conductive planes placed at $x = 0$ and $x = L$. Since the planes are perfectly conductive the tangent component of the electric field must be zero at the planes. Taking for the sake of simplicity a linearly polarised wave along y, the electric field must have the form

$$E_y(x,t) = E_{yo}(t)\sin(k_x x) \tag{4.29}$$

This solution satisfy the boundary condition at $x = 0$. In order to satisfy the boundary condition also at $x = L$ the quantity L must be equal to an integer or half integer number of wavelengths

$$L = \frac{l\pi}{k_x}, \tag{4.30}$$

with l an integer and $\mathbf{k}_x$ the wave vector. Only waves that satisfy Eq. (4.30) can exist in the domain and can be labelled with the index l. Thus, the electric and magnetic field in the domain can be expressed as the superposition of different standing waves as follows

$$E_y(x,t) = \sum_{l=1}^{\infty} E_l(t)\sin\left(l\pi\frac{x}{L}\right) \qquad B_z(x,t) = \sum_{l=1}^{\infty} B_l(t)\cos\left(l\pi\frac{x}{L}\right) \tag{4.31}$$

Each wave in Eq. (4.31) satisfies Maxwell's equation provided

$$\frac{dB_l}{dt} = -lk_o E_l \tag{4.32}$$

with $k_o = \pi/L$. Equation (4.32) can be substituted into the expression for the energy of the electromagnetic field, yielding

$$\begin{aligned} W = A\int_o^L dx\left(\frac{B^2}{2\mu_o} + \frac{\epsilon_o E^2}{2}\right) = \\ = A\int_o^L dx \sum_{l,m}\left(\frac{B_l B_m}{2\mu_o}\cos(lk_o x)\cos(mk_o x) + \frac{\epsilon_o E_l E_m}{2}\sin(lk_o x)\sin(mk_o x)\right) = \\ = \frac{AL}{2}\sum_{l=1}^{\infty}\frac{1}{l^2k_o^2}\left(\frac{\epsilon_o}{2}\left(\frac{dB_l}{dt}\right)^2 + \frac{l^2k_o^2}{2\mu_o}B_l^2\right) \end{aligned} \tag{4.33}$$

where the following identity has been taken into account

$$\int_0^L dx\cos(lk_o x)\cos(mk_o x) = \int_0^L dx\sin(lk_o x)\sin(mk_o x) = \frac{L\delta_{l,m}}{2} \tag{4.34}$$

Equation (4.33) suggests the identification of each standing wave with a harmonic oscillator with the energy being continuously exchanged between the electric and magnetic field similarly to what happens in a mechanical oscillator where the energy is continuously exchanged between kinetic and potential energy. We will come back to this analogy in connection with the discussion of the black body spectrum in Chap. 5.

4.4 Phase Velocity and Group Velocity

The wave-vector k and the angular frequency ω are related to the wavelength λ and the frequency ν as follows

$$\lambda = \frac{2\pi}{k} \qquad \nu = \frac{\omega}{2\pi} \tag{4.35}$$

The wave front is the locus of points that at a given time have the same phase. For example, the two-dimensional wave described by

$$A \ \cos\left(\omega t - k_x x - k_y y\right) \tag{4.36}$$

at time $t = 0$ is characterized by phase $\psi = t - k_x x - k_y y$ that is constant along the line $k_x x + k_y y = -\psi$ corresponding to a straight line orthogonal to **k** (Fig. 4.5). The wave propagation direction is the direction of **k**.

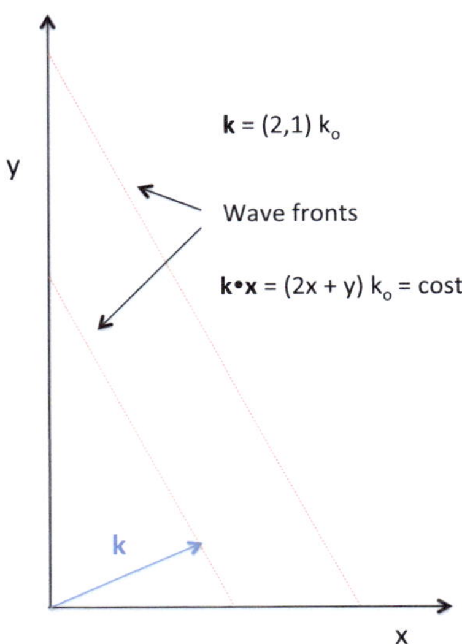

Fig. 4.5 Wave propagation. The direction of the wave vector **k** indicates the direction of wave propagation. The wave front—the locus of the point with the same phase- is orthogonal to **k**

Taking for the sake of simplicity the case of propagation along $x(k_y = 0)$, it is possible to define the velocity of a wave front defined by $\omega t - kx = \psi = cost$. The position is related to time through the relation

$$x_{\psi=const} = \frac{\omega}{k}t - \frac{\psi}{k} \tag{4.37}$$

Thus the velocity of the wave front, or *phase velocity*, is given by

$$v_{phase} = \frac{dx_{\psi=const}}{dt} = \frac{\omega}{k} \tag{4.38}$$

In wave propagation problems we are mainly interested in the velocity at which information can travel. However, a purely monochromatic wave carries no information. Therefore, it is not phase velocity the quantity of interest. If we want to send information we need to modulate the signal as for example made using the Morse alphabet composed by a combination of shorter and longer pulses. The simplest modulation is obtained through the superposition of two waves with slightly different frequencies and wave vectors (Fig. 4.6)

$$\begin{aligned} A\cos(\omega t - kx) + A\cos(\omega' t - k' x) = \\ = 2A\cos\left[\frac{(\omega - \omega')t}{2} - \frac{(k - k')x}{2}\right]\cos\left[\frac{(\omega + \omega')t}{2} - \frac{(k + k')x}{2}\right] \end{aligned} \tag{4.39}$$

This expression corresponds to a wave at frequency $(\omega + \omega')/2$ modulated with an amplitude at frequency $(\omega - \omega')/2$ (Fig. 4.8).

The velocity of the wave fronts of the rapidly varying part is simply the phase velocity

$$v_{phase} = lim_{\omega\to\omega'}\frac{(\omega + \omega')/2}{(k + k')/2} = \frac{\omega}{k} \tag{4.40}$$

Conversely, the velocity of the slowly varying amplitude is given by

$$v_{group} = lim_{\omega\to\omega'}\frac{(\omega - \omega')/2}{(k - k')/2} = \frac{\partial\omega}{\partial k} \tag{4.41}$$

and it is called *group velocity*. In the *specific* case of the propagation of electromagnetic waves in vacuum the group velocity and the phase velocity coincide and are both equal to the speed of light c. Note also that the phase velocity in this case is the same for all the frequencies, i.e. vacuum is a non-dispersive medium.

For the general case of a medium with dispersion relation $\omega = f(k)$, the phase velocity $\omega/k = f(k)/k$ is a function of k and the group velocity is determined from the derivative of the function $f(k)$.

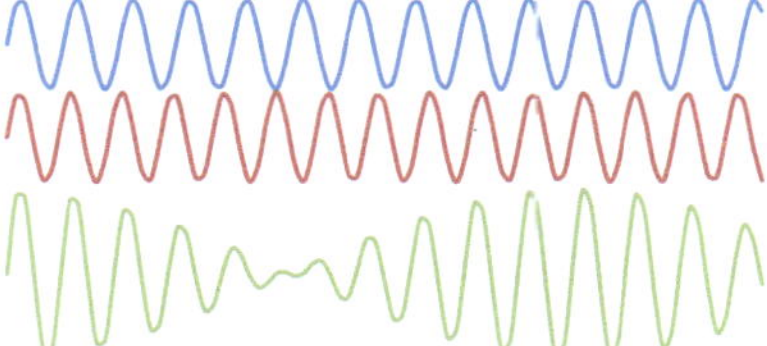

Fig. 4.6 A monochromatic wave does not carry any information. To send information the signal must be modulated e.g. by superimposing two waves with slightly different frequencies and wave vectors

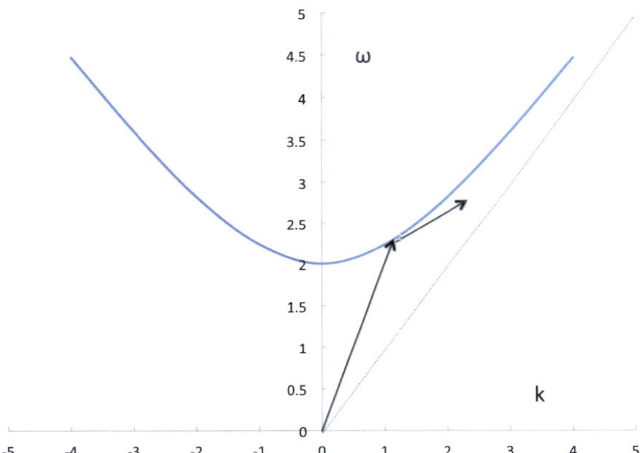

Fig. 4.7 The dispersion relation in a plasma is shown. No propagation is possible below the electron plasma frequency. For large values of the wave vector the dispersion relation approaches that of electromagnetic waves in vacuum. Note that the phase velocity is always larger than c whereas the group velocity (proportional to the slope of the local tangent to the curve) is always smaller than c

4.5 Electromagnetic Waves in the Ionosphere

As an example of propagation in a dispersive medium we consider the propagation of electromagnetic waves in the plasma of the ionosphere. Plasmas are a superposition of two gases of negatively charges electron and positively charged ions. At an height of about 100 km the Sun radiation ionise the atoms of the atmosphere producing a plasma.

The plasmas considered in this book are neutrals (the positive charge balances the negative charge). In the presence of an electromagnetic wave the charges are accelerated by the electric field and produce a current density that acts as a source in the Maxwell's equation (4.20) and (4.21). Thus, the wave no longer propagates in vacuum and the presence of the sources to determine the dispersion relation must be taken into account .

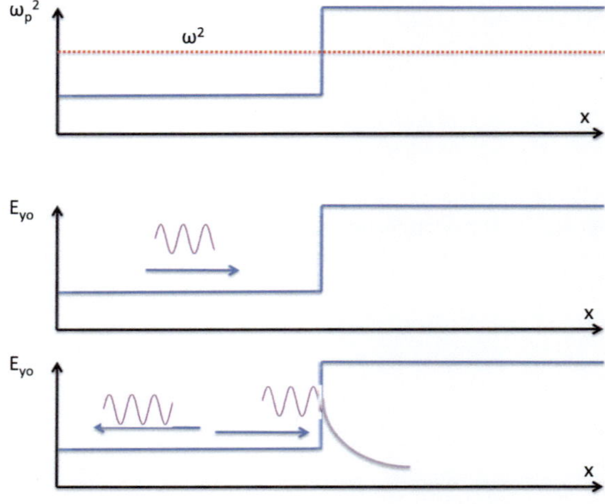

Fig. 4.8 An electromagnetic wave propagating in a plasma is reflected at the position where the local electron plasma frequency equals the wave frequency. Beyond this point the wave is evanescent

Since ions are much heavier than electrons it is possible to assume that they remain at rest. Electrons (charge $-e$) can be considered as a gas of free particles under the influence of an electromagnetic wave of the form

$$E = E_y \mathbf{y} = E_{yo}\, \mathbf{y}\, \cos(\omega t - kx) \tag{4.42}$$

The electrons will move according to the following equation[1]

$$m_e \frac{dv_y}{dt} = -e\, E_{yo}\, \cos(\omega t - kx) \tag{4.43}$$

with solution

$$v_y = -\frac{eE_{yo}}{m_e\omega}\, \sin(\omega t - kx) \tag{4.44}$$

and will produce a current density given by

$$J_y = -en_e v_y = \frac{n_e e^2 E_{yo}}{m_e\omega} \sin(\omega t - kx) \tag{4.45}$$

The source term Eq. (4.45) can be substituted into Eq. (4.20) for E_y yielding

$$\left(\frac{\partial^2}{\partial t^2} - c^2 \frac{\partial^2}{\partial x^2}\right) E_y = -c^2 \mu_o \frac{\partial J_y}{\partial t} = -\omega_p^2 E_y \tag{4.46}$$

[1] In the following the effect of the Lorentz force associated with the wave magnetic field B_z is neglected, see discussion in the exercises.

with $\omega_{pe} = (n_e e^2/(\epsilon_o m_e))^{1/2}$ the *electron plasma frequency* that is approximately given by $\omega_{pe}/(2\pi) \approx 8.98 n_e (m^{-3})^{1/2}$Hz. Taking into account Eq. (4.42) the dispersion relation for electromagnetic waves in a plasma is obtained (Fig. 4.7)

$$\omega^2 = \omega_{pe}^2 + k^2 c^2 \tag{4.47}$$

For frequencies well above the electron plasma frequencies ($\omega \gg \omega_{pe}$) the dispersion relation reduces to that of electromagnetic waves in vacuum. However for frequencies below ω_{pe} the change is dramatic because it is impossible to satisfy the dispersion relation unless k becomes an imaginary number. An imaginary value for k means that the wave amplitude is exponentially decreasing

$$e^{ikx} \approx e^{-|k|x} \tag{4.48}$$

and corresponds to an evanescent (non-propagating) wave. If an electromagnetic wave propagates in a region where the plasma density is (slowly) increasing as long as the density is such that $\omega \gg \omega_{pe}$ the wave will propagate as in vacuum. As the wave penetrates in regions with increasing density the wave vector decreases and vanishes at the point where $\omega = \omega_{pe}$. Beyond this point the wave cannot propagate and it is reflected back (Fig. 4.8).

The presence of the *ionosphere*, a layer of plasma 300 km thick starting at about 100 km above the level of the sea and produced by the Sun ultra violet radiation, was not known when Marconi started to work on radio waves at 800 kHz. When he tried to send a radio signal across the ocean, most of the people forecasted that the signal would have been lost in space. On the contrary, the signal arrived on the other side of the Atlantic thanks to the presence of the ionosphere. The density of the free electrons in the ionosphere is between 10^{10} m^{-3} and 10^{12} m^{-3} corresponding to frequencies between 1 MHz and 10 MHz. The Marconi signal was at sufficiently low frequency to undergo reflection by the ionosphere!

4.6 Emission from an Accelerated Point Charge

In this section the electromagnetic field produced by an accelerated charge is evaluated. The starting point is Eq. (4.16) that gives the vector potential for an arbitrary current density distribution. In many applications we are interested in the field far from the source. Thus, the denominator $|\mathbf{r} - \mathbf{r}\prime|$ can be replaced with $|\mathbf{r}|$, yielding

$$\mathbf{A}(\mathbf{r}, t) \approx \frac{\mu_o}{4\pi r} \int d^3\mathbf{r}\prime \mathbf{J}(\mathbf{r}\prime, t\prime_r) \tag{4.49}$$

with the retarded time approximated as $t\prime_r \approx t - r/c$. Assume that the electron is accelerated by an electric field of the form $\mathbf{E}_o \cos(\omega t)$

$$m\frac{d^2\mathbf{x}}{dt^2} = e\mathbf{E}_o \cos(\omega t) \tag{4.50}$$

The electron motion is an oscillation at frequency ω with velocity given by Eq. (4.44) (the electron located at $\mathbf{x} = 0$)

$$\frac{d\mathbf{x}}{dt} = \frac{e\mathbf{E}_o}{m\omega}\sin(\omega t) \tag{4.51}$$

This motion corresponds to a current density $\mathbf{J}(\mathbf{r}, t) = e(d\mathbf{x}/dt)\delta(\mathbf{r})$ that can be inserted into Eq. (4.49) yielding

$$\mathbf{A}(\mathbf{r}, t) \approx \frac{\mu_o}{4\pi r}\frac{e^2\mathbf{E}_o}{m\omega}\sin[\omega(t - r/c)] \tag{4.52}$$

This expression can be used to evaluate the magnetic field produced by the accelerated charge using Eq. (4.5), with the electric field in turn obtained using Faraday's law Eq. (4.3).

The space derivatives of the vector potential are associated with two kinds of contribution: the terms arising from the derivative of the amplitude and the terms arising from the derivative of the phase (through the factor r/c in the retarded time) with the latter providing a contribution $O(\omega/c)$ and the former being $O(1/r)$. Clearly, the contribution arising from the derivative of the phase is always dominant as long as the distance from the source is much larger than the wavelength in vacuum. Keeping only the leading contribution, the magnetic field produced by the moving charge can be determined

$$\mathbf{B} = \nabla \times \mathbf{A} \approx -\frac{\mu_o}{4\pi r}\frac{e^2}{mc}\cos[\omega(t - r/c)]\hat{\mathbf{r}} \times \mathbf{E}_o \tag{4.53}$$

with $\hat{\mathbf{r}} \equiv \mathbf{r}/r$. Ampére's law can be written as

$$\frac{\partial \mathbf{E}}{\partial t} = c^2 \nabla \times \mathbf{B} = -\frac{\mu_o}{4\pi r}\frac{\omega e^2}{m}\sin[\omega(t - r/c)]\hat{\mathbf{r}} \times (\hat{\mathbf{r}} \times \mathbf{E}_o) \tag{4.54}$$

which, in turn, yields the expression for the electric field

$$\mathbf{E} = \frac{\mu_o}{4\pi r}\frac{e^2}{m}\cos[\omega(t - r/c)]\hat{\mathbf{r}} \times (\hat{\mathbf{r}} \times \mathbf{E}_o) = -c\hat{\mathbf{r}} \times \mathbf{B} \tag{4.55}$$

Equations (4.54) and (4.55) describe an electromagnetic wave travelling from the source located at $r = 0$ in all the directions (spherical wave). Taking a propagation direction r, the electric field and the magnetic field are orthogonal to $\mathbf{r}$ (therefore they are tangent to the sphere $r = cost$) and orthogonal among themselves. The wave is transverse (as it should be since the propagation is in vacuum) therefore there are no component of $\mathbf{E}$ and $\mathbf{B}$ along $\mathbf{r}$. As to the relation between the direction of $\mathbf{E}$ and

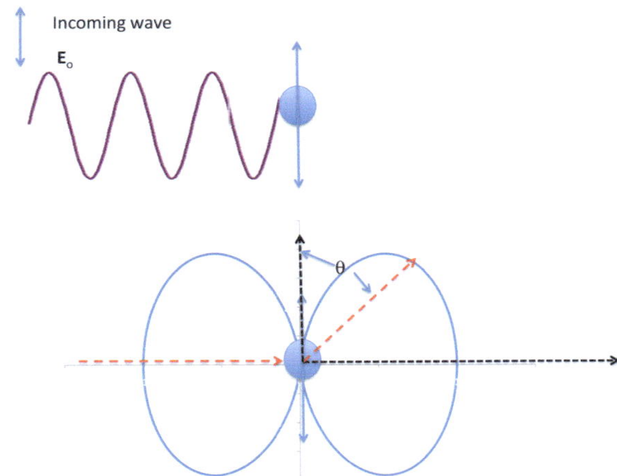

Fig. 4.9 Dipole radiation. The incoming electromagnetic wave is polarized with the electric field along y. The emitted radiation is in all the directions but its amplitude for emission at $\theta = 0$ is zero

that of the electric field $\mathbf{E}_o$ of the primary wave, Eq. (4.55) tells that $\mathbf{E}$ is in the plane defined by $\mathbf{E}_o$ and $\mathbf{r}$ (whereas $\mathbf{B}$ is orthogonal to the plane, Fig. 4.9).

The above expressions can be used to evaluate the Poynting vector $\mathbf{S} = \mathbf{E} \times \mathbf{B}/\mu_o$

$$\mathbf{S} = \frac{c}{\mu_o}|B|^2\hat{\mathbf{r}} = \frac{\mu_o}{c}\left(\frac{1}{4\pi r}\right)^2\left(\frac{e^2}{m}\right)^2 \cos^2[\omega(t - r/c)]|\hat{\mathbf{r}} \times \mathbf{E}_o|^2\hat{\mathbf{r}} \tag{4.56}$$

Denoting with θ the angle between the line of sight and the electric field $\mathbf{E}_o$, the distribution of the radiated power is characterised by a $\sin^2\theta$ dependence. No radiation is emitted in the direction of $\mathbf{E}_o$ (Fig. 4.9).

The total radiation P emitted by the electron is obtained by integrating the Poynting vector over the solid angle

$$P = \int r^2 d\Omega \mathbf{S} \cdot \mathbf{r} =$$

$$= \frac{\mu_o}{c}\left(\frac{1}{4\pi}\right)^2\left(e|\frac{d^2x}{dt^2}|\right)^2 \int_{-1}^{1} d\cos\theta \int_0^{2\pi} d\phi \, \sin^2\theta = \frac{(e|\frac{d^2x}{dt^2}|)^2}{6\pi\epsilon_o c^3} \tag{4.57}$$

this is the *Larmor formula* for the power emitted by an accelerated point charge in the non relativistic limit. It is valid for an arbitrary accelerations (not just the oscillatory motion that we have used to derive Eq. (4.49)).

4.7 Scattering of Electromagnetic Waves. Thomson Scattering

The amount of power emitted by an electron accelerated by an electromagnetic wave Eq. (4.57) can be expressed in a more intuitive way using the concept of cross section.

The phenomenon discussed in the previous section is the scattering by an electron of an electromagnetic wave with electric field $\mathbf{E}_o$ and magnetic field $\mathbf{B}_o$ propagating in the direction orthogonal to both $\mathbf{E}_o$ and $\mathbf{B}_o$. The Poynting vector of the incident electromagnetic wave is given by

$$\mathbf{S}_o = \frac{\mathbf{E}_o \times \mathbf{B}_o}{\mu_o} \tag{4.58}$$

The incident wave is a wave propagating in vacuum and is described by the dispersion relation Eq. (4.28). Using Faraday's law the Poynting vector can be expressed as follows

$$\mathbf{S}_o = \frac{\mathbf{E}_o \times (\mathbf{k} \times \mathbf{E}_o)}{\mu_o} = \epsilon_o c |E_o|^2 \mathbf{k} \tag{4.59}$$

The quantity $\mathbf{S}_o$ is a power per unit surface. The power scattered by the electron is given by Eq. (4.57) that can be written as

$$P = \frac{8\pi}{3} r_{cl}^2 S_o \tag{4.60}$$

with $r_{cl} = e^2/(4\pi\epsilon_o mc^2) \approx 2.8 \times 10^{-15}$m the *classical electron radius*.

The result of the each scattering event will be that energy is removed from the incident beam at a rate given by Eq. (4.60). If instead of a single electron we have a target of electrons with density n_e the incoming wave will undergo multiple scattering. The cumulative effect will be obtained by multiplying Eq. (4.60) by the number of scattering events. In a volume of infinitesimal area dS and height dx along the propagation direction the number of events is simply $n_e dS dx$. Since the quantity of interest is the change in the energy flux S_o we have to divide by the area dS yielding

$$dS_o = n_e dx P = n_e \frac{8\pi}{3} r_{cl}^2 S_o dx \tag{4.61}$$

Equation (4.61) is identical to Eq. (3.21). Thus the cross section for the scattering process can be defined as done in Chap. 3. The process takes the name of *Thomson scattering* and its cross section is given by

$$\sigma_c = \frac{8\pi}{3} r_{cl}^2 \tag{4.62}$$

The Thomson scattering cross section is about $0.66b$. We note that the frequency of the scattered radiation is the same of the incident radiation, a result that will be modified when we will consider quantum mechanics.

4.8 Electric Dipole Emission

The solutions for the vector and scalar potentials Eqs. (4.16) and (4.17) in many situations need to be evaluated far from the sources. This implies that the term $|\mathbf{r}-\mathbf{r}\prime|$ can be replaced as a first approximation with r (see e.g. Eq. (4.49)). More rigorously, far from the source, the term $|\mathbf{r}-\mathbf{r}\prime|$ can be replaced with an infinite series in $(r\prime/r)$ that produces the *multipolar expansion* of the two equations. Note that the infinite series is associated to both the $|\mathbf{r}-\mathbf{r}\prime|$ term in the denominator of the integrands and the analogous term in the retarded time. For the problem of radiation emission the relevant limit to be considered is $r \gg \lambda \gg d$ where d is the typical extension of the source. In this case only the contributions arising from the retarded time need to be retained while ignoring those from the denominator of the integrands. We refer to the discussion presented in Ref. [10] to cover all the possible regimes.

We go back to Eq. (4.49) and keep the first order correction to the retarded time. To maintain the discussion general, we will assume that the source is oscillating at a single frequency with time dependence of the form $\mathbf{J}(\mathbf{r},t)=\bar{\mathbf{J}}(\mathbf{r})e^{-i\omega t}$. Noting that $|\mathbf{r}-\mathbf{r}\prime|\approx r-\hat{\mathbf{r}}\cdot\mathbf{r}\prime$ (with $\hat{\mathbf{r}}\equiv\mathbf{r}/r$) and taking the limit $kr\prime \ll 1$, Eq. (4.52) is modified as follows

$$\mathbf{A}(\mathbf{r},t)\approx\frac{\mu_o}{4\pi r}\int d^3\mathbf{r}\prime\bar{\mathbf{J}}(\mathbf{r}\prime)e^{-i\omega(t-\frac{r}{c})}(1-i\frac{\omega\hat{\mathbf{r}}\cdot\mathbf{r}\prime}{c}) \tag{4.63}$$

The first term is what we have already considered in the previous section. It can be expressed in a more general form via integration by parts

$$\int d^3\mathbf{r}\prime\bar{\mathbf{J}}(\mathbf{r}\prime)=-\int d^3\mathbf{r}\prime\mathbf{r}\prime\nabla\cdot\bar{\mathbf{J}}(\mathbf{r}\prime)=-i\omega\int d^3\mathbf{r}\prime\mathbf{r}\prime\bar{\rho}(\mathbf{r}\prime,t\prime_r) \tag{4.64}$$

where use has been made of the continuity equation

$$\frac{\partial\rho}{\partial t}+\nabla\cdot\mathbf{J}=0 \tag{4.65}$$

and $\rho=\bar{\rho}e^{-i\omega t}$. It is convenient to define the *electric dipole* moment $\mathbf{p}$ as

$$\mathbf{p}\equiv\int d^3\mathbf{r}\prime\mathbf{r}\prime\bar{\rho} \tag{4.66}$$

In terms of the dipole moment the average power emitted by a point charge oscillating at frequency ω can be written as

$$P = \frac{(\omega^2 p)^2}{12\pi\epsilon_o c^3} \tag{4.67}$$

The next order contribution in Eq. (4.63) can be decomposed as follows

$$\hat{\mathbf{J}}\hat{\mathbf{r}} \cdot \mathbf{r}\prime = \frac{1}{2}(\hat{\mathbf{r}} \cdot \mathbf{r}\prime\bar{\mathbf{J}} + \hat{\mathbf{r}} \cdot \bar{\mathbf{J}}\mathbf{r}\prime) + \frac{1}{2}(\mathbf{r}\prime \times \bar{\mathbf{J}}) \times \hat{\mathbf{r}} \tag{4.68}$$

The first term in the r.h.s is related to the *electric quadrupole* moment whereas the second term (the magnetisation associated to the current **J**) is associated with the *magnetic dipole* moment **m** given by

$$\mathbf{m} = \frac{1}{2}\int d^3\mathbf{r}\prime(\mathbf{r}\prime \times \mathbf{J}) \tag{4.69}$$

taking a filament current flowing along a closed path it is easily shown that the magnetic moment is given by $m = IS$ with S being the area enclosed by the path. The contribution to the vector potential associated with the magnetic dipole moment

$$\mathbf{A}(\mathbf{r}, t) \approx \frac{\mu_o}{4\pi r}\frac{i\omega}{c}\hat{\mathbf{r}} \times \mathbf{m}e^{-i\omega(t-\frac{r}{c})} \tag{4.70}$$

The calculation of the magnetic field and of the electric field then follows the same path as in the previous section, yielding at the lowest order in $\omega r/c$

$$\mathbf{E} = -\frac{\mu_o}{4\pi r}\frac{\omega^2}{c}\hat{\mathbf{r}}\times\mathbf{m}e^{-i\omega(t-\frac{r}{c})}$$

$$\mathbf{B} = \frac{\mu_o}{4\pi r}\frac{\omega^2}{c^2}(\hat{\mathbf{r}} \times \mathbf{m}) \times \hat{\mathbf{r}}e^{-i\omega(t-\frac{r}{c})} \tag{4.71}$$

Equations 4.71 are formally identical to Eqs. (4.53) and (4.55) with the change $E \rightarrow B$, $B \rightarrow -E$ and $p \rightarrow m$, therefore the electromagnetic field pattern is similar with a rotation by $\pi/2$ of the polarisation plane.

4.9 Emission from a Damped Harmonic Oscillator

Consider an electron subject to an elastic force $-m\omega_o^2 x$ and a viscous force $-m\gamma dx/dt$. The damped harmonic oscillator is a good approximation to a body close to a stable equilibrium state.

The power scattered under the effect of a linearly polarised electromagnetic wave at frequency ω can be evaluated as above using the Larmor formula. The particle acceleration is obtained by solving the following equation

$$m_e \frac{d^2\mathbf{x}}{dt^2} = -m_e \omega_o^2 \mathbf{x} - m_e \gamma \frac{d\mathbf{x}}{dt} + e\mathbf{E}_o \cos(\omega t) \tag{4.72}$$

The solution can be easily obtained with the following position

$$\mathbf{x}(t) = \mathbf{X}(\omega) e^{i\omega t} + c.c. \tag{4.73}$$

yielding

$$\mathbf{X}(\omega) = \frac{e\mathbf{E}_o}{2m_e} \frac{1}{\omega_o^2 - \omega^2 + i\omega\gamma} \tag{4.74}$$

Upon substituting into Eq. (4.73) and taking the second derivative with respect to time the resulting expression for the acceleration is

$$\mathbf{a} = \frac{-\omega^2 e\mathbf{E}_o}{m_e} \frac{(\omega_o^2 - \omega^2)\cos\omega t + \gamma\omega \sin\omega t}{(\omega_o^2 - \omega^2)^2 + \gamma^2\omega^2} \tag{4.75}$$

Upon inserting into the Larmor formula and performing a time average, we obtain

$$P = \frac{e^4 E_o^2}{12\pi\epsilon_o m_e^2 c^3} \times \frac{\omega^4}{(\omega_o^2 - \omega^2)^2 + \gamma^2\omega^2} \tag{4.76}$$

The first fraction on the r.h.s. is equal to the average power emitted by a free electron. The second fraction provides the frequency dependent modifications due to the harmonic and the damping forces.The scattering cross section has a characteristic behaviour shown in Fig. 4.10.

For frequencies below the characteristic frequency ω_o the cross section increase as ω^4. This is the region of the so called *Rayleigh scattering*. There is a narrow peak of σ around $\omega = \omega_o$ in the so called *resonant region*. The width of the resonant region is measured by the damping rate γ of the harmonic oscillator. The lower is the damping rate the narrower is the resonance. In the region $\omega \gg \omega_o$ the scattering cross section attains a constant value given by Eq. (4.62) (Thomson scattering).

The behavior shown in Fig. 4.10 in the resonance region is the prototype of the resonance cross sections that we will find in many places in this book.

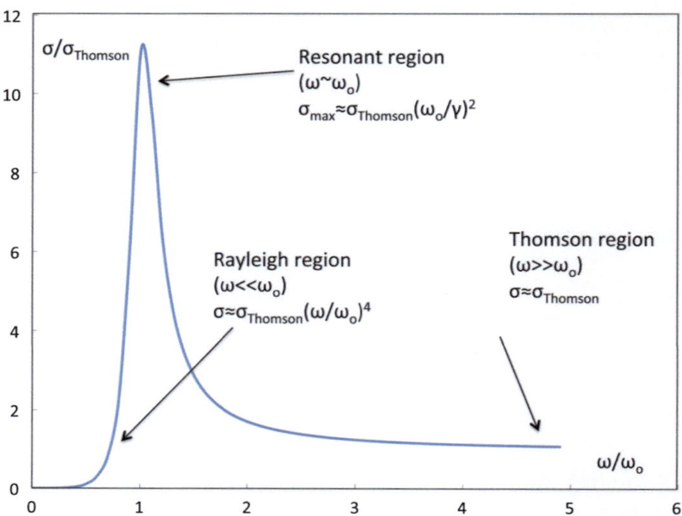

Fig. 4.10 Harmonic oscillator cross section

4.10 Interference and Diffraction of Electromagnetic Waves

The wavelike nature of light was experimentally demonstrated at the beginning of the 19th century by Young [5] and Fresnel [6]. The experimental arrangement is shown in Fig. 4.11. A plane wave passes through two slits and the intensity is measured on a screen at a distance $L \gg d$. If the corpuscular theory were correct, the figure on the screen should be that shown in (Fig. 4.11a). Instead, what appears is a series of lines as shown in (Fig. 4.11b).

This outcome can be understood on the basis of the Huygens-Fresnel principle. When the wave arrives at each of the two slits it excites spherical waves centered at the slit position. The electromagnetic field on the screen will be the superposition of the fields produced by both slits. The fields from the two slits will add in the points where their phase is the same whereas they will cancel each other in the points where the two phases differ by π. Let us consider the condition to have a positive interference. Looking at (Fig. 4.11b) the fields will add together provided the difference in their path is an integer multiple of the wavelength. In the limit $L \gg d$ the two paths are almost parallel lines and the difference in the optical path can be approximated as $d \ \sin\varphi$, therefore the condition for constructive interference is

$$d \ \sin\varphi = n \ \lambda \tag{4.77}$$

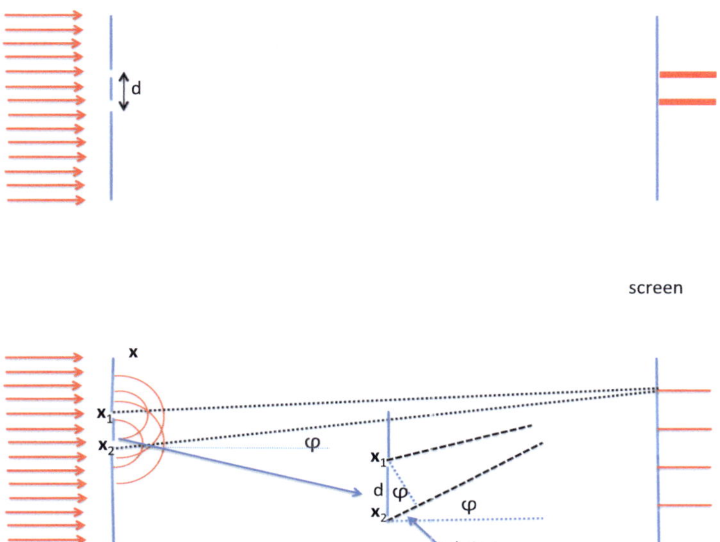

Fig. 4.11 Light passing through two slits. If corpuscular theory were correct the image of the screen would have reproduced the image of the two slits (**a**). The interference pattern (**b**) can be understood as a consequence of the Huygens-Fresnel principle. If the distance between the slits and the screen is much larger than the slits separation the waves emitted by the two slits are almost parallel and the difference in the optical path is $d \sin\varphi$. This difference must be equal to an integer number of wavelengths in order to obtain constructive interference

The image of a single slit of aperture $a \ll L$ can be determined in a similar way. If the corpuscular theory were correct the image would be that in Fig. 4.12a. The experimental result is instead shown in Fig. 4.12b.

In this case to understand the result it is easier to look at the condition for *destructive interference* by taking pairs of points at distance $a/2$. Destructive interference is obtained if the difference between the optical paths of each pair of waves is half wavelength. Thus, a minimum of the intensity will be observed at the points where the following condition is satisfied

$$\frac{a}{2}\sin\varphi = \frac{\lambda}{2} \tag{4.78}$$

or $a \sin\varphi = \lambda$. This reproduces the position of the first minimum. Upon repeating the argument for the pairs at a distance $a/4$ we can find the next order minimum, and so on.

It is important to note that by reducing the slit aperture the position of e.g. the first minimum increases. This means that the image on the screen becomes wider. On the basis of a corpuscular theory we would have expected the opposite!

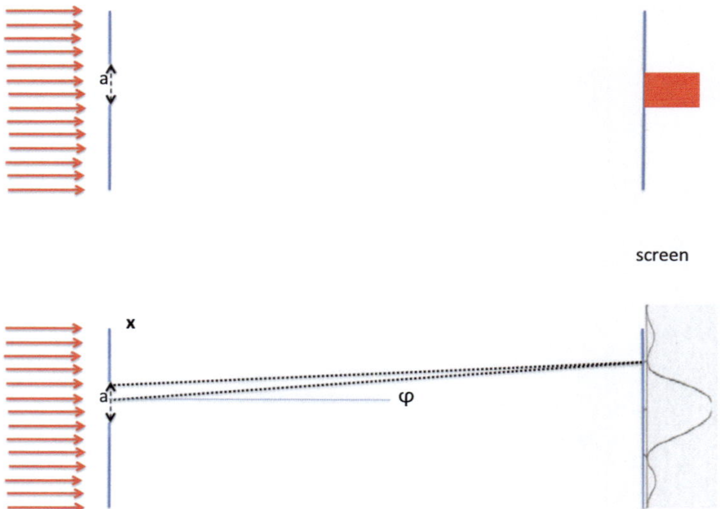

Fig. 4.12 Diffraction experience. The image expected on the basis of corpuscular theory is shown in (**a**). The diffraction figure (**b**) can again be understood on the basis of the Huygens-Fresnel principle by taking pairs of points at a distance $a/2$ inside the aperture and looking for the condition for destructive interference

4.11 The Bragg Diffraction Grating

As anticipated in Chap. 1 the Bragg diffraction grating has played a crucial role to progress spectroscopy. In this section we determine the condition for positive interference of the radiation reflected by the grating.

Making reference to Fig. 4.13 we assume that the grating is made of a lattice of atoms that absorb and re-emit radiation with the angle of incidence (the angle between the ray and the normal to the surface) equal to the angle of reflection (the angle between the reflected ray and the normal). The difference in the optical path depends on the lattice constant d and is given by $d \sin\theta$ with θ the complement to

Fig. 4.13 Diffraction grating. The incident ray is reflected at an angle identical to the angle of incidence

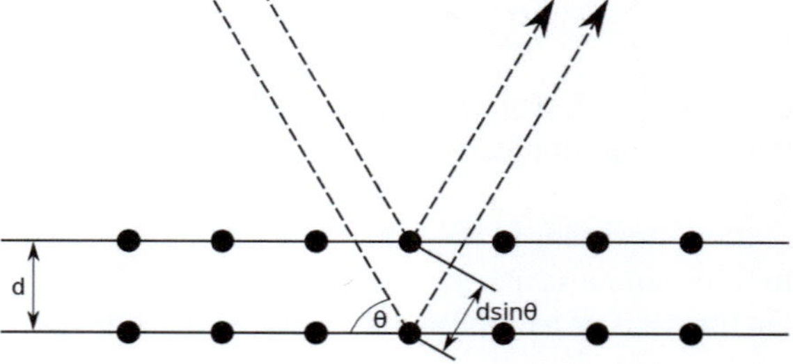

$\pi/2$ of the angle of incidence (see Fig. 4.13). Thus, the condition for constructive interference is

$$2d \quad \sin\theta = n\lambda \tag{4.79}$$

The Bragg diffraction grating allows testing the matter properties at distances comparable to that of radiation.

4.12 Galilean Relativity

The law of classical mechanics are the same for the reference frames moving at constant velocity (inertial frames). Therefore all the *inertial systems* are equivalent. This was realized by Galileo in the *Dialogo sui massimi sistemi* [11]. The trajectory of a body is different in the frame at rest with respect to an inertial frame moving along x at velocity v and the position in the moving frame ($x\prime$, $y\prime$, $z\prime$) are related to those in the frame at rest through the Galilean transformations

$$x\prime = x - vt \quad y\prime = y \quad z\prime = z \quad t\prime = t \tag{4.80}$$

but the equation of dynamics are identical in the two frames. The different trajectories are shown in Fig. 4.14 for two inertial frames represented by the station and a train moving at velocity v.

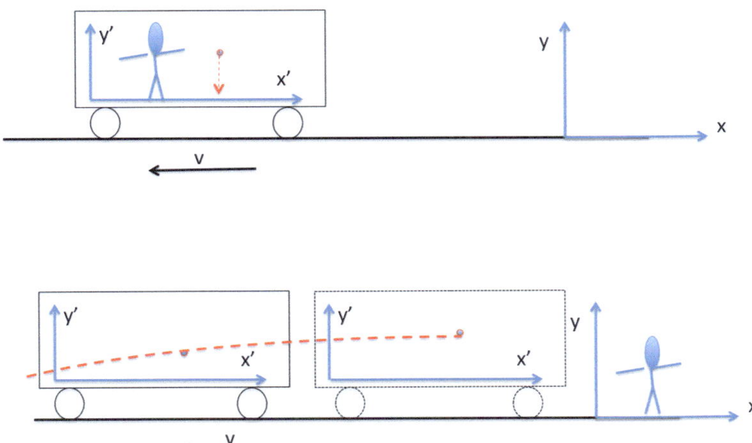

Fig. 4.14 Galilean relativity. Two inertial frames (the station at rest and the wagon moving at constant velocity) are equivalent. The law of dynamics are the same and the trajectory of an object can be obtained from the transformations in Eq. (4.80)

An object that is let falling on the train from the height $z = z_o$ will follow a vertical trajectory covered in a uniformly accelerated motion with acceleration $-g$ described by the parametric equations

$$x\prime = 0 \quad y\prime = 0 \quad z\prime = z_o - \frac{gt^2}{2} \tag{4.81}$$

In the station frame the trajectory will correspond to a parabola described by the parametric equation

$$x = vt \quad y = 0 \quad z = z_o - \frac{gt^2}{2} \tag{4.82}$$

or $z = z_o - g(x/v)^2/2$. However in both frames the equation of motion are

$$\frac{d^2x}{dt^2} = \frac{d^2x\prime}{dt^2} = 0 \quad \frac{d^2y}{dt^2} = \frac{d^2y\prime}{dt^2} = 0 \quad \frac{d^2z}{dt^2} = \frac{d^2z\prime}{dt^2} = -g \tag{4.83}$$

A consequence of the Galilean invariance is that the velocity of an object can only be measured with respect to a certain inertial system. Velocity cannot be defined in *absolute* terms.

However, Maxwell's equations pose a problem since they are not invariant under a Galilean transformation. The transformation that make Maxwell's equations invariant (the Lorentz transformations [12]) imply that in all the frames the speed of light is always the same. This has profound consequences on all the physical phenomena. The consequences are small if the velocity of the frame is small with respect to the speed of light, but, as the speed of light is approached, they become important.

4.13 The Michelson-Morley Experiment

To understand how the invariance of the velocity of light modifies the dynamics it is convenient to discuss the Michelson and Morley experiment [13], originally devised to demonstrate the relative motion of the Earth with respect to a fictitious medium called *ether* that at that time was supposed to be the fluid supporting the light oscillations.

The experiment made use of the Michelson interferometer, an extremely precise instrument to measure the difference in the optical path of two rays. The principle of the instrument is shown in Fig. 4.15.

Light is emitted in A. At point B it encounters a semi-reflective plate and is 50% reflected towards a mirror at point C and 50% transmitted towards a mirror at point D. After the mirror reflections the two rays recombine in E. If the optical paths are the same the two signals will produce constructive interference. If the

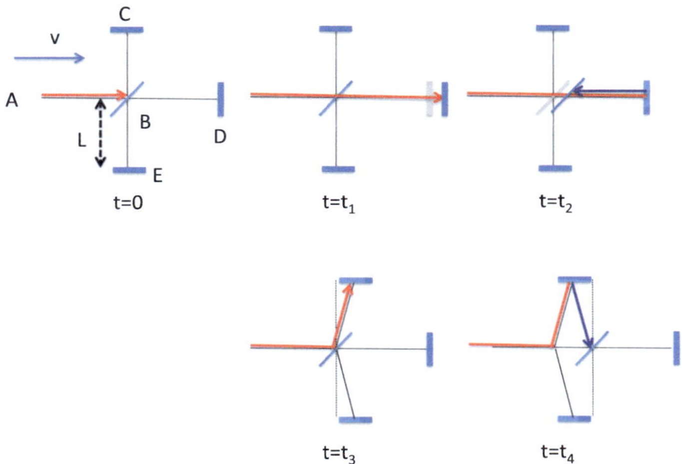

Fig. 4.15 Michelson-Morley experiment

optical paths differ by half wavelength, destructive interference results. Thus, the Michelson interferometer is a powerful tool to measure differences in the optical path.

The original idea behind the Michelson-Morley experiment is that if the interferometer is moving with velocity v with respect to the ether, a change in the optical path is expected. Thus, placing the instrument with the main arm parallel and perpendicular to the direction of motion it would be possible to measure the velocity of the inertial frame. In the original expectations such a velocity would have been the velocity of the interferometer with respect to the ether seen as a *privileged* frame.

Let us try to work out the details.

Assume that initially the arm ABC is placed along the direction of motion and evaluate the time that the ray reflected in C takes to go from the plate to the mirror and back to the plate (after that time the path of the two rays coincide). The rays arrive in C at a time t_1 such that the path travelled by the wave (ct_1) is equal to the interferometer length L plus the extra space travelled by the mirror in the same time (vt_1)

$$ct_1 = L + vt_1 \quad or \quad t_1 = \frac{L}{c - v} \tag{4.84}$$

On the way back the time t_2 needed to go from the mirror to the plate can be calculated similarly, taking into account that the space to be travelled is smaller because the ray is propagating opposite to v

$$ct_2 = L - vt_2 \quad or \quad t_2 = \frac{L}{c + v} \tag{4.85}$$

Now consider the optical path along the perpendicular arm. The time t_3 needed to go from B to D is given by Pythagoras theorem (see Fig. 4.15).

$$ct_3 = (L^2 + v^2 t_3^2)^{1/2} \quad or \quad t_3 = \frac{L}{(c^2 - v^2)^{1/2}} \tag{4.86}$$

The same amount of time is clearly needed to go back from D to B. If ω is the wave frequency, the phase difference between the two rays for $v \ll c$ is given by

$$\Delta\phi = \omega(t_1 + t_2 - 2t_3) \approx -\omega \frac{L}{c} \frac{v^2}{c^2} \tag{4.87}$$

After rotating the interferometer by 90° it easy to show that the phase difference has equal magnitude and opposite sign. The effect would be therefore clearly visible.

However, no phase difference was measured in the experiment!

The explanation is that, contrary to what is expected on the basis of Galilean invariance, the interferometer length in the direction of the velocity v as measured in the laboratory frame is reduced by a tiny but finite amount [14]. The length of the arm becomes

$$L\prime = L\left(1 - \frac{v^2}{c^2}\right)^{1/2} \tag{4.88}$$

If we now repeat the calculation, the time to perform the path BCB is given by

$$t_1 + t_2 = \frac{L\prime}{c - v} + \frac{L\prime}{c + v} = \frac{2L}{(c^2 - v^2)^{1/2}} \tag{4.89}$$

and is equal to the time needed to perform the path BDB that is given by the same expression as before ($2t_3$) since the length of the arm perpendicular to direction of motion is unaffected. Thus, no phase difference is expected.

We have introduced the contraction of lengths as an *ad hoc* hypothesis. As we will see later, it is a consequence of the Lorentz transformations.

4.14 Simultaneity of Events and Time Dilatation

One of the consequences of special relativity is that simultaneity of events is no longer an absolute concept. Events that occur at the same time in one frame may not be simultaneous in another frame.

To illustrate this, let us consider a train moving at constant speed v. Suppose that two light signals are emitted from the center of the wagon. The observer on the train will see that they reach the two ends at the same time $L/(2c)$. Let us now take the point of view of an observer in the station frame. He will see the light signal arriving in A (see Fig. 4.16) travelling a shorter distance than the signal arriving in B since

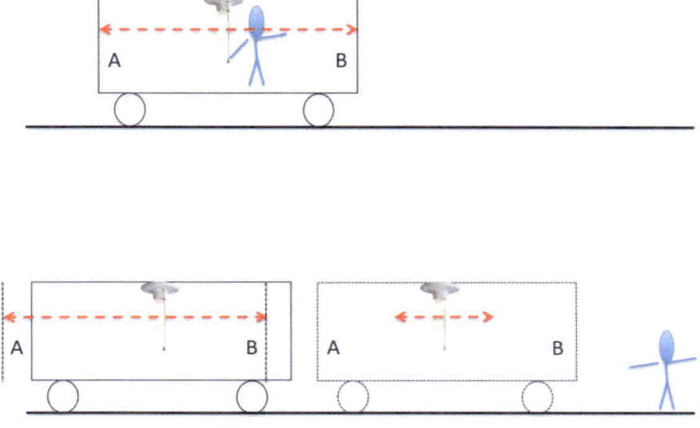

Fig. 4.16 Simultaneity of events. A light signal emitted in the center of the wagon will arrive at the two ends of the wagon at the same time for an observer on the train. An observer in the station will measure different arrival times. Two events (the arrival of the light signals at the ends of the wagon) that are simultaneous for an observer on the train are not simultaneous for an observer in the station

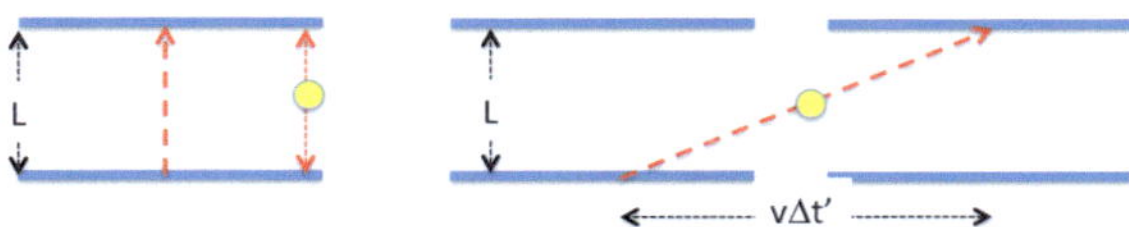

Fig. 4.17 Elementary clock. The clock is made by two ideally reflecting parallel mirrors and a bunch of light oscillating between the two. The time for the light to travel the distance between the two mirrors is different for an observer on the train and an observer in the station. For both the propagation velocity is c but the path made by the light as measured by the observer in the station is larger. Therefore, for the observer in the station the time on the train flows more slowly

we have to take into account that from the time of emission to the time of arrival the wagon has moved by an amount $v\Delta t$. However the speed of light is the same in the wagon and in the station. Thus, they will *not* be seen by the station observer to arrive at the same time.

The notion of time is different in the two frames. To measure time we need the simplest possible tool. For our purpose, let us use a clock made of a bunch of light bouncing back and forth between two perfectly reflecting mirrors. The time needed to go from one mirror to the other is the elementary time interval (Fig. 4.17).

The interval measured on the train is simply

$$\Delta t = \frac{L}{c} \tag{4.90}$$

In the rest frame the motion of light follows an oblique trajectory and the path is now given by $(L^2 + v^2\Delta t'^2)^{1/2}$. Thus, the time needed to go from one mirror to the other is

$$\Delta t' = \frac{L}{(c^2 - v^2)^{1/2}} \tag{4.91}$$

and is therefore *longer* than the time on the wagon. The time on the wagon *as observed from the station frame* flows more slowly. Note that the situation is symmetric as both frames are inertial frames and therefore equivalent. The observer in the wagon measures the time flowing more slowly in the station frame. What happens then if after some time the train stops and return back to the station of departure? Well, to do this experiment we have to accelerate and decelerate the train in such a way that the train is no longer an inertial system. So let us make the experiment in a different way. Two trains are both moving at constant velocity v but in opposite directions. When the first train passes through the station it sends a signal to synchronize its clocks with the clock in the station. After a time t_1(as measured in the station) it encounters the second train coming in the opposite direction and the observers in the two trains synchronize their clocks. The distance (as measured by the observer in the station) travelled by the first train at the time of the encounter is $L = vt_1$. The second train will pass through the station again after a time t_1 (again as measured in the station—the two trains have the same velocity v) and will send a third signal to the observer in the station.

How much time has passed for the observer in the station and for the two observers on the trains? For the observer in the station the time interval is simply $2t_1$. The time passed in both trains according to the observer in the station is however smaller by a factor $(1 - v^2/c^2)^{1/2}$. Let us now consider the point of view of the observers in the trains. After leaving the station the observer in the first train encounters the second train after a time $t_1{}' = L'/v$ where L' is the distance measured by the observer in the train. Since the train is moving, this length is contracted by a factor $(1 - v^2/c^2)^{1/2}$ with respect to L $L' = (1 - v^2/c^2)^{1/2} L$ and the time between the first two encounters is $t_1{}' = (1 - v^2/c^2)^{1/2} t_1$. A similar time interval is measured by the observer in the second train between the encounter with the first train and the encounter with the station. Thus we arrive at the conclusion that indeed time has passed more slowly on the trains than in the station. This is the famous *twin paradox*. The asymmetry between the observer in the station and the observers on the train is due to the fact that only the former can make the statement to be at rest. The observers on the train cannot be both at rest.

4.15 Lorentz Transformations

All the phenomena that we have described in the last two paragraphs can be accounted for if we replace the Galilean transformation Eq. (4.80) with the more general set of Lorentz transformations [12]

$$x' = \gamma(x - vt) \quad y' = y \quad z' = z \quad t' = \gamma\left(t - \frac{vx}{c^2}\right) \tag{4.92}$$

with $\gamma = (1 - v^2/c^2)^{-1/2}$. In order to familiarise with the Lorentz transformations a few examples are reported below.

- *Do the Lorentz transformations ensure that the velocity of light is the same in all the frames?* A ray of light travel in the unprimed system along the trajectory $x = ct\ \cos\varphi\quad y = ct\ \sin\varphi\quad z = 0$ where it has been assumed that light is emitted at $t = 0$ from the origin at an angle φ with respect to the x axis in the (x, y) plane. The trajectory observed on the train is

$$x\prime = \gamma(c\ \cos\varphi - v)t \quad y\prime = ct\ \sin\varphi \quad z\prime = 0 \quad t\prime = \gamma\left(1 - \frac{v\ \cos\varphi}{c}\right)t \tag{4.93}$$

The resulting trajectory in the primed frame is $x\prime = ct\prime\ \cos\varphi\prime\quad y\prime = ct\prime\ \sin\varphi\prime\quad z\prime = 0$ and corresponds again to the propagation at constant velocity c and at an angle $\varphi\prime$ with

$$\cos\varphi\prime = \frac{\cos\varphi - \frac{v}{c}}{1 - \frac{v}{c}\cos\varphi} \tag{4.94}$$

- *A ruler of length L is placed on the train (the frame with the primed coordinates). What is the length of the ruler for the observer in the station?* The end points of the ruler are placed at $x\prime = 0$ and $x\prime = L$. The corresponding positions in the station frame at a generic time $t = T$ are obtained from the first of Eq. (4.92) yielding $x = vT$ and $x = vT + L/\gamma$. Therefore the length measured in the station frame is contracted by a factor γ.
- *A particle is moving on the train at constant velocity $u\prime$ from $x\prime = 0$ to $x\prime = L\prime$. How long does it take for the observer in the station?* Again we have simply to apply the Lorentz transformations for the two events start of the path (corresponding to $x\prime = 0,\ t\prime = 0$) and end of the path (corresponding to $x\prime = L\prime,\ t\prime = L\prime/u\prime$). With an arbitrary choice of the origin in the station system we can assume that the first event corresponds to $x = 0,\ t = 0$. From Eq. (4.92) the second event corresponds to

$$x = \gamma L\prime\left(1 + \frac{v}{u\prime}\right) \quad t = \gamma\frac{L\prime}{u\prime}\left(1 + \frac{v\ u\prime}{c^2}\right) \tag{4.95}$$

Upon dividing the space by the time, it is possible to determine the velocity in the primed system

$$u = \frac{v + u'}{1 + \frac{v\ u\prime}{c^2}} \tag{4.96}$$

- *A particle is moving on the train at constant velocity $u\prime$ but now along $y\prime = 0$. How long does it take for the observer in the station?* In this case the particle velocity has no component along $x\prime$ in the primed system. The transit takes again a time $t\prime = L\prime/u\prime$. The particle remains at $x\prime = 0$ or $x = vt$. The last of Eq. (4.92) yields $t = \gamma L\prime/u\prime$ showing the time dilatation.

4.16 The Equivalence Between Mass and Energy

The most important consequence (for energy applications) of special relativity is the equivalence between mass and energy expressed by the celebrated Einstein equation

$$E = m_o \gamma c^2 \tag{4.97}$$

The quantity m_o is the mass of the particle (sometimes referred to as *rest mass*[2]) and the energy $E_o = m_o c^2$ is the *rest energy* of the particle (i.e. its energy when at rest).

The kinetic energy T is the difference between the total energy and the rest energy. For velocities much smaller than c we can expand the factor γ to obtain

$$T \equiv E - E_o \approx \frac{1}{2} m_o v^2 \tag{4.98}$$

This is the *non-relativistic limit* ($v \ll c$). The opposite limit ($v \to c$) is called the *ultra relativistic limit*. In this limit $\gamma \to \infty$ and the particle energy diverges. What happens is that we are accelerating the body by applying a force. However the velocity cannot exceed the speed of light, so as v approaches c we have to apply a larger and larger force. The work done goes in the energy of the object.

The total energy can be also expressed using the particle momentum

$$p = m_o \gamma v \tag{4.99}$$

By combining Eqs. (4.97) and (4.99) we have

$$E = (m_o^2 c^4 + p^2 c^2)^{1/2} \tag{4.100}$$

that relates E to p. If $E \gg E_o$ (*ultra-relativistic limit*) we have

$$E \simeq pc \tag{4.101}$$

This condition does not depend explicitly on the rest mass and is therefore exact for massless particle.

[2] Equation (4.97) is sometimes interpreted using the concept of *relativistic mass* $m \equiv m_o \gamma$. This is today no longer considered appropriate. The only mass of an object is its rest mass.

4.17 Suggestions for Further Readings

A full and rigorous treatment of electromagnetism can be found in the Jackson book [10]. For special relativity there are several introductory books. We suggest the presentation in Ref. [15].

4.18 Exercises

Problem 4.1 The Ge crystal lattice has a constant $d = 5.658\text{Å}$. The electromagnetic radiation is diffracted at an angle $\theta = 30^o$ at the second order. Evaluate the wavelength and the frequency of the electromagnetic radiation.

Problem 4.2 Consider the dispersion relation given in Eq. (4.47). Evaluate the phase and the group velocity for $\omega = 2 \times \omega_p$.

Problem 4.3 Evaluate the rest energy of the deuterium nucleus and compare it to the sum of the rest energy of the proton and the neutron.

Problem 4.4 Evaluate the velocity (as a fraction of the velocity of light of an electron and a proton both with energy $1GeV$.

References

1. I. Newton, Opticks: or, A Treatise of the Reflexions, Refractions, Inflexions and Colours of Light. (1704)
2. R. Hooke, *Micrographia: or, Some Physiological Descriptions of Minute Bodies Made by Magnifying Glasses*, ed. by J. Martyn, J. Allestry (London, 1665)
3. R. Descartes, Discours de la méthode, pour conduire la raison et chercher la vérité des sciences, plus La Dioptrique et Les Météores qui sont des essais sur cette méthode, Girard, 1642
4. C. Huygens, Traité de la Lumiére Chez Pieter van der Aa, marchand libraire, ed. by A. Leide (1690)
5. T. Young, The Bakerian lecture: on the theory of light and colours. Philos. Trans. R. Soc. Lond. **92**, 12–48 (1802)
6. A. Fresnel, Mémoire sur la diffraction de la lumiére (deposited 1818, "crowned" 1819), in Oeuvres complétes (Paris: Imprimerie impériale, 1866–70), vol. 1, pp. 247–363; partly translated as "Fresnel's prize memoir on the diffraction of light", in H. Crew (ed.), The Wave Theory of Light: Memoirs by Huygens, Young and Fresnel, American Book Co., 1900, pp. 81–14
7. J.C. Maxwell, *A Treatise on Electricity and Magnetism*, vol. 1 (Clarendon Press, Oxford, 1873)
8. A. Einstein, On a heuristic point of view concerning production and transformation of light. Ann. Phys. **17**, 132 (1905)
9. F.M. Grimaldi, Physico-mathesis de lumine, coloribus et iride aliisque adnexis, Girolamo Bernia, Johann Zieger, (1665)
10. J.D. Jackson, *Classical Electrodynamics*, 3rd edn. (John Wiley & Sons Inc, New York, 2021)
11. C. Galilei, Dialogo sopra i due massimi sistemi del mondo. Piovan Battista landing. Firenze (1632)

12. H.A. Lorentz, Simplified theory of electrical and optical phenomena in moving systems, in *Proceedings of the Royal Netherlands Academy of Arts and Sciences*, vol. 1 (1899), pp. 427–442. Electromagnetic phenomena in a system moving with any velocity smaller than that of light, in *Proceedings of the Royal Netherlands Academy of Arts and Sciences* vol. 6 (1904), pp. 809–831
13. A. Michelson, E. Morley, On the relative motion of the earth and the luminiferous ether. Am. J. Sci. **34**(203), 333–345 (1887)
14. G.F. FitzGerald, The ether and the Earth's atmosphere. Science **13**(328), 390 (1889)
15. B. Greene, *The Elegant Universe: Superstrings, Hidden Dimensions, and the Quest for the Ultimate Theory* (Random House Inc., Vintage Series, 2000)

Chapter 5
The Experimental Evidence of Quantum Mechanics

Abstract *The experimental evidence that led to the formulation of quantum mechanics is reviewed. Starting from the blackbody problem and the introduction of the concept of energy quanta the photoelectric effect, the Bohr-Rutherford model of the atoms and the Compton effect are discussed. The de Broglie relations for the description of matter as waves are introduced. The Heisenberg uncertainty principle is presented.*

In this chapter we review the main experimental evidence that at the beginning of the 20th century led physicists to propose quantum mechanics. The formulation of the new theory brought a significant change in the way we describe reality and indeed Albert Einstein always refused to accept that the universe could behave in a non deterministic way (summarised in the sentence "God does not play dice"). Nevertheless quantum mechanics is today a powerful tool that has several practical applications from quantum computing to metrology and cryptography.

5.1 The Black-Body Problem

Historically the first problem that required for its solution the introduction of quantisation was the black body problem. A black body is a system in which the electromagnetic radiation is in thermal equilibrium with the matter the body is made of. Radiation is continuously emitted and reabsorbed by matter. Such a strong coupling leads to thermal equilibrium. An example of a black body is a pizza oven. Radiation is produced by heating the oven wall and is continuously absorbed and re-emitted by the wall.

The electromagnetic field inside the oven is made of standing waves (see Fig. 5.1) corresponding to the natural modes of oscillation inside the cavity. For the sake of illustration we assume that the walls of the cavity are made of perfectly conducting materials, therefore the tangent component of the wave electric field must vanish at the wall. Assuming, for the sake of simplicity a cubic oven with edge of length L, each standing wave must have a wavelength such that only an integer or a half integer

F. Romanelli, *Physics of Nuclear Energy*, Springer Series in Plasma Science and Technology, https://doi.org/10.1007/978-981-97-9609-0_5

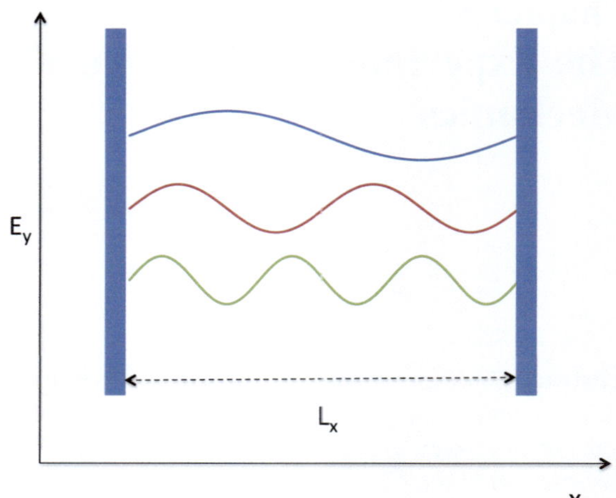

Fig. 5.1 A black body made of a cavity with perfectly conducting walls is characterised by a superposition of standing waves with the wave vector component in each direction being an integer multiple of the lowest wave vector. Each standing wave corresponds to an harmonic oscillator in which there is a continuous exchange of energy between the magnetic and the electric component of the wave

number of wavelengths is contained between two opposed walls (see Eq. 4.30)

$$L = \frac{l\pi}{k_x}, \quad L = \frac{m\pi}{k_y}, \quad L = \frac{n\pi}{k_z} \tag{5.1}$$

with l, m, n integers and $\mathbf{k}$ the wave vector. Thus, the electromagnetic field inside the oven consists of an infinite number of waves (Fig. 5.1b) with different l, m, n numbers. We have seen in Chap. 4 that a standing electromagnetic wave is equivalent to a harmonic oscillator with the energy being continuously exchanged between the electric and the magnetic field of the wave, similarly to the exchange of kinetic and potential energy in a mechanical oscillator. Therefore the electromagnetic field in the oven is equivalent to an infinite number of harmonic oscillators.

The spectral density of modes (the number of standing waves with frequency between ω and $\omega + d\omega$) can be evaluated from the dispersion relation of the electromagnetic waves in vacuum

$$\frac{\omega^2}{c^2} = k^2 = \left(\frac{\pi}{L}\right)^2 (l^2 + m^2 + n^2) \equiv \left(\frac{\pi}{L}\right)^2 p^2 \tag{5.2}$$

Since we are interested in large values of l, m, n we can consider the limit in which p can be regarded as a continuous quantity. The number $d\hat{N}$ of modes between ω and $\omega + d\omega$ will be the volume of a spherical layer octant of radius between p and $p + dp$ (Fig. 5.2)

$$d\hat{N} = \frac{1}{8} 4\pi p^2 dp = \frac{1}{8} 4\pi \left(\frac{L}{\pi c}\right)^3 \omega^2 d\omega = 4\pi \frac{V}{c^3} \nu^2 d\nu \tag{5.3}$$

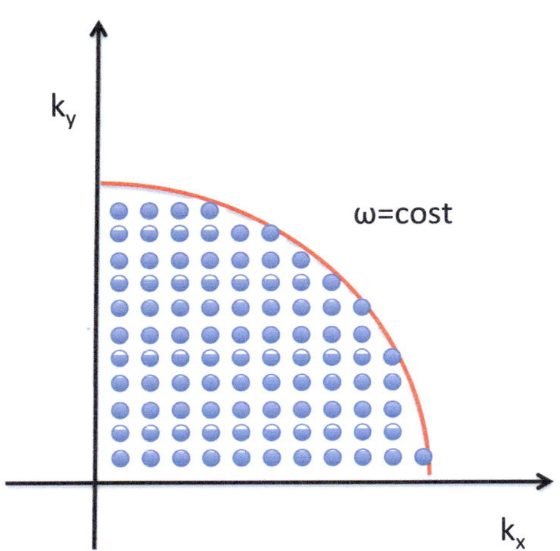

Fig. 5.2 The number of discrete modes inside an octant in **k** space can be determined by taking the continuous limit valid for large values of k

with $\omega = 2\pi\nu$. The number of modes per unit volume will be, taking into account that for each mode there are two independent polarizations

$$dN = \frac{2d\hat{N}}{V} = \frac{8\pi}{c^3}\nu^2 d\nu \tag{5.4}$$

The average energy in each mode $< E_\nu >$ is a function of the frequency only. Thus the energy density per unit frequency $u(\nu)$ of the electromagnetic field can be written as

$$dU = u(\nu)d\nu =< E_\nu > dN =< E_\nu > \frac{8\pi}{c^3}\nu^2 d\nu \tag{5.5}$$

To progress further the quantity $< E_\nu >$ must be evaluated on the basis of thermodynamic arguments.

To determine $< E_\nu >$ it is possible to exploit the equivalence of each wave to a harmonic oscillator. In conditions of thermodynamic equilibrium a harmonic oscillator has an average energy T. Since there is an infinite number of waves, if the average energy of each wave is the same the total energy would (badly) diverge. This problem was realised at the end of the 19th century under the name of *ultraviolet catastrophe*.

We know that this result is meaningless. The energy in a pizza oven is finite! How should we solve the problem? The right idea came to Max Planck [1] who argued that each standing wave is made by quanta of energy $h\nu$ with h a constant to be determined from the experiment. According to the Planck hypothesis the energy in each mode must be an *integer* multiple of $h\nu$. As we will see in a moment this hypothesis has a substantial impact in the evaluation of the average energy in each mode of oscillation.

If $h\nu \ll T$, the classical equipartition principle suggests that the average number of quanta at frequency ν will be $< n > \approx T/(h\nu)$. In this limit the mode is made by a *large* number of *small* quanta and the difference between the classical continuum approach and the Planck hypothesis will be minimal. The number of quanta will fluctuate (with a distribution that we do not know yet) but on average it will be such to satisfy the equipartition principle.

The impact of the Planck hypothesis becomes apparent for modes with frequency $h\nu \gg T$. In this case if the number of quanta is zero the energy in the mode is zero. However, already for a number of quanta equal to *one* the energy exceeds the average energy expected on the basis of classical thermodynamics. Thus, we may expect that the probability of having a finite number of quanta will be small at high frequency.

In order to evaluate the probability of having n quanta we have to go back to the statistical approach outlined in Chap. 2. The probability will follow a Boltzmann distribution

$$p(n) = \frac{e^{-\frac{nh\nu}{T}}}{\sum_{n=0}^{\infty} e^{-\frac{nh\nu}{T}}} \tag{5.6}$$

The normalisation factor at the denominator of Eq. (5.6) can be evaluated from the identity $\sum_{n=0}^{\infty} x^n = (1-x)^{-1}$ valid for $|x| < 1$. The average energy by summing the energy of n quanta averaged using the probability given in Eq. (5.6)

$$< E_\nu > = \sum_{n=0}^{\infty} n\ h\nu\ p(n) = \frac{h\nu}{e^{\frac{h\nu}{T}} - 1} \tag{5.7}$$

In the limit $h\nu \ll T$ the exponential in the denominator can be expanded for small arguments to recover the classical result $< E_\nu > \approx T$, whereas for $h\nu \gg T$ the average energy is exponentially small. Note that the classical limit is also recovered if we take $h \to 0$. Since $h\nu$ is the quantum energy Eq. (5.7) can also be interpreted by saying that the average number of quanta $< n_\nu >$ at frequency ν is

$$< n_\nu > = \frac{1}{e^{\frac{h\nu}{T}} - 1} \tag{5.8}$$

again, for $h\nu \ll T$ we recover the classical result whereas for $h\nu \gg T$ the average number of quanta is exponentially small. Equation (5.8) is the distribution function in energy of a $Bose-Einstein$ gas that we have already found in Chap. 2.

Using the expression for $< E_\nu >$ into Eq. (5.5) it is possible to obtain the spectral density

$$u(\nu) = < E_\nu > \frac{8\pi}{c^3}\nu^2 = \frac{8\pi h}{c^3} \frac{\nu^3}{e^{\frac{h\nu}{T}} - 1} \tag{5.9}$$

Equation 5.9 exactly reproduces the experimental data provided the value $6.6261 \times 10^{-34}\,Js$ is chosen for the Planck constant h (Fig. 5.3). It also shows that

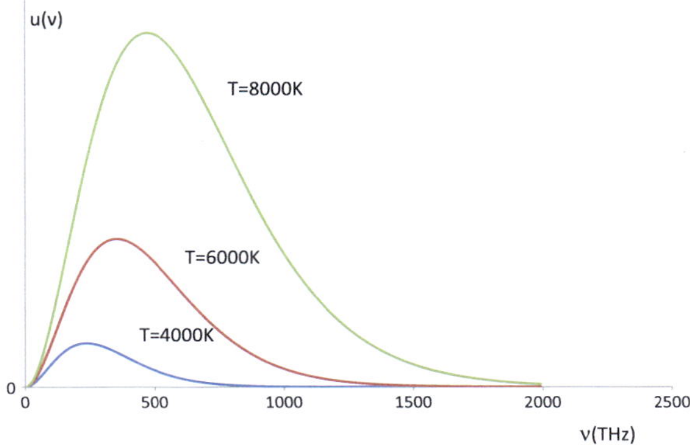

Fig. 5.3 The energy density of the electromagnetic field per unit frequency increases as ν^2 at low frequency as expected by classical physics arguments. However at high frequency $u(\nu)$ decreases exponentially preventing the ultraviolet catastrophe. The frequency corresponding to the maximum of the spectrum depends on the blackbody temperature and follows the Wien law Eq. 5.10

the maximum spectral density of a black body ($du(\nu)/d\nu = 0$) occurs for

$$\lambda_{max} T = 0.0029\,\mathrm{mK} \quad (Wien\ law) \tag{5.10}$$

The total energy density is obtained by integrating Eq. (5.9) over all the frequencies

$$U = \int_0^\infty u(\nu)d\nu = \frac{\pi^4}{15}\frac{8\pi h}{c^3}\left(\frac{T}{h}\right)^4 \tag{5.11}$$

The total energy density U can be also interpreted as the average energy of quanta E_f times their density n_f. Using this interpretation, an expression for the power radiated per unit surface by a black body can be obtained by evaluating the outgoing flux of quanta (Fig. 5.4).

The flux is obtained by integrating the normal component of the velocity of each quantum ($c\ \cos\theta$) over the fraction of solid angle that corresponds to quanta leaving the surface (corresponding to $0 \le \theta \le \pi/2$ in Fig.5.4)

$$dP = n_f E_f dS \int \frac{d\Omega}{4\pi} c\ \cos\theta = dS\frac{c}{4}U = dS\sigma T(K)^4 \tag{5.12}$$

with

$$\sigma = \frac{2\pi^5 k_B^4}{15h^3c^2} = 5.67 \times 10^{-8} W/(m^2K^4) \tag{5.13}$$

being the *Stefan-Boltzmann constant* and $k_B = 1.3807 \times 10^{-23} J/K$ the Boltzmann constant. The Planck result, originally presented as a mathematical expedient to solve the ultraviolet catastrophe problem, proved to be extremely fruitful.

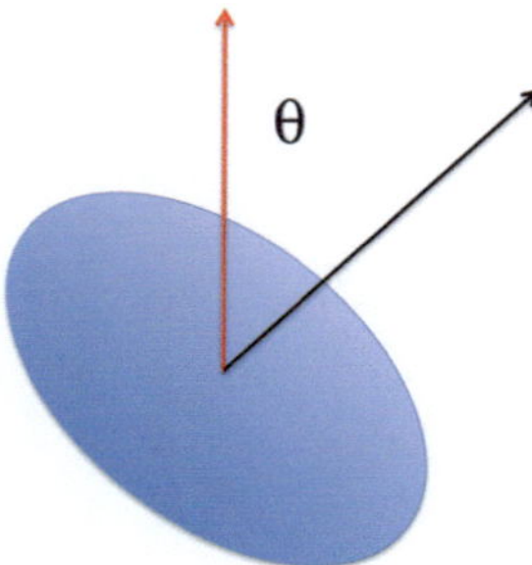

Fig. 5.4 The radiation per unit surface emitted by a blackbody can be evaluated by integrating the flux of photons through the surface. The flux of photons in the direction normal to the surface is simply $n_f c \cos\theta$ with n_f the photon density. The average energy of the photons times the photon density is given by Eq. 5.11. The power emitted per unit surface follows the Stefan Boltzmann law

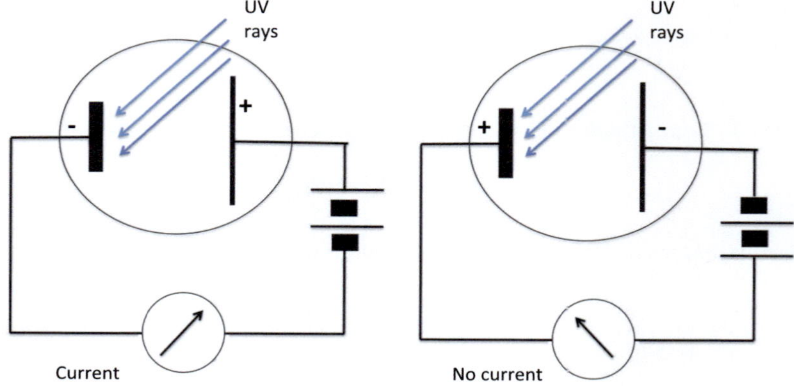

Fig. 5.5 Two electrodes are inserted in a glass tube under vacuum. When the electrode made of an alkali metal is irradiated by UV radiation electrons are emitted. If the alkali metal is negatively biased the electron are accelerated towards the other electrode and a current is detected. If the alkali metal is positively biased the other electrode tends to repel the electrons and a current is detected only if the potential difference is less that the energy of the emitted electrons

5.2 The Photoelectric Effect

In 1887 Hertz [2] discovered the photoelectric effect by using an experimental apparatus similar to that shown in Fig. 5.5. Two electrodes are inserted in a glass tube under high vacuum conditions. One electrode is an alkali metal and can be irradiated by electromagnetic waves in the ultraviolet range. Initially, the alkali metal electrode is at negative potential and the other electrode at positive potential. When the metal is irradiated a current starts to flow through the circuit. If the polarity is reversed, above a certain potential the current stops.

What is happening is that electrons are emitted from the metal at a certain energy and they flow towards the other electrode. In the experiment with reversed polarity we can stop the current if the potential difference is larger than the energy at which electrons are emitted. The experiment can be repeated by varying the intensity and the frequency of the wave. The result is the following:

- The energy at which electrons are emitted is independent of the intensity and increases linearly with the wave frequency.
- A threshold in frequency exists below which no electrons are ejected.
- The number of emitted electrons is proportional to the wave intensity.

This result cannot be explained by classical physics. Classically, we would expect that electrons are ejected when they absorb sufficient energy from the wave independently of the frequency at which absorption takes place. Thus, no frequency threshold is expected. We simply have to wait enough time for the electrons to absorb energy. The larger the wave intensity the shorter is the waiting time. However, in the experiment the emission is immediate even at very low intensity provided we are above the frequency threshold.

The photoelectric effect was explained by Einstein in [3] on the basis of the idea of quanta of light (*photons*). Each photon with energy $h\nu$ colliding with an electron of the metal transfers its energy to the electron. If the energy is larger than the extraction potential A of the electron a current is generated. The number of electrons that are extracted is proportional to the number of photons, i.e. to the wave intensity. The electron energy is given by

$$E = h\nu - A \tag{5.14}$$

The photoelectric effect was the demonstration that the Planck hypothesis was describing the true nature of radiation. For the explanation of the photoelectric effect Einstein was awarded the 1921 Nobel prize.

5.3 Atomic Spectra and Bohr Model

Thanks to the use of the diffraction grating, in the 19th century it was shown that atoms emit and absorb radiation at well defined frequencies and not on a continuous spectrum. Balmer [4] had shown that some of the hydrogen lines could be described by the expression

$$\lambda = 364.6nm \frac{n^2}{n^2 - 4} \tag{5.15}$$

with n integer larger than two. This series includes the well known H_α line at $656nm$ ($n = 3$) that is one of the most widely used lines to detect hydrogen in the universe. In 1889 Rydberg proposed a general expression of the form [5]

$$\frac{1}{\lambda} = R\left(\frac{1}{n_1^2} - \frac{1}{n_2^2}\right) \tag{5.16}$$

with n_1 and n_2 integer numbers and $R = 1.097 \times 10^7 m^{-1}$ being the *Rydberg constant*. The Rydberg expression reproduces the Lyman series ($n_1 = 1$)[6], the Balmer series ($n_1 = 2$), the Paschen series ($n_1 = 3$) [7] and the Brackett series ($n_1 = 4$) [8].

The understanding of this structure was a major contribution by Niels Bohr [9]. When Bohr proposed his theory Rutherford had already shown that the atom was made of a central positive nucleus with the electrons orbiting around. However, this description was incomplete.

Let us start by considering the atom a classical system and, for the sake of simplicity, assume that electrons follow circular orbits of radius r defined by the balance between the centrifugal force and the Coulomb force

$$\frac{m_e v^2}{r} = \frac{Ze^2}{4\pi\epsilon_o r^2} \tag{5.17}$$

The energy is given by the sum of kinetic and potential energy

$$E = \frac{m_e v^2}{2} - \frac{Ze^2}{4\pi\epsilon_o r} \tag{5.18}$$

Equations (5.17) and (5.18) can be combined to obtain a relation between the energy and the radius of the orbit

$$E = -\frac{Ze^2}{8\pi\epsilon_o r} \tag{5.19}$$

We know from Chap. 4 that an accelerated electric charge emits energy. Using the Larmor formula Eq. (4.57) the power radiated by an electron moving on its orbit can be estimated

$$P = \frac{e^2 a^2}{6\pi\epsilon_o c^3} = \frac{e^2}{6\pi\epsilon_o c^3}\left(\frac{Ze^2}{4\pi\epsilon_o m_e r^2}\right)^2 \tag{5.20}$$

The time needed to lose all its energy can be estimated by dividing the kinetic energy obtained from Eq. (5.18) by Eq. (5.20) yielding

$$\tau \approx \frac{3}{4Z}\frac{r}{c}\left(\frac{r}{r_{cl}}\right)^2 \tag{5.21}$$

We know that atoms have a dimension of the order $r \approx 5 \times 10^{-11} m$. Thus, within $10 ps$ the electron would fall on the nucleus!

Bohr proposed to overcome the problem by assuming that the electron does not radiate energy provided it is on a special orbit defined by the condition that its orbital angular momentum is an integer multiple of $\hbar \equiv h/(2\pi)$ (Fig. 5.6)

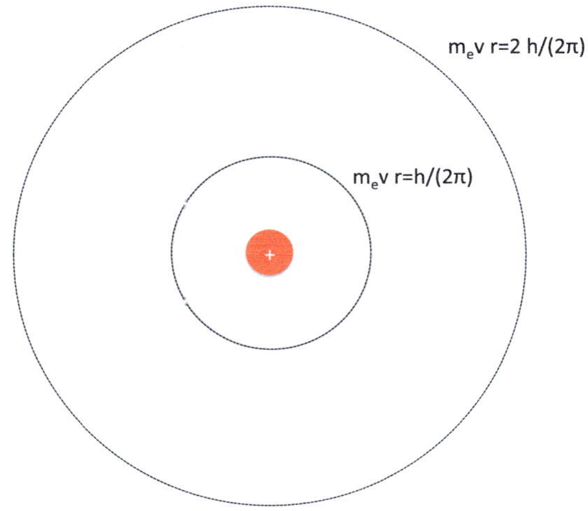

Fig. 5.6 The Bohr hypothesis identified priviledged orbits those for which the angular momentum is an integer multiple of $h/2\pi$

$$m_e v r = n\hbar \tag{5.22}$$

Upon using Eq. (5.17) we can express the Bohr condition as

$$r = a_o \frac{n^2}{Z} \tag{5.23}$$

with $a_o = \epsilon_o h^2/(\pi m_e e^2)$ the *Bohr radius*. Upon inserting Eq. (5.23) into Eq. (5.19) we obtain

$$E = -\frac{Z^2 m_e e^4}{8\epsilon_o^2 h^2} \frac{1}{n^2} \tag{5.24}$$

Equation (5.24) shows that the electron energy can take only well defined values corresponding to the radius of the orbit.

If we now assume that the emission and absorption of radiation takes place when the electron moves between two orbits characterised by orbital numbers n_1 and n_2, the energy conservation will relate the energy of the radiation quantum $h\nu$ to the energy difference of the electron in the two orbits

$$\frac{1}{\lambda} = \frac{\nu}{c} = \frac{Z^2 m_e e^4}{8\epsilon_o^2 h^3 c}\left(\frac{1}{n_1^2} - \frac{1}{n_2^2}\right) \tag{5.25}$$

Taking $Z = 1$ for hydrogen and substituting the values of the various constants we obtain $m_e e^4/(8\epsilon_o^2 h^3 c) = 1.097 \times 10^{-7} m^{-1}$ i.e. the exact value of the Rydberg constant!

Finally, in concluding this section we note that Eq. (5.24) for the ground state of the hydrogen atom $Z = 1$ and $n = 1$ yields the hydrogen ionisation energy $E_{ionisation} = 13.6\,\text{eV}$.

5.4 The Compton Effect

In Chap. 4 we have discussed the Thomson scattering on the basis of classical electromagnetism. An electron accelerated by the electric field of an electromagnetic wave emits a wave at the same frequency. This diffusion process has a cross section given by $8\pi r_{cl}^2/3$.

Classical electromagnetism is in good agreement with experiments as long as the photon energy is much smaller than 1 MeV. However, at higher frequencies the re-emitted wave is not at the same frequency. This phenomenon, discovered by Compton in [10] is a further evidence of quantum mechanics.

To understand why, let us take a corpuscular approach and consider the incoming wave as a flux of photons. Each photon has an energy $h\nu$ and a momentum that we can determine from the energy-momentum relation for a massless particle Eq. (4.100) $p_{photon} = E_{photon}/c = h\nu/c$. The momentum is directed along the propagation direction. Taking into account the dispersion relation of electromagnetic waves in vacuum we have

$$\mathbf{p}_{photon} = \hbar\mathbf{k} \tag{5.26}$$

Assume that the electron is at rest in the laboratory frame and that the incoming photon is moving in the x direction (Fig. 5.7).

The energy and momentum after the collision can be determined from the conservation of energy and momentum. Since the photon energy may be comparable to the electron rest energy we need the relativistic expressions. Denoting with a prime

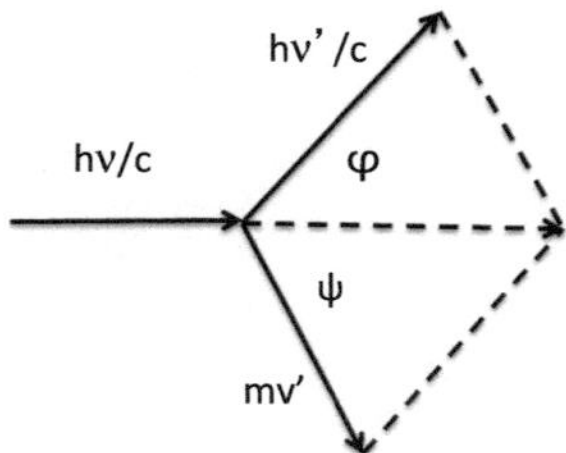

Fig. 5.7 In the Compton effect a photon of energy $h\nu$ and momentum $\hbar\mathbf{k}$ collide with an electron initially at rest. The energy and momentum of the photon and of the electron after the collision are determined by the energy and momentum conservation laws. Since part of the energy has been transferred to the electron the photon decreases its energy (and therefore its frequency). This is in contrast with classical Thomson scattering where the frequency of the wave is unchanged

the quantities after the collision, we have

$$h\nu + m_o c^2 = h\nu\prime + (p_e\prime^2 c^2 + m_o^2 c^4)^{1/2} \tag{5.27}$$

$$\hbar \mathbf{k} = \hbar \mathbf{k}\prime + \mathbf{p}_e{}' \tag{5.28}$$

It is convenient to determine the electron momentum after the collision from Eq. (5.28) and then substitute its expression in Eq. (5.27) to obtain

$$\begin{aligned}(h\nu + m_o c^2 - h\nu\prime)^2 &= \hbar^2 |\mathbf{k} - \mathbf{k}\prime|^2 c^2 + m_o^2 c^4 = \\ &= (h\nu)^2 + (h\nu\prime)^2 - 2(h\nu)(h\nu\prime)\cos\varphi + m_o^2 c^4\end{aligned} \tag{5.29}$$

that can be solved for $\nu\prime$ yielding

$$\frac{c}{\nu\prime} - \frac{c}{\nu} = \frac{h}{m_o c}(1 - \cos\varphi) \tag{5.30}$$

The quantity $h/(m_o c)$ is *Compton wavelength* of the electron $\approx 2.4263 \times 10^{-12} m$. Again, for $h \rightarrow 0$ the effect disappears and the frequency of the emitted wave would be the same as the frequency of the incident wave. Equation (5.30) can also be written as follows

$$\nu\prime = \frac{\nu}{1 + \frac{h\nu}{m_o c^2}(1 - \cos\varphi)} \tag{5.31}$$

that shows that the Compton effect becomes visible for photons of wavelengths of the order of the Compton wavelength, i.e. for photons with energies of the order of the rest energy of the particle.

5.5 How the Neutron Was Discovered

In 1930 Bothe and Becker (see Ref. [11]) while investigating the reaction between beryllium and the 5.3 MeV alpha particles produced by Po decay observed a penetrating radiation. Two years later Joliot and Curie [12] observed that the radiation passing through a material rich in hydrogen produced $\approx$5 MeV protons and interpreted the phenomenon as Compton scattering. In their opinion the radiation emitted in the $\alpha - Be$ reaction was electromagnetic radiation emitted by the excited ^{13}C nucleus formed via the capture of the alpha particle by beryllium. The Compton scattering of the gamma ray on hydrogen would have produced the high energy protons.

In 1932 Chadwick [13] pointed out that this was inconsistent with the conservation of energy and momentum and proposed that the radiation was an entirely new particle, the neutron.

To understand the Chadwick argument let us consider Eq. (5.29) written for a proton instead of an electron. Since the kinetic energy of the proton is much smaller than its rest energy we can use the non-relativistic limit. The proton kinetic energy T_p is given by

$$T_p = \frac{p_{p'}^2}{2m_p} = \frac{(h\nu)^2}{m_p c^2}(1 - \cos\varphi) \tag{5.32}$$

If the scattered proton has a energy of 5 MeV it means that the photon energy must be at least

$$h\nu = \left(\frac{5\,\text{MeV} \times 938\,\text{MeV}}{2}\right)^{1/2} \approx 50\,\text{MeV} \tag{5.33}$$

However the amount of energy available in the reaction $\alpha +^9 Be \rightarrow^{13} C + \gamma$ (i.e. the difference between the rest energy of ^{13}C and the sum of the rest energies of the α particle and of ^9Be) is only 10 MeV (that becomes 15 MeV if the alpha particle has an energy of 5 MeV), much less than the 50 MeV needed to explain the phenomenon as a Compton effect. The only possible way to explain the $\alpha +^9 Be$ reaction was to assume that the reaction products were a ^{12}C nucleus and a new particle, the neutron. In this case the energy available for the neutron would have been 5.7 MeV and could be transmitted to protons via elastic scattering.

5.6 The Wave Properties of Matter and the de Broglie Relations

After Einstein showed that light can have both a wavelike and corpuscular nature it was natural to ask whether matter could exhibit a similar dualism. In 1925 Louis de Broglie [14] proposed that matter could behave as a wave with wavelength and frequency given by

$$p = \frac{h}{\lambda} \quad E = h\nu \tag{5.34}$$

If the relation between E and p is known a dispersion relation can be derived that determine the group velocity of such a wave. Taking for example the relativistic equation Eq. (4.100) and the definition of the group velocity Eq. (4.41), from Eq. (5.34) we have

$$\mathbf{v}_g = \frac{\partial E}{\partial \mathbf{p}} = \frac{m_o \gamma \mathbf{v} c^2}{m_o \gamma c^2} = \mathbf{v} \tag{5.35}$$

Thus, the wave group velocity coincides with the particle velocity. This is a strong indication that we are on the right track, but would it be possible to demonstrate that

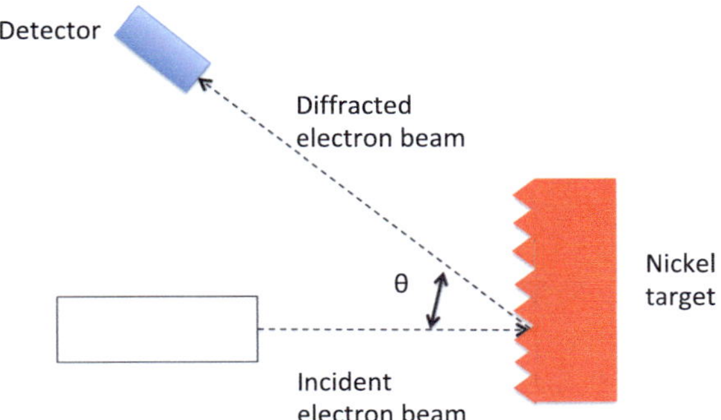

Fig. 5.8 Davisson and Germer experiment. The experimental setup consisted in bombarding a corrugated Nickel surface and to detect the diffracted beam

the main features of wavelike behaviour (diffraction and interference) also apply to matter waves?

The de Broglie idea had an immediate demonstration by Davisson and Germer [15] while studying the effect of the corrugation of a nickel surface by bombarding the surface with an electron beam and observing the beam reflection (Fig. 5.8). They discovered the diffraction pattern typical of a diffraction grating with the position of the maxima in agreement with the Bragg condition.

The result is shown in Fig. 5.9 from Ref. [15]. The angle of incidence is fixed at $\theta = 50^{o}$[1] and the accelerating potential Φ of the electron beam is varied. The wavelength is related to the accelerating potential by the following equation

$$\lambda = \frac{h}{(2m_e e\Phi)^{1/2}} \tag{5.36}$$

The maxima are expected when $2d \sin\theta = n\lambda$, with the nickel lattice constant being $0.091 nm$, thus for

$$\Phi^{1/2} = \frac{nh}{(2m_e e)^{1/2} 2d\cos(\theta/2)} \approx n7.54(V^{1/2}) \tag{5.37}$$

Equation (5.37) indeed reproduces the first, third and fifth maxima.

The wavelike nature of electrons has been used to develop the Transmission Electron Microscope. The advantage of TEM is the possibility of focussing a beam of charged particles using electric and magnetic field and of changing the wavelength continuously by changing the beam energy. The maximum theoretical resolution is high. The limitations come from the aberrations of the focalization optics, the interaction of the beam and the sample surface and limited resolution of the detectors.

[1] Note that in Ref. [15] the incidence angle is defined as the complement to 90° of the usual definition.

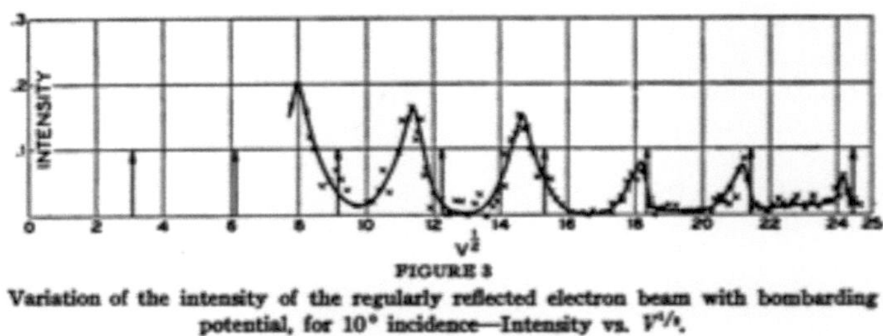

Fig. 5.9 Davisson and Germer experiment. The scattering of electrons from the surface of a nickel crystal shows the characteristic features of wavelike phenomena (from Ref. [15])

5.7 The Heisenberg Uncertainty Principle

Perhaps the most revolutionary aspect of quantum physics is the uncertainty principle. To understand its origin, let us consider a beam of electrons with velocity along y passing through a narrow slit of aperture a. This reproduces the Fresnel experiment using electrons instead of electromagnetic waves. When the electron passes through the slit we can state that we know its position along x with an uncertainty given by

$$\Delta x \approx a \tag{5.38}$$

The beam produces a diffraction pattern on the screen meaning that also regions not in front of the screen can be reached by the electron. We can interpret this effect by saying that in passing through the slit the electron takes a component of momentum along x given by

$$\Delta p_x \approx p \sin \phi \tag{5.39}$$

Taking for ϕ the value corresponding to the first diffraction minimum we have $a \sin \phi = \lambda$ with λ given by the de Broglie relation Eq. (5.34). Upon combining these relations we obtain a condition on the product of the two uncertainties

$$\Delta x \Delta p_x = ap \sin \phi = p\lambda = h \tag{5.40}$$

A proper analysis provides a better estimate that leads to the uncertainty principle [16]

$$\Delta x \Delta p_x \geq \frac{\hbar}{2} \tag{5.41}$$

Thus we cannot simultaneously measure with arbitrary accuracy the position and the velocity of a particle.

5.8 Exercises

Problem 5.1 Rutherford had made the hypothesis that the neutron was a proton-electron pair. What would be the electron energy if it had to be confined within a nuclear radius ($R \approx 10^{-15}m$)?

Problem 5.2 Evaluate the energy variation of a photon scattered at 30° by an electron initially at rest. Assume a photon in the UV rage and in the gamma range.

Problem 5.3 Evaluate the de Broglie wavelength of an electron of energy 100 keV and 1 MeV.

References

1. M. Planck, On the theory of the energy distribution law of the normal spectrum. Verhandl. Dtsch. Phys. Ges. **2**, 237
2. H. Hertz, Untersuchungen über die Ausbreitung der elektrischen Kraft Ein-leitung (Sándig Reprint Verlag, Vaduz, 1993) Ed. (1894)
3. A. Einstein, On a heuristic point of view concerning production and transformation of light. Ann. Physic **17**, 132 (1905)
4. J.J. Balmer, Notiz uber die Spectrallinien des Wasserstoffs. Annalen der Physik **261**(5), 80–87 (1885)
5. J.R. Rydberg, Researches sur la constitution des spectres d'émission des éléments chimiques. [Investigations of the composition of the emission spectra of chemical elements]. Kongliga Svenska Vetenskaps-Akademiens Handlingar [Proceedings of the Royal Swedish Academy of Science]. 2nd series (in French) **23**(11), 1–177 (1889)
6. T. Lyman, The spectrum of hydrogen in the region of extremely short wave-length. Memoirs Am. Acad. Arts Sci., New Ser. **13**(3), 125–146 (1906)
7. F. Paschen, Zur Kenntnis ultraroter Linienspektra. I. (Normalwellenlängen bis 27000 A.-E.). Annalen der Physik **332**(13), 537–570 (1908)
8. F.S. Brackett, Visible and infra-red radiation of hydrogen. Astrophys. J. **56**, 154 (1922)
9. N. Bohr, On the constitution of atoms and molecules, part I. Philos. Mag. **26**, 1–25 (1913)
10. A.H. Compton, A quantum theory of the scattering of X-rays by light elements. Phys. Rev. **21**(5), 483–502 (1923)
11. W. Bothe, H. Becker, Künstliche Erregung von Kern-γ-Strahlen [Artificial excitation of nuclear γ rays]. Z. Phys. **66**, 289 (1930)
12. I. Curie, F. Joliot, Émission de protons á grande vitesse par les substances hydrogénées sous l'influence des rayons γ trés pénétrants. C. r. hebd. séances Acad. Sci. Paris **194**, 273 (1932)
13. J. Chadwick, Possible existence of a neutron. Nature **129**, 312 (1932)
14. L. de Broglie, Recherches sur la théorie des quanta (Researches on the quantum theory). Thesis (Paris), 1924; de Broglie, L. Ann. Phys. (Paris) **3**, 22 (1925)
15. C.J. Davisson, L.H. Germer, Reflection of electrons by a crystal of nickel. Proc. Natl. Acad. Sci. United States of America **14**(4), 317–322 (1928)
16. W. Heisenberg, Uber den anschaulichen Inhalt der quantentheoretischen Kinematik und Mechanik. Zeitschrift für Physik (in German). **43**(3–4), 172–198 (1927)

Chapter 6
The Theoretical Basis of Quantum Mechanics

Abstract *The Schrödinger equation is introduced and discussed in few special cases: free particle, particle in a potential well and particle impinging on a barrier. The tunnel effect is discussed first for a rectangular barrier and then applied to the Coulomb barrier of nucleus to discuss the alpha decay and the fusion process.*

This chapter provides the background information on the Schrödinger equation and its solution in a few cases of relevance for the nuclear phenomena. The solution of the Schrödinger equation is illustrated for one dimensional problems. The probability of penetration through a potential barrier is discussed and applied to the description of alpha decay and fusion cross sections.

6.1 The Schrödinger Equation

We have seen in the previous chapter that matter can exhibit wave-like features that are described by the De Broglie relations. In this chapter we show how to describe the wave dynamics by introducing the Schrödinger equation [1].

A wave can always be expressed as a superposition of plane waves of the form $\hat{\psi} \propto \exp(-i(\omega t - \mathbf{k}\cdot\mathbf{r}))$. The de Broglie relations Eq. (5.34) on the other hand suggest that the energy and the momentum of a particle can be obtained by applying the following operators to the wave function

$$E\hat{\psi} = \hbar\omega\,\hat{\psi} = i\hbar\frac{\partial\hat{\psi}}{\partial t}, \quad \mathbf{p}\hat{\psi} = \hbar\mathbf{k}\hat{\psi} = -i\hbar\nabla\hat{\psi}, \tag{6.1}$$

with $\hbar = h/(2\pi)$. The relation between energy and momentum in non-relativistic theory provides the desired equation

$$i\hbar\frac{\partial\hat{\psi}}{\partial t} = -\frac{\hbar^2}{2\,m}\Delta\hat{\psi} + V(\mathbf{r})\hat{\psi} \tag{6.2}$$

with $V(\mathbf{r})$ being the interaction potential.

F. Romanelli, *Physics of Nuclear Energy*, Springer Series in Plasma Science and Technology, https://doi.org/10.1007/978-981-97-9609-0_6

The solution of Eq. (6.2) with appropriate initial and boundary conditions is the wave function. The quantity $|\hat{\psi}|^2 dx$ is the probability of finding the particle between x and $x + dx$ (see discussion on the interpretation of the wave function later in this chapter).

The Schrödinger equation is a wave equation and can be solved either as an initial value problem (e.g. for the propagation of a wave packet in space, knowing $\hat{\psi}(\mathbf{r}, t = 0)$ determine $\hat{\psi}$ at an arbitrary time t) or as an eigenvalue problem (e.g. determine the characteristic modes of oscillation of the system—the standing wave problem as discussed in Chap. 4) with the energy E being the eigenvalue. In the latter case we will consider solutions of the Schrödinger equation of the form $\hat{\psi}(\mathbf{r}, t) = \exp(iEt/\hbar)\psi(\mathbf{r})$ and solve Eq. (6.2) with appropriate boundary conditions. Equation (6.2) reduces to

$$E\psi = -\frac{\hbar^2}{2m}\Delta\psi + V(\mathbf{r})\psi \tag{6.3}$$

This is the *time-independent Schrödinger equation.*

6.2 The Free Particle

If the interaction potential is zero (no forces acting on the particles) the solution of Eq. (6.3) can be written in terms of plane waves. For the sake of simplicity let us assume that the solution is independent of y and z in such a way that Eq. (6.3) reduces to an ordinary differential equation

$$\frac{\hbar^2}{2m}\frac{d^2\psi}{dx^2} + E\psi = 0 \tag{6.4}$$

The classical analogous of this problem is a free particle moving along x with momentum p_x constant in time and energy $E = p_x^2/2m > 0$. The solution of the Schrödinger equation is

$$\psi = Ae^{-ikx} + Be^{ikx} \tag{6.5}$$

with

$$\frac{\hbar^2 k^2}{2m} = E \tag{6.6}$$

The solution is the combination of two waves, one propagating along the positive x axis (the solution with amplitude A) and the other propagating along the negative x axis (the solution with amplitude B)—see Chap. 4 for the discussion of electromagnetic waves propagation.

The boundary conditions correspond to the choice of the propagation direction. The problem with the particle coming from $x = -\infty$ and going to $x = +\infty$ will

Fig. 6.1 Free particle. A free particle can be represented as a plane wave propagating along a fixed direction in space (the x direction in the figure). The sign of the wave vector determine if the particle moves to positive or negative values of x

correspond to the choice $B = 0$ and vice-versa, if the particle is moving from $x = +\infty$ to $x = -\infty$ the choice will be $A = 0$ (Fig. 6.1).

Any value of energy $E \geq 0$ is allowed. Thus the spectrum of eigenvalues is continuous.

6.3 One-Dimensional Potential Well

Let us now consider a one-dimensional problem in which the potential is zero everywhere except that for $-a \leq x \leq a$ where it takes the (negative) value $V = -V_o$. The potential (see Fig. 6.2) takes the form of a rectangular well.

Before entering in the details of the solution of the Schrödinger equation it is instructive to consider the classical analogous of this problem. The force can be written in terms of the potential as

$$\mathbf{F} = -\nabla V \tag{6.7}$$

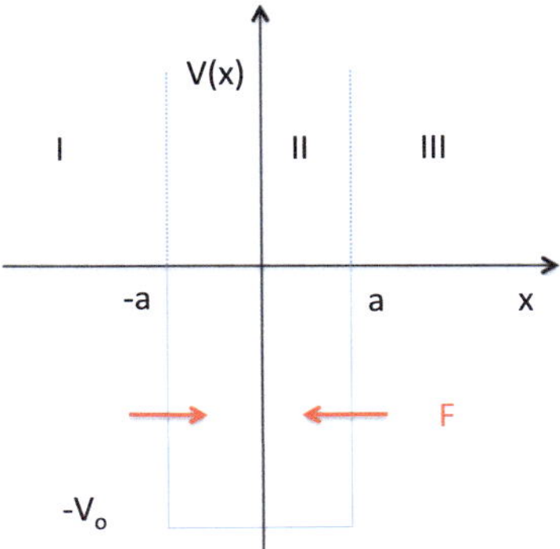

Fig. 6.2 Potential well. A bound state can be represented as a potential well. A particle with energy below the well maximum is trapped in the well. When the particle hits the boundaries of the well it feels a force that pushes it back in the well

and in this specific case corresponds to a force that it is zero both inside and outside the potential well and infinite at the well boundary $x = \pm a$. To be specific, the force is directed along the *positive* x direction at $x = -a$ and along the *negative* x direction for $x = +a$. To check the sign of the force let us replace the sharp boundary with a gentler behavior than the one shown in Fig. 6.2 for example by taking $V(x)$ a linear function of x between $-a - \epsilon$ and $-a + \epsilon$, and then taking the limit $\epsilon \to 0$. Similarly for $x = +a$.

The particle motion will depend on the particle energy. If the energy E satisfies the condition

$$0 \geq E \geq -V_o \tag{6.8}$$

the particle will perform an oscillatory motion between $x = -a$ and $x = +a$ and it will be trapped in the potential well. Inside the potential well the particle will have a kinetic energy $T = E + V_o \geq 0$ whereas it is not allowed outside the potential well since its kinetic energy would be negative if Eq. (6.8) is satisfied.

For $E > 0$ the particle can travel in all the three regions shown in Fig. 6.2 because its kinetic energy is always positive. However, its motion will be perturbed every time it arrives at the well boundaries where it will receive a kick. We expect that a particle traveling from $x = +\infty$ with negative velocity will be accelerated by the force at $x = a$ and then decelerated by the same amount at $x = -a$ continuing up to $x = -\infty$. Similar considerations can be made for a particle moving with positive velocity from $x = -\infty$ to $x = +\infty$.

The classical analogous suggests that different boundary conditions must be applied for the case $E < 0$ (particle bound to stay inside the well) and the case $E > 0$ (particle moving everywhere in x). Therefore we will treat these two cases separately.

For both cases the solution of the time-independent Schrödinger equation is made simple by noting that in all the three regions of Fig. 6.2 the equation reduces to a second order ordinary differential equation with constant coefficients and therefore can be written as in Eq. (6.5) with k given by

$$\frac{\hbar^2 k^2}{2m} = E + V \tag{6.9}$$

However, since V is a function of x we have to determine the solution separately in the three regions and then impose the continuity of the solution and its derivative at the well boundaries.

6.3.1 *Bound States* $0 \geq E \geq -V_o$

Inside the well (region II) the quantity $(E + V_o)$ is positive and the wave function will be the superposition of two plane waves propagating in the positive and negative directions

$$k_{II} = \pm\frac{[2m(E + V_o)]^{1/2}}{\hbar} \tag{6.10}$$

The two waves correspond to the particle oscillating back and forth in the well. Thus, in region II the solution will be

$$\psi_{II} = A_{II}e^{ik_{II}x} + B_{II}e^{-ik_{II}x} \tag{6.11}$$

Outside the well $V_o = 0$ and the wave vector is a purely imaginary values

$$k_I = k_{III} = \pm i\frac{2m(-E)^{1/2}}{\hbar} \equiv \pm ik_{ext} \tag{6.12}$$

Imaginary values for k correspond to solutions that grow or decay exponentially in space. This is again reminiscent of the problem of propagation of electromagnetic waves in the ionosphere that we have discussed in Chap. 4. The solution in regions I and III will be of the form

$$\psi_I = A_I e^{-k_{ext}x} + B_I e^{k_{ext}x} \tag{6.13}$$

$$\psi_{III} = A_{III}e^{-k_{ext}x} + B_{III}e^{k_{ext}x} \tag{6.14}$$

The solutions that are exponentially growing for $x \to \infty$ and $x \to -\infty$ have no physical meaning. Only the solutions that are exponentially decreasing as $|x| \to \infty$ are acceptable. Therefore the boundary conditions for the bound state problem are (Fig. 6.3)

$$A_I = 0 \qquad B_{III} = 0 \tag{6.15}$$

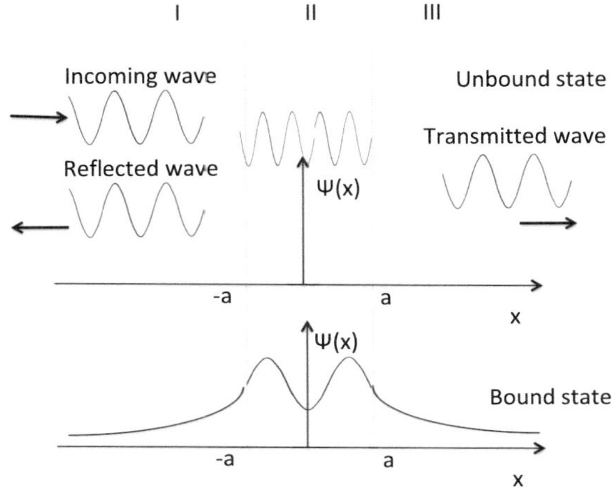

Fig. 6.3 In the case of the potential well we have to distinguish between unbound states (top part of the figure) and bound states (bottom part of the figure). For unbound states the wave can travel over the entire domain. Bound states have evanescent waves outside the potential well

Note that the boundary conditions determine two of the six unknown amplitudes. The remaining four unknowns are determined by matching the wave function and its derivative for $x = -a$ and $x = a$, yielding

$$B_I e^{-k_{ext} a} = A_{II} e^{-ik_{II} a} + B_{II} e^{ik_{II} a} \tag{6.16}$$

$$k_{ext} B_I e^{-k_{ext} a} = ik_{II} A_{II} e^{-ik_{II} a} - ik_{II} B_{II} e^{ik_{II} a} \tag{6.17}$$

$$A_{III} e^{-k_{ext} a} = A_{II} e^{ik_{II} a} + B_{II} e^{-ik_{II} a} \tag{6.18}$$

$$-k_{ext} A_{III} e^{-k_{ext} a} = ik_{II} A_{II} e^{ik_{II} a} - ik_{II} B_{II} e^{-ik_{II} a} \tag{6.19}$$

This is a system of four linear homogeneous equations that has non trivial solutions only if the determinant of the equations vanishes

$$\det \begin{pmatrix} e^{-k_{ext} a} & -e^{-ik_{II} a} & -e^{ik_{II} a} & 0 \\ k_{ext} e^{-k_{ext} a} & -ik_{II} e^{-ik_{II} a} & ik_{II} e^{ik_{II} a} & 0 \\ 0 & -e^{ik_{II} a} & -e^{-ik_{II} a} & e^{-k_{ext} a} \\ 0 & -ik_{II} e^{ik_{II} a} & ik_{II} e^{-ik_{II} a} & -k_{ext} e^{-k_{ext} a} \end{pmatrix} = 0 \tag{6.20}$$

that can be rearranged to give the following equations

$$tg(k_{II} a) = \frac{k_{ext}}{k_{II}} \tag{6.21}$$

$$cotg(k_{II} a) = -\frac{k_{ext}}{k_{II}} \tag{6.22}$$

corresponding respectively to solutions with even and odd parity for reflections at $x = 0$.

Equations (6.21)–(6.22) require a numerical solution. In the case of an infinite potential well ($V_o \rightarrow \infty$) the solution can be found analytically and corresponds to

$$k_{II} a = (2n + 1)\frac{\pi}{2} \quad n \geq 0 \tag{6.23}$$

$$k_{II} a = n\pi \quad n \geq 1 \tag{6.24}$$

Note that the wave vector inside the well never vanishes. This means that the particle inside the potential well has always a finite kinetic energy.

6.3.2 Unbound States

Particles with $E > 0$ are described by the superposition of propagating waves in all the three regions of Fig. 6.2 with the wave vector k given by

$$k_I = k_{III} = \pm\frac{(2mE)^{1/2}}{\hbar} \quad k_{II} = \pm\frac{(2m(E + Vo))^{1/2}}{\hbar} \tag{6.25}$$

For unbound states the boundary conditions provide less restrictions than in the bound case. The problem that we are dealing with can be formulated as follows. A particle is coming from $x = -\infty$ with positive velocity. The associated wave function is given by $\psi = A_I \exp(ik_I x)$. When the particle hits the left well boundary ($x = -a$) it is partially transmitted to $x > -a$ (with a modified wave vector) and partially reflected towards $x = -\infty$ (corresponding to a wave function $\psi = B_I \exp(-ik_I x)$). A second transmission/reflection takes place at $x = a$. Beyond this point the particle can travel undisturbed towards $x = +\infty$. Thus, the boundary conditions in this case reduce to

$$B_{III} = 0 \tag{6.26}$$

Obviously, we can equally consider the symmetric problem in which the particle is moving from $x = +\infty$ with negative velocity and after the transmission/reflection at $x = a$ and $x = -a$ continues undisturbed to $x = -\infty$ to conclude that in this case the only boundary condition would be $A_I = 0$. In both cases we have *five* unknowns but only *four* conditions (the matching of ψ and its derivative at $x = -a$ and $x = a$). Four of the unknowns can be determined in terms e.g. of the amplitude of the incoming particle (i.e. in terms of A_I for the problem with boundary condition $B_{III} = 0$ and in terms of B_{III} for the problem with boundary condition $A_I = 0$). The spectrum of eigenvalues is again continuous (all the values with $E > 0$ are allowed).

So, what is the information obtained by solving this problem? The information is the probability of transmission of the particle through the well and the associated probability for the particle to be reflected. Since in our problem particles are neither created nor destroyed, the sum of the probabilities of reflection and transmission must be 100%.

To make the discussion mathematically easier, we consider a simplified version of this problem in which we have a potential step, rather than a potential well, with the step located at $x = 0$ (see Fig. 6.4).

We will label the two regions $x < 0$ and $x > 0$ again as region I and region III (we do not have region II any longer). The boundary conditions are identical to those just discussed but the wave vector in region III is given by

$$k_{III} = \pm\frac{(2m(E + Vo))^{1/2}}{\hbar} \tag{6.27}$$

Taking for example the problem with boundary condition $B_{III} = 0$, the matching of ψ and its derivative in $x = 0$ yields

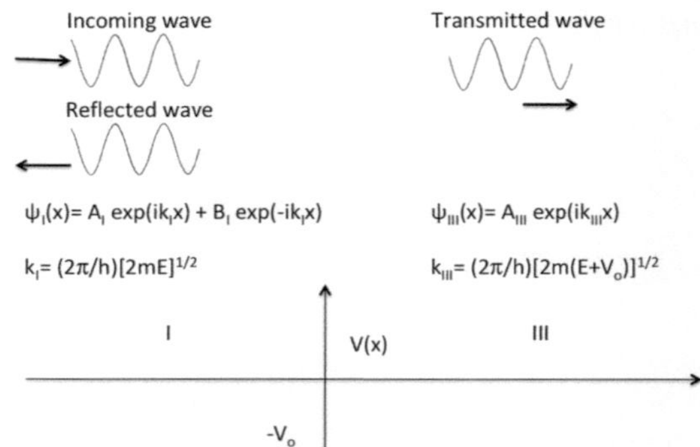

Fig. 6.4 Potential step. A particle with energy above the step in the potential is propagating over the entire x domain. When the particle arrives at the discontinuity it is partially reflected and partially transmitted. Since the process does not lead to absorption the same of the probability of transmission and the probability of reflection is 100%

$$A_I + B_I = A_{III} \tag{6.28}$$

$$k_I A_I - k_I B_I = k_{III} A_{III} \tag{6.29}$$

That can be easily solved to give

$$A_{III} = \frac{2k_I}{(k_I + k_{III})} A_I \tag{6.30}$$

$$B_I = \frac{(k_I - k_{III})}{(k_I + k_{III})} A_I \tag{6.31}$$

The transmission coefficient is obtained from Eq. (6.30)

$$T = \frac{|A_{III}|^2}{|A_I|^2} = \frac{4k_I^2}{(k_I + k_{III})^2} \tag{6.32}$$

It is possible to show that this solution indeed satisfies the condition of matching between the incoming particle flux and the transmitted plus reflected particle flux. The flux is given by the particle velocity times the modulus square of the amplitude. The incoming flux is therefore given by

$$\Gamma_{in} = \frac{hk_I}{m} |A_I|^2 \tag{6.33}$$

whereas the transmitted and the reflected flux are given respectively by

$$\Gamma_{tr} = \frac{hk_{III}}{m}|A_{III}|^2 \tag{6.34}$$

$$\Gamma_{ref} = -\frac{hk_I}{m}|B_I|^2 \tag{6.35}$$

Upon substituting Eqs. (6.27)–(6.31) into Eqs. (6.33)–(6.35) it is easy to show that

$$\Gamma_{in} = \Gamma_{tr} + \Gamma_{ref} \tag{6.36}$$

6.4 The Physical Meaning of the Wave Function

The physical meaning of the wave function has been the subject of discussion among the fathers of quantum mechanics. The interpretation that eventually emerged is due to Max Born who proposed to consider the wave function a *probability amplitude* and was the basis of the Copenhagen interpretation of quantum mechanics. Within this approach a quantum state is the superposition of different eigenstates of the system. Each eigenstate in the wave function will be characterised by a certain amplitude. When the measurement of a specific physical quantity (*observable*) is made, the system jumps on a specific eigenstate and the result of the measurement will be the corresponding eigenvalue. The square of the amplitude of the resulting eigenstate gives the probability density of measuring the corresponding physical value. If the wave function is composed by a single eigenstate, then only one result is possible. Otherwise, all the values corresponding to the various eigenstates can result, with a probability associated with the amplitude of the corresponding eigenstate. The probabilistic approach did not satisfy some of the physicists of the time (in particular Albert Einstein who criticised the interpretation with the famous phrase "God does not play dice") but was the best compromise to avoid contradictions between the corpuscular and the wave description of matter.

6.5 One Dimensional Potential Barrier and the Tunnel Effect

Let us now consider the case of a potential that is zero everywhere except between $x = 0$ and $x = a$ where it has a positive value V_o. Looking again at the classical limit it is possible to realise that this situation corresponds to a potential barrier. If the particle energy is lower than V_o a particle arriving from $x = -\infty$ will be reflected back. Quantum mechanics allows for an entirely new phenomenon.

Again we can solve the Schrödinger equation in the three regions shown in Fig. 6.5. If $E < V_o$ the solution will correspond to oscillations for $x \leq 0$ and $x \geq a$ and to increasing and decreasing exponential functions inside the barrier

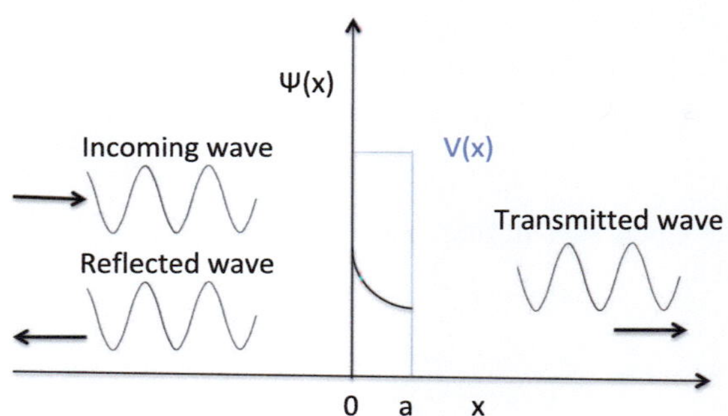

Fig. 6.5 Potential barrier. A particle moving from the left with an energy below the barrier energy would be reflected on the basis of classical physics. In quantum physics the particle has a finite probability to tunnel through the barrier

$$\psi_I = A_I e^{ik_I x} + B_I e^{-ik_I x} \tag{6.37}$$

$$\psi_{II} = A_{II} e^{-k_{II} x} + B_{II} e^{k_{II} x} \tag{6.38}$$

$$\psi_{III} = A_{III} e^{ik_{III} x} + B_{III} e^{-ik_{III} x} \tag{6.39}$$

with

$$k_I = k_{III} = k = \pm\frac{(2mE)^{1/2}}{\hbar} \quad k_{II} = \pm\frac{(2m(V_o - E))^{1/2}}{\hbar} = k_o \tag{6.40}$$

Using the same argument that we have employed to discuss the case of unbound states for the potential well we can assume as boundary condition $B_{III} = 0$ and determine the amplitudes in region I in terms of A_{III}.

$$A_I = \frac{k_o}{4k}\left(\left(1 - i\frac{k}{k_o}\right)^2 e^{k_o a} - \left(1 + i\frac{k}{k_o}e^{-k_o a}\right)\right)e^{ika}A_{III} \tag{6.41}$$

The interesting result is that the amplitude of the wave function in region III is different from zero: there is a finite probability for the particle to pass through the potential barrier (*tunnel effect*)! The probability can be obtained from Eq. (6.41)

$$T = \frac{|A_{III}|^2}{|A_I|^2} = \frac{16\frac{k^2}{k_o^2}}{(1 - \frac{k^2}{k_o^2})^2(e^{2k_o a} + e^{-2k_o a} - 2) + \frac{4k^2}{k_o^2}(e^{2k_o a} + e^{-2k_o a} + 2)} \tag{6.42}$$

The interesting limit is when the barrier is high enough or wide enough (or both) that the probability of tunnel effect is small. In this limit Eq. (6.42) yields

$$T \approx 16\frac{k^2 k_o^2}{(k^2 + k_o^2)^2}e^{-2k_o a} \tag{6.43}$$

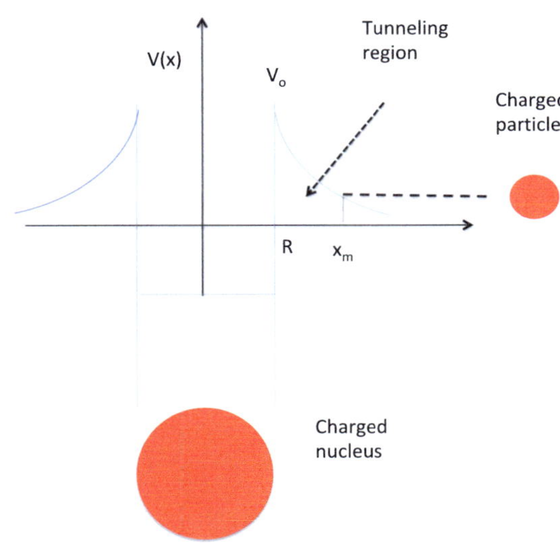

Fig. 6.6 Nuclear potential. The nuclear potential can be modelled as a square well with the dimension of the nuclear radius. Inside the well the proton and the neutrons are held together by the nuclear forces. Outside the nucleus the Coulomb potential repels positively charged particles

which shows that the transmission coefficient is exponentially small.

Note that the tunnel effect is a quantum mechanical effect. If we take the limit $h \to 0$ the exponential function in Eq. (6.41) vanishes.

6.6 The Nuclear Potential and the Gamow Factor

A first application of the tunnel effect is the determination of the probability for two charged nuclei to fuse. If we take two light nuclei (for example deuterium and tritium) the interaction potential is repulsive at large distances, as both nuclei are positively charged, and attractive at small distances, with the two fusing nuclei forming a ^5He nucleus that in a very short time disintegrate in an α-particle and a neutron. This potential can be modelled, as shown in Fig. 6.6, as a potential well for a distance $r \leq R$ with R being the nuclear radius and a Coulomb potential for $r \geq R$. A proper solution should be found by solving the Schrödinger equation in three dimensions. However, it is possible to show that for our purposes a one-dimensional treatment is justified.

In the centre of mass system the problem is equivalent to the penetration of a particle of mass equal to the reduced mass of the two particles through a potential barrier. Classically, the reduced mass particle would approach the origin up to the point at which its kinetic energy equals the repulsive Coulomb potential. At this point the particle would be reflected back. However in quantum mechanics we know that there is a finite probability for the particle to penetrate through the barrier and we want to quantify such a probability (Fig. 6.7). Unfortunately, the potential barrier is not a square barrier as in paragraph 6.5 but rather a space dependent barrier. However this problem can be overcome as we will see in a moment.

In a potential constant in space the wavelength of the wave function is constant. If the potential is not constant the wavelength will change with x. However, if the spatial scale of variation of the wavelength is much longer than the wavelength itself we can construct a solution using the WKB approximation [2]. To illustrate the method let us consider the following equation

$$\frac{d^2 f}{dx^2} + Q(x) f = 0 \tag{6.44}$$

Within the WKB approximation in each point a *local* wave vector $k(x)$ is defined from the relation

$$\frac{df}{dx} = ikf \tag{6.45}$$

The wave vector can be expressed in terms of $Q(x)$ as

$$k(x) = Q(x)^{1/2} \tag{6.46}$$

and the function f can be expressed, as a first approximation, as

$$f \approx f(0) e^{i \int Q(x')^{1/2} dx'} \tag{6.47}$$

The approximation is valid as long as the scale of variation of $k(x)$ is longer than $1/k$, i.e. for

$$k(x) \gg \frac{\frac{dk}{dx}}{k} \tag{6.48}$$

or

$$Q(x)^{1/2} \gg \frac{Q\prime}{2Q} \tag{6.49}$$

For our problem

$$Q(x) = \frac{2m}{\hbar^2} (\frac{Z_1 Z_2 e^2}{4\pi\epsilon_o x} - E) \tag{6.50}$$

Thus the WKB approximation is satisfied far from the point $x = x_m$ where $k(x) = 0$ at which, on the basis of classical physics, the particle would be reflected back

$$x_m = \frac{Z_1 Z_2 e^2}{4\pi\epsilon_o E} \tag{6.51}$$

For our purpose here we need simply to replace the exponential tunnelling factor given in Eq. (6.42) with its generalisation using the WKB approximation

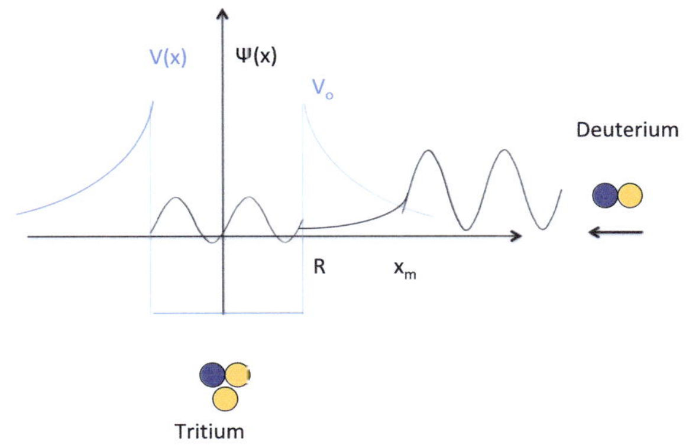

Fig. 6.7 In the fusion between deuterium and tritium the deuterium nucleus has to pass through the Coulomb barrier of the tritium nucleus if its energy is below the maximum of the potential well. Classically, it would be reflected at $x = x_m$. Quantum mechanics allow the deuterium nucleus to propagate inside the well with a finite probability

$$e^{-2k_o a} \rightarrow e^{-2\int_R^{x_m} k_o(x)dx} \equiv e^{-2G} \tag{6.52}$$

with G being the *Gamow factor* [3]. The Gamow factor can be evaluated explicitly yielding

$$G = \frac{(2m_r E)^{1/2}}{\hbar} x_m(\arccos(y) - y(1 - y^2)^{1/2}) \tag{6.53}$$

and $y = (R/x_m)^{1/2}$. In the limit $y \ll 1$ the Gamow factor takes the following asymptotic form

$$G \approx \frac{\pi}{2} \frac{Z_1 Z_2 e^2}{2\epsilon_o h} (\frac{2m_r}{E})^{1/2} \tag{6.54}$$

6.7 Fusion Cross Sections

A simple estimate for the fusion cross sections can be obtained by assuming that the typical size of the fusing nuclei is of the order of the de Broglie wavelength, at least for sufficiently low energies. At high energy the de Broglie wavelength becomes smaller than the nuclear radius and the cross section must be evaluated using the nuclear radius itself.

The use of the de Broglie wavelength to estimate the cross section can be rigorously justified by solving the scattering problem through an expansion in partial waves. Here we limit the discussion to qualitative arguments. It is intuitive that the largest probability for fusion reaction will be associated to head-on collisions, i.e. collisions with the lowest possible impact parameter b (see Chap. 3 for the definition of the impact parameter). The impact parameter is associated with the total angular momentum L in the centre of mass system through the relation $L = m_r b v_{-\infty}$ with $v_{-\infty}$ being the velocity before the collision. As the reduced mass particle approaches the scattering centre, its velocity is deflected but the angular momentum is conserved because the Coulomb interaction is described by a central potential. The effective potential will be the sum of the Coulomb potential and the centrifugal potential given by $L^2/(2m_r r^2)$.

At small distances the centrifugal potential is larger than the Coulomb potential and makes the penetration even more difficult unless $L = 0$. Thus, the question becomes of how large is the amplitude of the incoming particle wave function corresponding to $L = 0$. In quantum theory the incoming particle is described by a plane wave with given momentum $\hbar k$. A plane wave is not a state with defined angular momentum but rather the superposition of many waves with different L values. In the classical limit we may assume that the wave component with the lowest angular momentum corresponds to particles with $L < \hbar$ or, taking into account that $L = m_r b v_{-\infty}$, to

$$b_{min} \approx \frac{\hbar}{m_r v_{-\infty}} \approx \frac{\lambda_{deBroglie}}{2\pi} \tag{6.55}$$

On the basis of the arguments presented in Chap. 3 the cross section can be estimated as $\sigma \approx \pi b_{min}^2$. Such an estimate would apply in the absence of the Coulomb interaction. Thus, in the case of charged nuclei, the cross section must be weighted for the probability of tunnelling across the Coulomb barrier. Taking Eq. (6.54), the fusion cross section can be approximately written as

$$\sigma_{fusion} \approx \pi b_{min}^2 e^{-2G} \propto \frac{e^{-\frac{B_G}{E^{1/2}}}}{E} \tag{6.56}$$

with $B_G = \pi Z_1 Z_2 e^2/(2\epsilon_o h)(2m_r)^{1/2}$. This expression captures the main features of the fusion cross sections: an exponential decrease at low energies and a slow decrease at high energies. In general, fusion cross sections are expressed in the following form

$$\sigma_{fusion} = S(E)\frac{e^{-\frac{B_G}{E^{1/2}}}}{E} \tag{6.57}$$

where $S(E)$ is a slow function of E called the *astrophysical factor*. We will come back to the fusion cross section in Chap. 8.

6.8 Alpha Decay

Alpha decay consists in the emission of an alpha particle by a heavy nucleus. The number of nuclei N undergoing an alpha decay follows an exponential law

$$\frac{dN}{dt} = -\lambda_{alpha} N \tag{6.58}$$

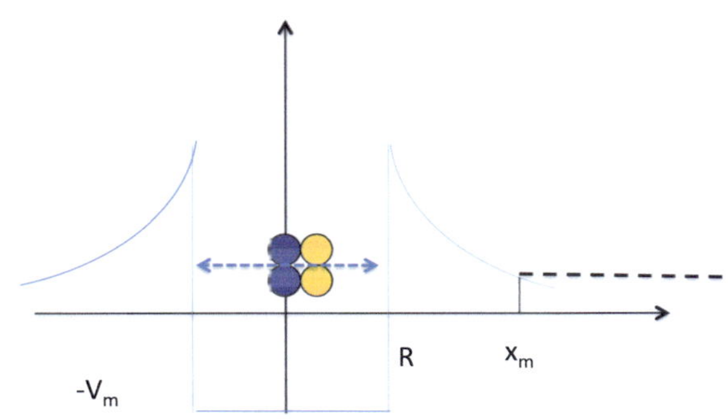

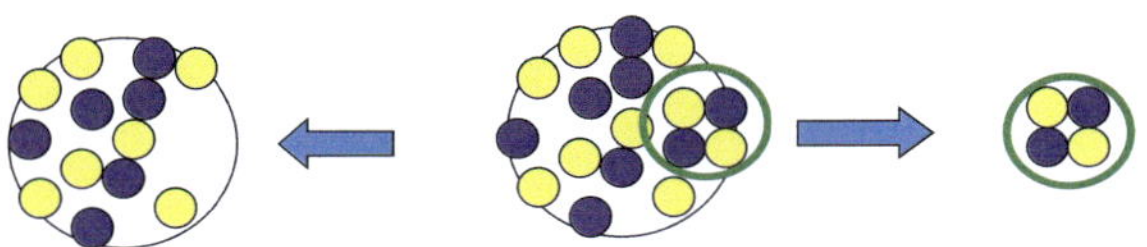

Fig. 6.8 Alpha particles are tightly bound systems and tend to preserve their identity inside the nucleus. They oscillate back and forth between the boundary of the nucleus. When they hit the boundary they have a finite probability to tunnel through the barrier. This model of the alpha decay reproduces the empirical Geiger-Nuttal law for the decay constant as a function of the energy and atomic number of the decaying nucleus

with λ_{alpha} the decay constant that is experimentally well described by the *Geiger-Nuttal* law [4]

$$\ln(\lambda_{alpha}) = -a_1 \frac{Z}{E^{1/2}} + a_2 \tag{6.59}$$

We can make the following picture of the decay. Alpha particles (that are among the more tightly bound nuclear objects and therefore tend to maintain their identity also when they are inside a complex nucleus) are confined inside the nuclear potential well (corresponding to a potential $-V_m$), oscillating between the well boundaries with a velocity that depends on their energy. Classically, their velocity will be given by (Fig. 6.8)

$$v_{alpha} = \left(\frac{2(E + V_m)}{m}\right)^{1/2} \tag{6.60}$$

The frequency of collisions with the wall will be

$$\nu_{coll} = \frac{v_{alpha}}{2R} \tag{6.61}$$

For each collision there is a finite probability that the alpha particle will tunnel through the Coulomb barrier. The probability can be again estimated from Eq. (6.43) and Eq. (6.52) to be $4e^{-2G}$. Therefore, we expect that the number of alpha decays per unit time λ_{alpha} will be given by

$$\lambda_{alpha} \approx \frac{v_{alpha}}{2R} 4e^{-2G} \tag{6.62}$$

Taking the asymptotic expression for the Gamow factor Eq. (6.54)

$$\ln \lambda_{alpha} \approx -\frac{\pi Z_1 Z_2 e^2}{2\epsilon_o h}\left(\frac{2m}{E}\right)^{1/2} + \ln\left(\frac{2v_{alpha}}{R}\right) \tag{6.63}$$

This expression reproduces closely the Geiger-Nuttal law Eq. (6.59) and confirms the simple picture of α decay.

6.9 Suggestions for Further Readings

The aim of the present chapter was only to give the minimal theoretical background for the interpretation of nuclear phenomena. A number of textbooks exist where the interested reader can find a detailed introduction to quantum mechanics. The text of Max Born [5] provides a very intuitive approach to this field.

6.10 Exercises

Problem 6.1 Evaluate the energy of the bound states of an electron in an infinite potential well of width $2a_o$ with a_o the Bohr radius.

Problem 6.2 Evaluate the transmission coefficient of an electron of energy equal to the thermal energy at $0\,°C$ through a barrier of width a_o and height 10 eV.

Problem 6.3 On the basis of the general expression for the alpha decay constant evaluate the value for the decay of a ^{238}U nucleus with the emission of a 4.2 MeV alpha particle. Assume that the well width is the sum of the radius of the alpha particle and that of the nucleus after the decay using the estimate $R = 1.2 \times 10^{-15} m A^{1/3}$ and that the well corresponds to a potential of -15 MeV.

References

1. E. Schrödinger, Quantization as an Eigenvalue Problem. Annalen der Physik **79**, 361-376; 489-527 (1926)
2. J. Heading, *An Introduction to Phase-Integral Methods* (Metheun and Co., Ltd., London; Wiley, New York, 1962)
3. G. Gamow, Quantum theory of the atomic nucleus. ZP **51**, 204 (1928)
4. H. Geiger, J.M. Nuttall, The ranges of the α particles from various radioactive substances and a relation between range and period of transformation. Philos. Mag., Series 6, **22**(130), 613–621 (1911)
5. M. Born, Atomic Physics .Dover (1990) Fisica atomica. Boringhieri (1976)

Chapter 7
The Structure of the Nucleus and the Radioactive Decay Processes

Abstract *The chart of radionuclides is introduced. The liquid drop model is discussed and it is shown how to reproduce the main characteristics of the radionuclide chart. The equations describing multiple radioactive decays are discussed and the concept of secular equilibrium is introduced. The energetics of alpha and beta decays is discussed. The final part of the chapter is devoted to the illustration of natural radioactivity. A mention to the radioactive dating techniques is also provided. The radiometric units are also introduced and some notion of the effect of radioactivity on human health is provided.*

A nuclide is a species of atom characterised by the atomic number Z, the mass number A and the energy state of the nucleus. There are about 4400 [1] known nuclides of which about 250 are stable. In this chapter a classification of radio-nuclides and a model to determine their mass will be presented first. All the elements up to lead ($Z = 82$) have at least one stable isotope with the exception of technetium ($Z = 43$) and promethium ($Z = 61$). The unstable nuclides decay through various radioactive processes that will be examined in detail.

7.1 The Chart of Nuclides

When protons and neutrons form a bound state the total mass of the bound system is smaller than the sum of the mass of the constituents. Such a *mass defect* is associated, on the basis of the mass-energy equivalence, with the binding energy that produces a negative contribution to the total energy of the system: in order to separate the bound state energy must be provided. Examples are listed below:

- The proton and the neutron have a rest energy of 939.565 MeV and 939.566 MeV, respectively. The rest energy of the deuteron (made of one proton and one neutron) is 1875.613 MeV. Thus the binding energy is

$$938.272\,\text{MeV} + 939.565\,\text{MeV} - 1875.613\,\text{MeV} = 2.224\,\text{MeV}$$

F. Romanelli, *Physics of Nuclear Energy*, Springer Series in Plasma Science and Technology, https://doi.org/10.1007/978-981-97-9609-0_7

- The rest energy of an alpha particle (made of two protons and two neutrons) is 3727.379 MeV, corresponding to a binding energy of 28.3 MeV.

The mass of a nucleus is related to the mass of its constituents and the binding energy[1] via the following equation

$$m(A, Z) = Zm_p + (A - Z)m_n - \frac{E_{binding}}{c^2} \tag{7.1}$$

From Eq. (7.1), upon adding and subtracting a term Am_{AMU} with $m_{AMU} \equiv 1AMU$ corresponding to an energy $m_{AMU}c^2 = 931.495\,\text{MeV}$, the binding energy per nucleon can be written as

$$\frac{E_{binding}}{A} = (m_n - m_{AMU})c^2 - \frac{(m(A, Z) - Am_{AMU})c^2}{A} + (m_p - m_n)c^2\frac{Z}{A} = \approx 8.071\,\text{MeV} - \frac{(m(A, Z) - Am_{AMU})c^2}{A} - 1.293\,\text{MeV}\frac{Z}{A} \tag{7.2}$$

The first term provides a contribution approximately 8 MeV. Since the AMU is defined as 1/12 of the ^{12}C isotope (6 protons and 6 neutrons), for ^{12}C the second term identically vanishes and the binding energy per nucleon of the ^{12}C isotope is given by the sum of the first and the third term yielding a binding energy per nucleon for this nuclide equal to 7.424 MeV. The experimentally determined values for the second term in the R.H.S of Eq. (7.2) varies in a limited range (between $\pm 0.9\,\text{MeV}$). The last term in Eq. (7.2) is almost constant since ratio Z/A for stable nuclides ranges between 0.5 and 0.38: its contribution to Eq. (7.2) varies in the range $0.575\,\text{MeV} \pm 0.08\,\text{MeV}$. As the second and third term in Eq. (7.2) are much smaller than the first, the binding energy per nucleon is approximately independent of A. Nevertheless, it is the small deviation from a constant that matters for the application to energy production (see Fig. 1.7)!

The chart of nuclides in the plan (Z, N), with $N \equiv A - Z$, is given in Fig. 7.1 [1]. It has a characteristic cigar shape. The general features can be summarised as follows:

- There is a narrow band (corresponding to the black points in Fig. 7.1) of nuclides that are stable or have a very long half-life. This band corresponds to the condition $Z \approx N$ for low to intermediate mass numbers.
- Most of the stable nuclides (156) have both Z and N even. There are 50 stable nuclides with Z odd and N even and 48 with Z even and N odd. There are only five stable nuclides with both Z and N odd: D, ^6Li, ^{10}B, ^{14}N and ^{180}Ta.
- Below and above the line of stable nuclides there are two large regions of nuclides that decay via beta emission (negative beta emission below and positive beta emission above).

[1] Note that in the following we will speak of the binding energy as a positive quantity as we have included its contribution in Eq. (7.1) with a negative sign.

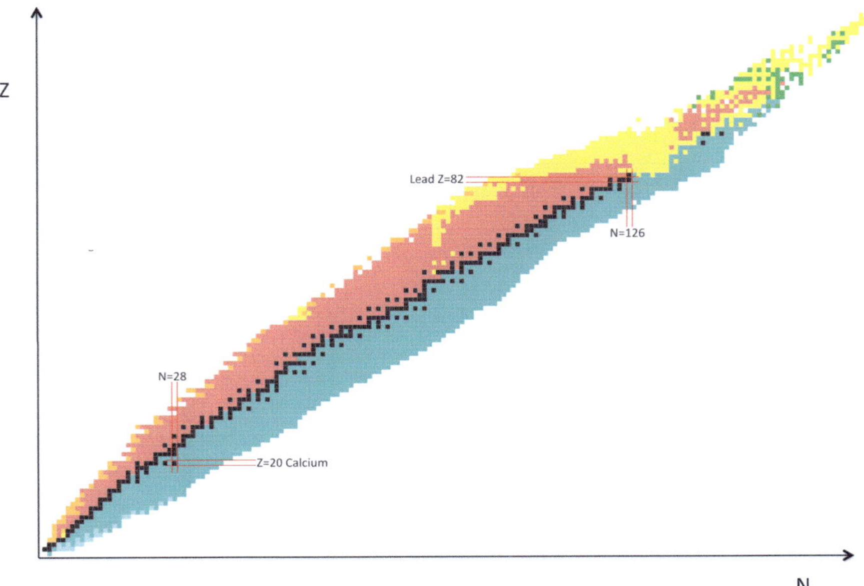

Fig. 7.1 Chart of nuclides. The chart in the (Z, N) plane from Ref. [1] exhibits a characteristic cigar shape. The black points correspond to stable nuclides. Above the stability line nuclei decay via positive beta emission (pink region) up to the proton drip line that represents the upper boundary. Below the stability line nuclei decay via negative beta emission (blue region) down to the lower boundary corresponding to the neutron drip line. For large A values nuclei can decay via alpha emission (yellow region). For even larger mass numbers spontaneous fission can occur (green points)

- The negative beta emission region (blue) is bounded from below by the neutron drip line. The positive beta emission region (pink) is bounded from above by the proton drip line.
- For large values of both Z and N (i.e. for large values of A) there are radioactive nuclides undergoing alpha emission (the yellow region). Low Z element do not undergo alpha decay with the exception of ^6Be and ^8Be.
- For even larger values of A there are radio-nuclides that undergo spontaneous fission (green points).

In the next section a semi phenomenological model will be derived to account for these observations. Data for the stable and unstable nuclides (life-time, decay mode, etc.) can be found at [2].

7.2 The Liquid Drop Model

The simplest description of the nucleus is that of a liquid drop [3]. High-energy scattering experiments show that the nucleus can be approximated as a spherical object made of A closely packed nucleons. This implies that the nuclear radius R must be proportional to $A^{1/3}$ (Fig. 7.2a)

$$R(m) \approx r_{nucleus} A^{1/3} \tag{7.3}$$

with $r_{nucleus} \approx 1.2 \times 10^{-15} m$. An expression for the binding energy can be derived by combining the observations listed in Sect. 7.1 [4, 5].

The first contribution is associated with the force that keeps the nucleus together. If each nucleon interacted with *all* the other nucleons via this force this term would be proportional to $A(A-1) \approx A^2$ and the energy per nucleon would not be a constant value, contradicting the experimental evidence. Therefore we must assume that this interaction involves each nucleon and *its closest neighbours only*. If K is the number of nucleons surrounding each nucleon, the contribution of this term to the binding

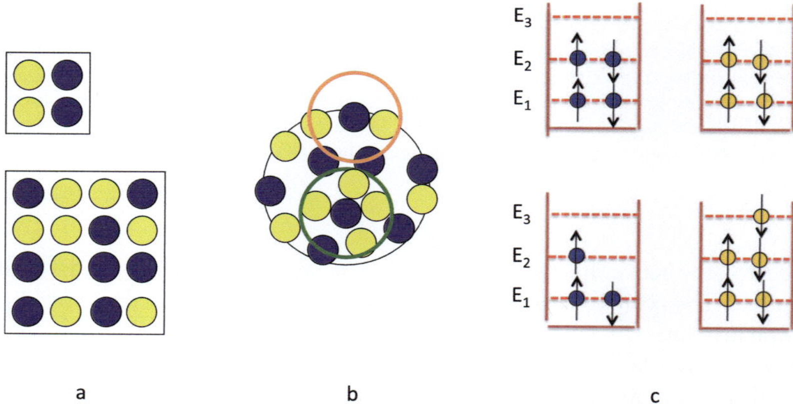

Fig. 7.2 The nucleus can be modelled as a liquid drop. As the nucleons are tightly packed the volume is proportional to the number of nucleons. Thus the nuclear radius scale as $A^{1/3}$ as illustrated in (**a**) for a 2D case (in which the linear size scale as $A^{1/2}$). The volume contribution to the binding energy depends only on the interaction with the closest neighbours as shown in (**b**) (green circle). However, particles at the nuclear surface provide a lower contribution since the interaction takes place with a lower number of particles (orange circle). This effect is responsible for the surface term that reduces the binding energy. The symmetry term **c** can be explained on the basis of the Pauli exclusion principle. The top and bottom configuration have the same mass number ($A = 8$). The top configuration has been obtained by populating the energy levels by progressively accumulating two protons (blu) and two neutrons (yellow) in each level (one with spin up and the other with spin down). If now a proton is exchanged with a neutron as done in the bottom configuration it has to go to the third energy level decreasing the binding energy of the nucleus since it takes less energy to promote the additional neutron to free particle. Therefore the top configuration is energetically favoured

energy will be proportional to $K\ A$ and it will provide a constant contribution to the energy per nucleon. On the basis of these arguments the volume contribution to the binding energy is given by

$$E_{volume} = b_1 A \tag{7.4}$$

Note that this term is positive so its contribution to the mass in Eq. (7.1) is negative as it should be since it corresponds to a binding force.

The nucleons that stay on the surface of the nucleus feel only the attractive force from the nucleons closer to the centre. This means that the binding energy is reduced by an amount proportional to the nuclear surface. This additional contribution to the binding energy is given by

$$E_{surface} = -b_2 A^{2/3} \tag{7.5}$$

Note the negative sign.

A third contribution, again to reduce the binding energy, is associated with the repulsive Coulomb force of the protons. Taking the nuclear radius as the typical distance between protons and taking into account that there are Z protons in the nucleus, the contribution of the Coulomb energy will be

$$E_{Coulomb} = -b_3 \frac{Z^2}{A^{1/3}} \tag{7.6}$$

Stable nuclei tend to have equal number of protons and neutrons for low mass numbers (this symmetry tends to be weakened in high mass number nuclei). This symmetry can be interpreted as a manifestation of the Pauli exclusion principle. The nucleus can be described as a potential well with discrete energy levels that must be filled with protons and neutrons. Both protons and neutrons are fermions and obey the Pauli exclusion principle: the same state cannot be occupied by more than one particle. Proton and neutrons have very similar energy levels (see Fig. 7.2c). The minimum energy state corresponds to an equal number of protons and neutrons that have filled the energy levels up to certain energy. If the energy levels have been completely filled and e.g. a proton is replaced with a neutron, the additional neutron cannot be placed at the same energy level of the original proton because such a level is already filled and it must be placed at the next energy level. The resulting configuration has a lower binding energy because it takes less energy to promote the additional neutron to a free particle ($E \geq 0$). To account for the exclusion principle a symmetry term is introduced in the binding energy given by

$$E_{symmetry} = -b_4 \frac{(Z-N)^2}{A} \tag{7.7}$$

Finally, a phenomenological constraint must be introduced to account for the different number of nuclei with (Z, N) (even, even) and (odd, odd). To this aim an *ad hoc* pairing term is introduced in the binding energy given by

Table 7.1 Constants in the binding energy expressions

b_1(MeV)	b_2(MeV)	b_3(MeV)	b_4(MeV)	b_5(MeV)
15.6	17.2	0.70	23.3	12

$$E_{pairing} = b_5 \frac{\delta(Z, A)}{A^{1/2}} \tag{7.8}$$

with $\delta = +1$ for (Z, N) even/even, $\delta = 0$ for (Z, N) even/odd or odd/even and $\delta = -1$ for (Z, N) odd/odd. With such a choice even/even nuclei are more bound (and so more likely to occur) than odd/odd nuclei.

The five constants $b_1 - b_5$ can be fitted using the experimental results of isotope mass spectroscopy. The fitting constants are given in Table 7.1 [6].

The final expression for the nuclear mass (also known as Weizsäcker formula) is

$$m(A, Z)c^2 = Zm_pc^2 + (A - Z)m_nc^2 + \\ - b_1 A + b_2 A^{2/3} + b_3 \frac{Z^2}{A^{1/3}} + b_4 \frac{(A - 2Z)^2}{A} - b_5 \frac{\delta}{A^{1/2}} \tag{7.9}$$

We can test some of the predictions of Eq. (7.9). The nuclear mass vs. Z at constant A is a parabola. The minimum of the parabola corresponds to the lowest energy state for a given A. This condition identifies the locus of stable isotopes (it will be shown below that this condition can be expressed as the stability for beta decay). A trivial calculation shows that the condition $\partial m(Z, A)/\partial Z = 0$ at fixed A is satisfied for

$$0 = (m_p - m_n)c^2 + 2b_3 \frac{Z}{A^{1/3}} + 4b_4 \frac{2Z - A}{A} \tag{7.10}$$

or

$$Z = Z_{stable} = \frac{A}{2} \frac{1 + \frac{(m_n - m_p)c^2}{4b_4}}{1 + \frac{b_3}{(4b_4)} A^{2/3}} \approx \frac{1.014}{1 + 0.0076A^{2/3}} \frac{A}{2} \tag{7.11}$$

The stability curve is plotted in Fig. 7.3 and is in good agreement with the position of the stable nuclides. This is not surprising since Eq. (7.9) has been constructed in order to fit the data. However, it shows that Eq. (7.9) can be used to predict with good confidence the change in the rest energy of the nuclei that take part in a nuclear process.

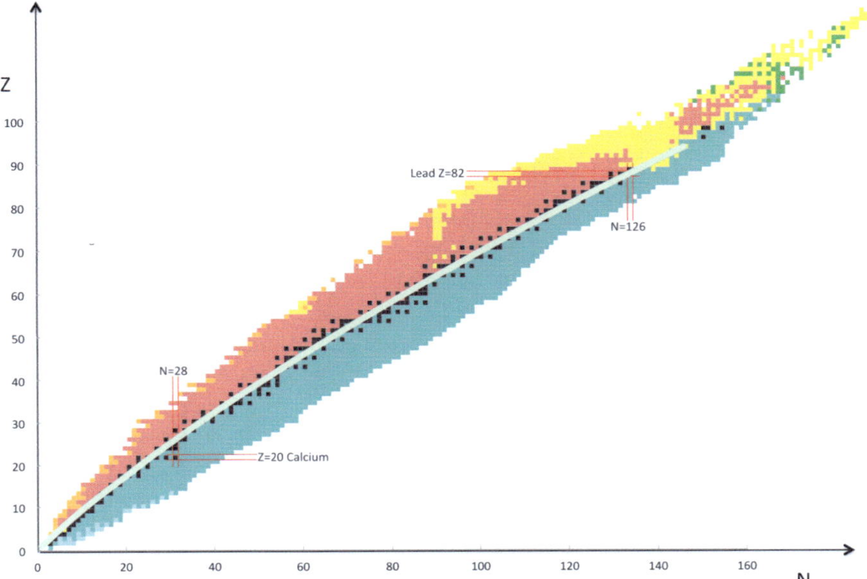

Fig. 7.3 Stable nuclides. The curve given by Eq. (7.11) is compared with the position of the stable nuclides in the (Z, N) plane

7.3 Radioactive Decays

Radioactive decays are spontaneous processes that transform a radionuclide into a different radionuclide or produce a transition of the nucleus from one energy state to another. There are three main decay processes: alpha, beta and gamma decay. At the boundary of the nuclide charts two more decay processes can occur: spontaneous proton or neutron emission. A sixth decay process, namely spontaneous fission, will be discussed in the next chapter. Decay processes are characterised by an exponential variation in time defined by the following equation for the number of nuclides N

$$\frac{dN}{dt} = -\lambda N \tag{7.12}$$

with λ the *decay constant* of the process. Equation (7.12) yields an exponential decay in time

$$N(t) = N(0)e^{-\lambda t} \tag{7.13}$$

The activity A is defined as the number of decay events per unit time

$$A(t) = \lambda N(t) \tag{7.14}$$

The characteristic time of variation of the radio nuclide population is usually expressed in terms of the *half-life* $\tau_{1/2}$, the time needed to decrease the initial population by a factor two. From Eq. (7.13) it can be related to the decay constant as follows

$$\tau_{1/2} = \frac{\ln 2}{\lambda} \tag{7.15}$$

The activity is often measured in *Bequerel (Bq)*, with $1Bq$ corresponding to one decay per second, but the SI unit is the *Curie (Ci)*, with $1Ci = 3.7 \times 10^{10} Bq$ being the approximate value of the activity of $1g$ of ^{226}Ra.

If the product of a radioactive decay is itself a radioactive nuclide a radioactive decay chain is initiated. In this case the number of parent nuclides N_o and the number of child radio nuclides N_1 are described by the following equations

$$\frac{dN_o}{dt} = -\lambda_o N_o \tag{7.16}$$

$$\frac{dN_1}{dt} = -\lambda_1 N_1 + \lambda_o N_o \tag{7.17}$$

that can be easily solved with initial conditions $N_o(0) = N_{oi}$ and $N_1(0) = 0$. The solution is shown in Fig. 7.4 for the case $\lambda_o \ll \lambda_1$.

If the parent nuclide decays on a time scale much longer than the child nuclide the latter initially increases almost linearly but on a time scale of the order λ_1^{-1} it reaches a maximum and on a longer time scale it follows the time variation of the parent nuclide with a constant ratio $N_1(t)/N_o(t)$: a so-called *secular equilibrium* is established

$$N_1(t) \approx \frac{\lambda_o}{\lambda_1} N_o(t) \tag{7.18}$$

with the ratio between the two nuclides populations being constant in time. In the opposite case $\lambda_o \gg \lambda_1$ the parent nuclide population is depleted on a very fast time scale in favour of the child nuclide. Note that Eq. (7.18) makes it apparent that when a secular equilibrium is established the activities of the parent and child nuclide are the same.

The child nuclide can in turn undergo a transmutation into a different radionuclide. In this case we speak of a *radioactive chain*. The evolution of the chain is governed by the parent nuclide with the lowest decay constant. The intermediate nuclides with much faster decay represent intermediate steps in the chain that is concluded by a stable isotope (or an isotope with very long half life). On the basis of what we just noted, each step of the decay process contribute for the same amount to the total activity.

The half life of a nuclide can be found in various public data repositories accessible online. The International Atomic Energy Agency provides simple tools and its libraries of nuclear data are freely available online. Data on radioactive decays can e.g. be found in [7].

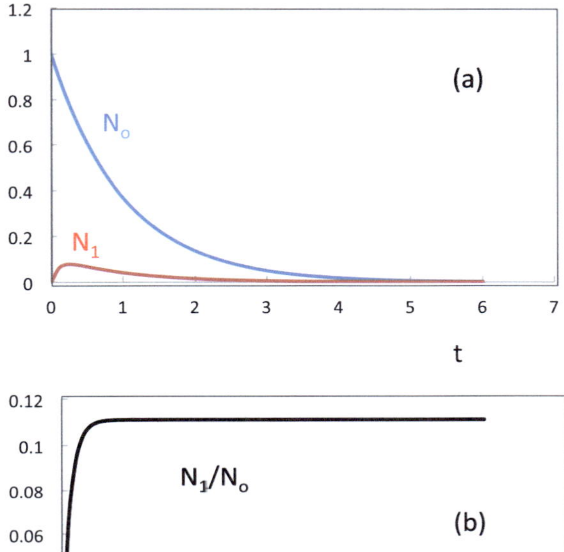

Fig. 7.4 Radioactive decay chain. The time evolution of the populations of the parent (blue) and child (red) nuclides is shown for the case when the parent decay constant is much smaller than the child decay constant (**a**). In this case a secular equilibrium is established. Looking at the ratio between the child and the parent nuclides (**b**) after a transient phase a constant value is approached that depends only on the ratio of the decay constants

7.4 Alpha Decay

Alpha decay consists in the emission by a nucleus of an alpha particle

$$ {}^{A}_{Z}X \rightarrow {}^{A-4}_{Z-2}X + \alpha \tag{7.19}$$

It is a two-body decay process. Thus, knowing the energy Q released in the decay, the kinetic energy T_α of the alpha is given by (see Chap. 1)

$$T_\alpha = \frac{m_{X-4}}{m_X} Q \approx (1 - \frac{4}{A}) Q \tag{7.20}$$

Since the decay involves nuclei with $A \geq 200$[2] almost all the energy of the decay is converted into the alpha particle energy.

[2] The only low mass number nuclide that decays in this way is ^{8}Be.

In order for alpha decay to occur, the sum of the masses of the products needs to be smaller than the mass of the original decaying nucleus

$$\frac{Q}{c^2} = m(Z, A) - [m(Z-2, A-4) + m_\alpha] > 0 \tag{7.21}$$

Using the Weizsäcker formula it is possible to estimate Q. Since Eq. (7.9) is not a good approximation for low mass number nuclides, the alpha particle mass will be computed using the experimental value for the binding energy of the alpha particle $E_{b\alpha} = 28.3\,\text{MeV}$. In substituting Eq. (7.9) into Eq. (7.21) the contribution arising from the mass of protons and neutrons exactly vanishes since their number is unchanged in the process

$$Q = 28.3\,\text{MeV} + E_{binding}(Z-2, A-4) - E_{binding}(Z, A) \tag{7.22}$$

In evaluating the difference of the binding energy it is convenient to make use of the fact that the relative change in both Z and A is small for the mass numbers of interest to obtain

$$Q = 28.3\,\text{MeV} - 2\frac{\partial E_{binding}(Z, A)}{\partial Z} - 4\frac{\partial E_{binding}(Z, A)}{\partial A} \tag{7.23}$$

Using Eqs. (7.9), (7.23) can be written as

$$Q = 28.3\,\text{MeV} - 4b_1 + 4\frac{\frac{2b_2}{3} + b_3 Z(1 - \frac{Z}{3A})}{A^{1/3}} - 4b_4\left(1 - \frac{2Z}{A}\right)^2 \tag{7.24}$$

The right hand side of this equation is a parabola in Z with a maximum for

$$Z = Z_{max} = \frac{A}{2}\frac{1 + \frac{b_3 A^{2/3}}{4b_4}}{1 + \frac{b_3 A^{2/3}}{12b_4}} \tag{7.25}$$

The parabola has a downward concavity. If for $Z = Z_{max}$ the condition $Q > 0$ is satisfied, then the decay can take place in the range of Z defined by the roots of the parabola. If for $Z = Z_{max}$ $Q < 0$ the decay is energetically forbidden.

Note that Z_{max} is above the stable nuclide curve given by Eq. (7.11) that, as we have anticipated, corresponds to the boundary for beta decay. Of the two roots of the equation $Q = 0$, we focus on the lowest root $Z = Z_-$. If for a certain A the lowest root is below the stability line, the nuclides are in reality unstable for alpha decay. The condition $Z_- = Z_{stable}$ sets therefore a threshold above which no stable nuclei can be found. Figure 7.5 compares the position of the stable curve in the (Z, N) plane with the position of the lowest root of Eq. (7.24) for various values of Q. It is possible to see that the above $A \approx 140$ the alpha decay is always energetically favorable.

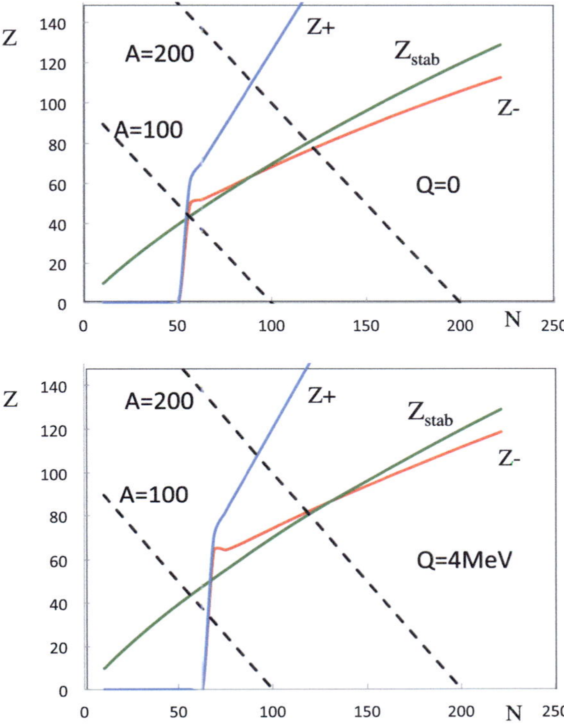

Fig. 7.5 The two roots of Eq. (7.24) (red and blue curves) are shown in the (Z, N) plane together with the line $Z = Z_{stable}$ (green line) for two values of Q. If the stability line enters the region between the two roots the corresponding nuclides results unstable for alpha decay. The lines $A = 100$ and $A = 200$ are also shown (dashed lines)

For $Q > 0$ the alpha decay is possible but, as discussed in Chap. 6, the decay constant has an exponential dependence on the energy of the emitted alpha particle that, in turn is approximately equal to Q. If $Q < 4\,\text{MeV}$, the Geiger-Nuttal law (Fig. 7.6) predicts half-lives above $10^{17}s(\approx 3Gy)$ that reduces to about $10^{12}s(30My)$ for $Q = 5\,\text{MeV}$. Thus, it is convenient to define an effective threshold for alpha decay by restricting the conditions to the decays with $Q \geq 4\,\text{MeV}$. The lowest root intercepts the stability curve for $A = 200$ (see Fig. 7.5) that can be taken as representative of the effective threshold for alpha decay.

Conversely, alpha decays with $Q > 8\,\text{MeV}$ lead to half-lives below $1\mu s$, i.e. to very short lived nuclides. Thus, a reasonable range for the energy released in alpha decay is 4–8 MeV.

The amount of energy released in alpha decay increases with the mass number of the decaying nucleus. In Fig. 7.7 it is shown the dependence of Q on A for three different choices of the atomic number: $Z = Z_{stable}$, $Z = Z_{stable} \pm 5$.

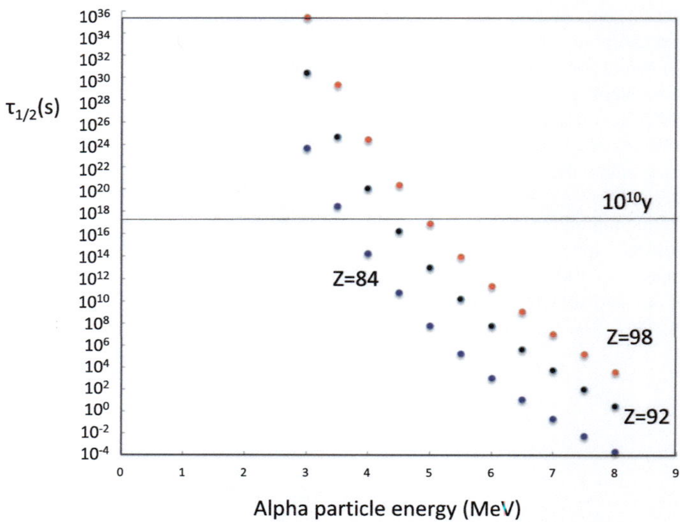

Fig. 7.6 Half-life for α decay. The half-life is shown versus the alpha particle energy for three elements. If the particle is emitted with energy below 4 MeV the half-life is longer than $3Gy$

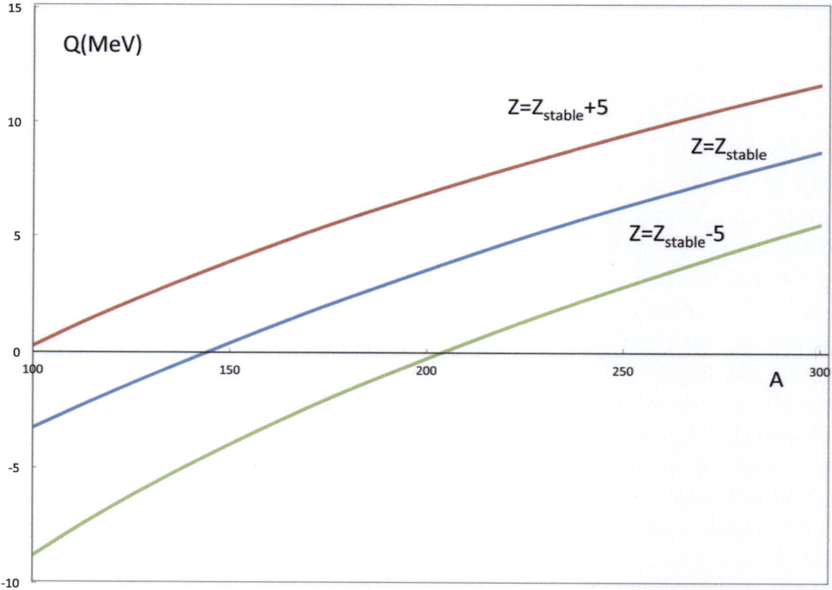

Fig. 7.7 The energy released in alpha decay is an increasing function of the mass number A

7.5 Beta Decay

There are three types of beta decay. The β_- *decay* of a nucleus consists in the emission of an electron and of an antineutrino with the transformation of a neutron into a proton

$$ {}^{A}_{Z}X \rightarrow {}^{A}_{Z+1}X + e + \bar{\nu}_e \tag{7.26}$$

The β_+ *decay* consists in the emission of a positron (i.e. an antielectron) and a neutrino with the transformation of a proton into a neutron

$$ {}^{A}_{Z}X \rightarrow {}^{A}_{Z-1}X + \bar{e} + \nu_e \tag{7.27}$$

Electron capture consists in the capture of an electron by the nucleus with the transformation of a proton into a neutron and the emission of a neutrino. It can be considered formally similar to the β_+ decay.

$$ {}^{A}_{Z}X + e \rightarrow {}^{A}_{Z-1}X + \nu_e \tag{7.28}$$

Beta decay is the result of the weak force. The number of baryons is conserved but the individual number of protons and neutrons is not. It is a three body process. Thus, the energy of the products is not defined and there is a spectrum of energies that can be possessed e.g. by the electron, with the maximum energy being the Q of the reaction that will be estimated below

$$\frac{Q}{c^2} = m(Z, A) - [m(Z \pm 1, A) + m_e] \tag{7.29}$$

with the upper sign referring to β_-decay and the lower to β_+ decay. From the Weizsäcker formula

$$Q = \pm(m_n - m_p)c^2 - m_e c^2 - b_3 \frac{\pm 2Z + 1}{A^{1/3}} - 4b_4 \frac{\pm(2Z - A) + 1}{A} - 2b_5 \frac{\delta(Z, A)}{A^{1/2}} \tag{7.30}$$

again with the upper sign for the β_- decay. Upon comparing this expression for Q with Eq. (7.10) it is apparent that the condition for stable nuclides is identical to the condition $Q = 0$ for beta decay provided the contributions of the electron mass and of the pairing term are neglected. The latter is about ± 2.4 MeV for $A \approx 100$ whereas the contribution of the Coulomb term (taking $Z = A/2$) is about 15 MeV. To be precise, $Q > 0$ for the β_- decay if Z is below the stability line and for the β_+ decay if Z is above the stability line.

Taking a nuclide close to the stability line ($Z = Z_{stable} + M$ with M a positive or negative integer and $|M| << Z_{stable}$)

$$Q = -m_e c^2 - (\pm 2M + 1)(\frac{b_3}{A^{1/3}} + \frac{4b_4}{A}) - 2b_5 \frac{\delta(Z, A)}{A^{1/2}} \tag{7.31}$$

Thus, for the β_- decay (the + sign in Eq. (7.31)) it is possible to have $Q > 0$ if $M < 0$ (i.e. if the nuclide is below the stability line) whereas for the β_+ decay (the − sign in Eq. (7.31)) $Q > 0$ implies $M > 0$ (nuclides above the stability line).

The above discussion shows that the majority of radionuclides in the radionuclide chart undergo a β decay.

7.6 Neutron and Proton Drip Lines

The proton and the neutron drip lines can be obtained by looking at the condition $Q > 0$ for the decays

$${}^{A}_{Z}X \rightarrow {}^{A-1}_{Z-1}X + p \quad (proton\ emission) \tag{7.32}$$

and

$${}^{A}_{Z}X \rightarrow {}^{A-1}_{Z}X + n \quad (neutron\ emission) \tag{7.33}$$

The application of the Weizsäcker formula yields for the neutron emission

$$Q = -b_1 + \frac{2b_2}{3A^{1/3}} - \frac{b_3 Z^2}{3A^{4/3}} + b_4 \left(1 - \frac{4Z^2}{A^2}\right) - b_5 \frac{\delta(Z, A) - \delta(Z, A-1)}{A^{1/2}} \tag{7.34}$$

whereas for the proton emission

$$Q = -b_1 + \frac{2b_2}{3A^{1/3}} - \frac{b_3 Z^2}{3A^{4/3}} + b_4 \left(1 - \frac{4Z^2}{A^2}\right) - b_5 \frac{\delta(Z, A) - \delta(Z-1, A-1)}{A^{1/2}} + \\ + \frac{2b_3 Z}{A^{1/3}} - 4b_4 \left(1 - \frac{2Z}{A}\right) \tag{7.35}$$

The proton and neutron drip lines are shown in Fig. 7.8.

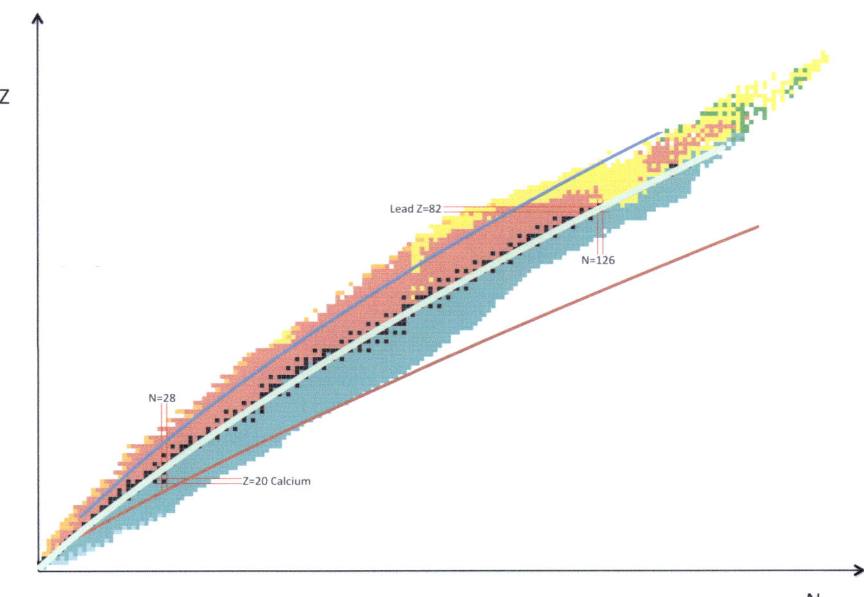

Fig. 7.8 Proton (blu) and neutron (red) drip lines compared with the stability line (pale green) in the (Z, N) plane

7.7 Gamma Decay

Gamma decay consists in the emission of an energetic photon when the nucleus relaxes from an excited state to a lower energy state. The gamma energy depends on the detail of the nuclear structure. It ranges between 0.1 and 10 MeV. The emission produces a recoil of the nucleus (from energy and momentum conservation).

Even if energetically possible, a gamma decay can be forbidden by other conservation laws such as the conservation of angular momentum.

A simple estimate for the decay constant can be given if the emission can be assumed to be that of an oscillating dipole (see Chap. 4). If P is the power emitted by the nuclear dipole and ν is the frequency of the emitted gamma, the decay constant is given by

$$\lambda = \frac{P}{h\nu} \tag{7.36}$$

The power emitted by an oscillating dipole is given by Eq. (4.67). The electric dipole moment of the nucleus can be estimated as $p \approx \frac{3}{4} e r_{nucleus} A^{1/3}$. Then, the decay constant for the electric dipole transition can be written as

$$\lambda(E_1) \propto \frac{e^2}{12\pi\epsilon_o\hbar}(\frac{E}{\hbar c})^3(\frac{3r_{nucleus}A^{1/3}}{4})^2 \propto E^3\, A^{2/3} \tag{7.37}$$

Table 7.2 Weisskopf estimates of the gamma decay constant. The notation (E_n) and (M_n) indicate the $2n - pole$ electric and magnetic emission. Photon energy E expressed in MeV

Electric transition	Decay constant (s^{-1})	Magnetic transition	Decay constant (s^{-1})
$\lambda(E_1)$	$1.0 \times 10^{14} A^{2/3} E^3$	$\lambda(M_1)$	$5.5 \times 10^{13} E^3$
$\lambda(E_2)$	$7.3 \times 10^7 A^{4/3} E^5$	$\lambda(M_2)$	$3.5 \times 10^7 A^{2/3} E^5$
$\lambda(E_3)$	$34 A^2 E^7$	$\lambda(M_3)$	$16 A^{4/3} E^7$
$\lambda(E_4)$	$1.1 \times 10^{-5} A^{8/3} E^9$	$\lambda(M_4)$	$4.5 \times 10^{-6} A^2 E^9$

Similarly, using the expressions derived in Chap. 4 it is possible to determine the decay constant for electric quadrupole emission and magnetic dipole emission. These simple arguments lead to the *Weisskopf estimates* of the gamma decay rate [8] with E_n denoting the electric transitions and M_n the magnetic transitions (n=1 dipole, n = 2 quadrupole, etc.) (Table 7.2).

The Weisskopf estimates are not very accurate but give the correct dependence of the decay constant with photon energy and nuclear mass.

7.8 Natural Radioactivity. Radioactive Decay Chains

Natural radioactivity is mainly due to the decay of the radio nuclides present in the Earth interior since its formation and, on a lower extent, to the radio nuclides that are continuously produced by the bombardment of cosmic rays.

The radio nuclides in the Earth interior belongs to four distinct families. Each family is made of the elements that have been produced through successive radioactive decays. If we express the mass number as $A = M + 4p$ with p an integer and $M = 0, 1, 2, 3$, each decay produces a nucleus with the same value of M since an alpha decay results in a mass number $A = M + 4(p - 1)$ and a beta decay does not change the value of A.

Each family is designated on the basis of the parent nuclide. The three families still existing on Earth are:

- the ^{238}U family $A = 4p + 2$ parent nuclide half-life $\tau_{1/2} = 4.5Gy$
- the ^{232}Th family $A = 4p$ parent nuclide half-life $\tau_{1/2} = 14Gy$
- the ^{235}U family $A = 4p + 3$ parent nuclide half-life $\tau_{1/2} = 0.7Gy$

The fourth family (the ^{237}Np family $A = 4p + 1$) is extinct since the parent nuclide has a short half-life (only $2.14My$) and what is left is only ^{209}Bi slowly decaying into ^{205}Tl ($\tau_{1/2} = 1.9 \times 10^{19} y$).

The decay chains of the four families are shown in Figs. 7.9 and 7.10.

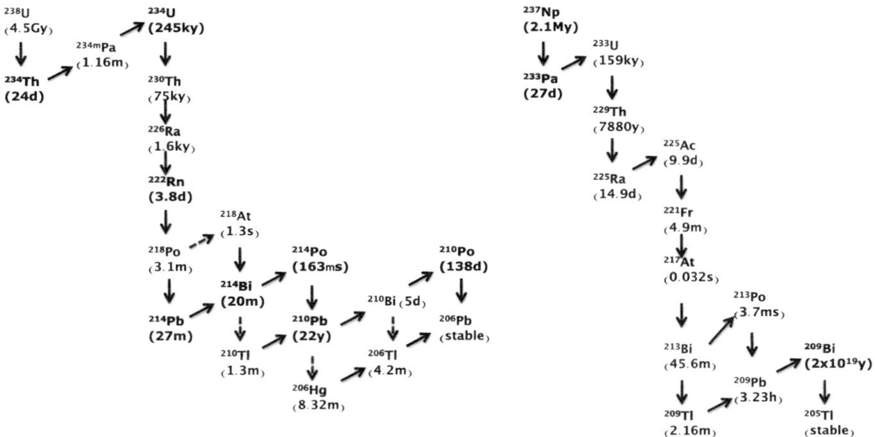

Fig. 7.9 Radioactive families. ^{237}Np and ^{238}U families. The ^{237}Np family is no longer active

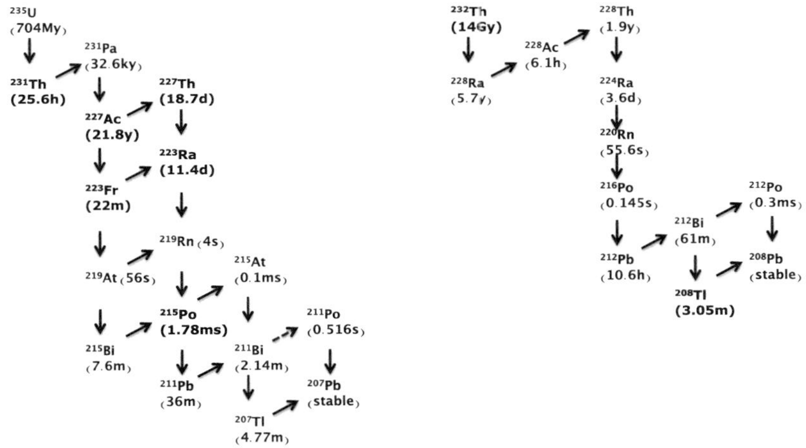

Fig. 7.10 Radioactive families. ^{232}Th and ^{235}U families

7.9 Radioactive Dating

The property of a radionuclide to decay on a certain time scale can be used to determine the age of an object. This technique is called *radioactive dating*. Looking at Eqs. (7.16)–(7.17) and assuming that the child nuclide is stable ($\lambda_1 = 0$) the equations for N_o and N_1 can be easily solved to give

$$N_o(t) = N_o(0)e^{-\lambda_o t} \tag{7.38}$$

$$N_1(t) = N_1(0) + N_o(0)(1 - e^{-\lambda_o t}) \tag{7.39}$$

The initial value of the parent nuclide $N_o(0)$ is not known (otherwise we could simply use the first of Eq. (7.38)). Thus, it is convenient to eliminate $N_o(0)$ yielding

$$N_1(t) = N_1(0) + N_o(t)(e^{\lambda_o t} - 1) \tag{7.40}$$

Thus, plotting the number of child and parent nuclei as measured at the time t for different specimens of the same object, Eq. (7.40) predicts a linear dependence with the slope of the line being related to the product $\lambda_o t$. Then, knowing the decay constant of the parent isotope, it is possible to determine the age t of the object.

The number of parent nuclei at the time t must be sufficiently large. This implies that a radioactive decay with $\lambda_o t \approx 1$ must be chosen for the dating. In the case of rocks uranium is used since its half life is of the order of the age of Earth. In the case of biological tissues a powerful method is the radiocarbon dating that uses the ^{14}C isotope that is continuously formed by the interaction of cosmic rays with the upper atmosphere and has a half life of about $5000y$. The radiocarbon enters the food cycle and is absorbed by the body of the living object until it dies and the intake stops.

7.10 Radiometric Units

The information provided by the activity of an element is not enough to determine its potential effect since it does not take into account the nature and energy of the emitted particles. These effects can be accounted for by introducing the notion of *dose*, i.e. the energy per unit mass deposited in a certain object. The dose is measured in *gray (Gy)* with $1Gy = 1$ J/kg. The dose measures an effect integrated over time. The effect of radiation on biological tissues depends on the nature of the absorbed particles. To account for this, the dose is weighted with appropriate coefficients that measure the biological damage. The biologically weighted dose is measured in *Sievert (Sv)*. In the case of electromagnetic radiation (gamma rays) the weighting factor is one, thus $1Gy$ of gammas produces an equivalent dose of $1Sv$. The same weighting factor applies to beta radiation. Protons have a weighting factor 2 and alphas a weighting factor 20. Thus, the biological damage of a beam of alpha particles at fixed absorbed dose is much larger than that of gammas. In the case of neutrons the weighting factor depends upon the neutron energy. These numbers are shown in Table 7.3.

Thus the biologically weighted dose is given by

$$Equivalent \quad dose(Sv) = \sum weight \quad \times \quad Physical\, dose(Gy) \tag{7.41}$$

The *dose rate* is the dose absorbed per unit time. A dose rate of $10\mu Sv/h$ means that after 20 hours the equivalent dose is $200\mu Sv$. The *contact dose rate D* is a measure of the hazard posed by a radioactive material. The γ dose rate can be evaluated e.g. in the approximation of a semi-infinite solid from the following expression

Table 7.3 Weights to convert the physical dose to the biologically weighted dose

Radiaton	Energy	Weight
Gammas and leptons	*any*	1
Protons	*any*	2
Alphas and heavy nuclei	*any*	20
Neutrons	<1 MeV	$2.5 + 18.2e^{-(\ln(E))^2/6}$
	1–50 MeV	$5.0 + 17.0e^{-(\ln(2E))^2/6}$
	>50 MeV	$2.5 + 3.25e^{-(\ln(0.04E))^2/6}$

$$D(Sv/h) = 5.76 \times 10^{-10} \frac{B}{2} \sum_i \frac{\mu_a(E_i)}{\mu_m(E_i)} E_{i,tot}(\text{MeV}) A_{sp,i} \tag{7.42}$$

with $E_{i,tot}$ the energy released in gamma rays in the decay, $A_{sp,i}$ the activity per unit mass (in Bq/kg), μ_a is the mass energy absorption coefficient of air and μ_m the mass energy attenuation coefficient in (m^2/kg) for the material, respectively. The summation is made over each photon energy group. The *build-up factor B* accounts for multiple scattering of the radiation. Codes such as $FISPACT$ [9] are available to evaluate the activity of a material due to the combined effect of radioisotope decay and of transmutations.

Equation 7.42 can be easily obtained in the case of a semi-infinite solid material by determining first the amount of energy per unit area at the solid surface by integrating over the contribution of all the volume elements

$$\begin{aligned} \Phi_i(W/m^2) &= 1.6 \times 10^{-13} B \int_0^1 2\pi d(cos\theta) \\ &\times \int_0^\infty r^2 dr \rho_m \frac{e^{-\mu_m \rho_m r}}{4\pi r^2} E_{i,tot}(\text{MeV}) A_{sp,i} \\ &= 1.6 \times 10^{-13} \frac{B E_{i,tot}(\text{MeV}) A_{sp,i}}{2\mu_m} \end{aligned} \tag{7.43}$$

From Eq. (7.43) the dose absorbed in a volume of area S and thickness Δ of air can be calculated

$$D_i(Sv/h) = 3600\mu_a(\rho_a S\Delta)\Phi_i = 3600\mu_a(\rho_a S\Delta)1.6 \times 10^{-13} \frac{B E_{i,tot}(\text{MeV}) A_{sp,i}}{2\mu_m} \tag{7.44}$$

Since the dose is the energy absorbed per unit mass the quantity $(\rho_a \Delta S)$ must be taken equal to $1kg$ and after summing over al the energy groups Eq. (7.42) is recovered.

The crucial problem is the effect of radioactivity on human health. The effect at high dose is well known whereas the effect of low doses is still debated. As a result, guidelines are continuously provided and updated by international bodies such as the International Commission on Radiological Protection (ICRP) [10]. At the moment, the maximum dose for exposed workers is $100mSv$ in five years with a maximum of $50mSv$ per year. National states set their own limits. Italy, for example, has a maximum annual dose limit of $20mSv$. The level of $20mSv$, taking into account a typical working time of 2000 h per year, corresponds to the exposure to radioactive sources producing a dose rate of $10\mu Sv/h$. This is the *limit for hands-on activities.*

It is interesting to compare these numbers with the values of natural radioactivity. On average the natural background corresponds to $2mSv$ per year. However there are significant differences in the natural background from place to place, depending e.g. to the proximity of volcanic zone [11] that can lead to a much larger annual dose to the population.

7.11 Suggestions for Further Readings

An old but still valid monograph for nuclear phenomena is the book of E. Segré [6].

7.12 Exercises

Problem 7.1 Using the Weizsäcker formula evaluate the binding energy of the following nuclei indicating the most stable nucleus and discuss which of the contributions in the mass formula make the other nuclei less stable
$^{27}Mg(Z = 12)$ $^{27}Al(Z = 13)$ $^{27}Si(Z = 14)$

Problem 7.2 Find the solution of Eqs. (7.16)–(7.17) with initial conditions $N_1(0) = N_o$ $N_2(0) = 0$. Plot the time variation of the activities of the two nuclei for $\lambda_o = 0.1$ $\lambda_1 = 1$ and $\lambda_o = 1$ $\lambda_1 = 0.1$.

Problem 7.3 If $1g$ of ^{226}Ra has an activity of $1Ci(= 3.7 \times 10^{10}Bq)$ evaluate the half life of ^{226}Ra.

Problem 7.4 Using the expression for Q for the alpha decay plot in the (Z, N) plane the boundaries of the intervals in Z for which $Q > 4$ MeV and $Q > 8$ MeV as A is varied between $A = 60$ and $A = 300$. Compare the result with the nuclide chart.

Problem 7.5 Plot in the (Z, N) plane the curves corresponding to the stable nuclei and the proton/neutron drip lines. Compare the result with the nuclide chart.

References

1. Z. Sóti, et al., Karlsruhe nuclide chart—new 10th edition 2018. EPJ Nucl. Sci. Technol. **5**, 6 (2019)
2. International Atomic Energy Agency (IAEA) at https://nds.iaea.org/relnsd/vcharthtml/VChartHTML.html
3. G. Gamow, Mass defect curve and nuclear constitution. Proc. R. Soc. A. **126**(803), 632–644 (1930)
4. C.F. von Weizsäcker, Zur Theorie der Kernmassen. Zeitschrift für Physik (in German). **96**(7–8), 431–458 (1935)
5. H.A. Bethe, R.F. Bacher, Nuclear physics A. Stationary states of nuclei. Rev. Mod. Phys. **8**, 82 (1936)
6. E. Segre', *Nuclei and Particles* (Benjamin/Cummings Pub. Co., Menlo park (CA), 1982)
7. Evaluated Nuclear Data Files (ENDF) https://www-nds.iaea.org/exfor/endf.htm
8. V.F. Weisskopf, Radiative transition probabilities in nuclei. Phys. Rev. **83**, 1073 (1951)
9. R.A. Forrest, M.R. Gilbert, FISPACT-2005: user manual. Report UKAEA FUS 514 (2005)
10. International Commission on Radiological Protection, General Principles for the Radiation Protection of Workers, ICRP Publication 75, Annals of the ICRP **27**(1) (1997)
11. United Nations Scientific Committee on the Effects of Atomic Radiation, Sources and Effects of Ionising Radiation (United Nations, New York, 2000)

Chapter 8
Nuclear Reactions

Abstract *The general nomenclature of nuclear reaction is introduced and a classification of the main types of reactions is presented. The energetics of fission reactions is discussed on the basis of the liquid drop model. The range in mass number where spontaneous fission is possible is derived. The last part of the chapter focuses on fusion reactions. The main reactions and their relative merits are discussed.*

8.1 Conservation Laws and Cross Sections

Nuclear reactions between two particles A and b leading to particles c and D can be indicated either as

$$A + b \rightarrow c + D \tag{8.1}$$

or

$$A(b, c)D \tag{8.2}$$

In Eq. (8.2) usually the lighter particles are those in lowercase letters. Most of the reactions are classified according to the particles within brackets. Examples are:

- Neutron elastic diffusion (n, n);
- Neutron inelastic diffusion (n, n'); the neutron transfers part of its energy to the internal degrees of freedom of nucleus A;
- Radiative capture (n, γ); the neutron absorption is followed by the emission of a photon;
- Charged particle emission (n, p), (n, d) or (n, α); the neutron absorption is followed by the emission of a charged particle;
- Neutron multiplication $(n, 2n)$; the neutron absorption is followed by the emission of two neutrons.

Indicating with Q the difference in the rest energy of the reactants and of the products, reactions are defined *exoergic* if $Q > 0$ and *endoergic* if $Q < 0$. Exoergic reactions can take place even if the kinetic energy of the reactants is zero. Endoergic

F. Romanelli, *Physics of Nuclear Energy*, Springer Series in Plasma Science and Technology, https://doi.org/10.1007/978-981-97-9609-0_8

reaction can take place only if the reactants have a finite kinetic energy. In general, for a reaction to take place the following inequality must be satisfied

$$Q + E^{(com)}_{kin(A)} + E^{(com)}_{kin(b)} > 0 \qquad (8.3)$$

with the kinetic energy evaluated in the centre of mass frame. It should be noted that Eq. (8.3) is necessary but not sufficient condition for a reaction to occur. Recalling the relation between the total kinetic energy in the laboratory frame and in the c.d.m. frame (see Chap. 3), the threshold energy in the laboratory frame of the reaction is given by

$$Q + E^{(lab)}_{kin(A)} + E^{(lab)}_{kin(b)} - \frac{(m_A + m_b)V^2_{com}}{2} > 0 \qquad (8.4)$$

Assuming the A particle at rest in the laboratory frame, the following equation for the threshold energy is obtained

$$E^{(lab)}_{kin(b)} > -Q(1 + \frac{m_b}{m_A}) \equiv E_{threshold} \qquad (8.5)$$

Reactions must satisfy the following conservation laws in addition to the conservation of energy, momentum and angular momentum:

- conservation of charge
- conservation of baryonic number (see Chap. 1)
- conservation of leptonic number (see Chap. 1)

The conservation of baryonic number implies that the sum of the number of neutrons and protons must be conserved. As we will see the number of neutrons and of protons is *separately* conserved except for the reactions involving the weak force.

For example, the beta decay of the neutron ($n \rightarrow p + e + \bar{\nu}$) conserves the charge (zero before the reactions and zero after) the baryonic number (1 before and 1 after) the leptonic number (zero before and zero after with the electron having a leptonic number +1 and the antineutrino a leptonic number −1). However, since the beta decay is a manifestation of the weak force it does not conserve separately the number of protons and neutrons but only the baryonic number.

A number of important reactions are listed below

- $^{14}N(\alpha, p)^{17}O$ (Q = −1.19 MeV)—the first reaction to be produced in a laboratory by Rutherford in 1919 using the alphas generated in the Po decay.
- $^{7}Li(p, \alpha)\alpha$ (Q = 17.27 MeV)—the first reaction obtained with an accelerator (Cockroft and Walton in 1931).
- $^{9}Be(\alpha, n)^{12}C$ (Q = 5.71 MeV)—the reaction that led to the discovery of the neutron (Bothe and Becker 1931)
- $^{14}N(n, p)^{14}C$ (Q = 0.625 MeV)—the reaction in the upper atmosphere that produces the radioactive isotope used for dating.
- $n(p, \gamma)D$ (Q = 2.22 MeV)—the radiative capture of neutron by protons.
- $p(p, \bar{e}\nu)D$ (Q = 0.42 MeV)—the reaction that burns hydrogen in the stars.

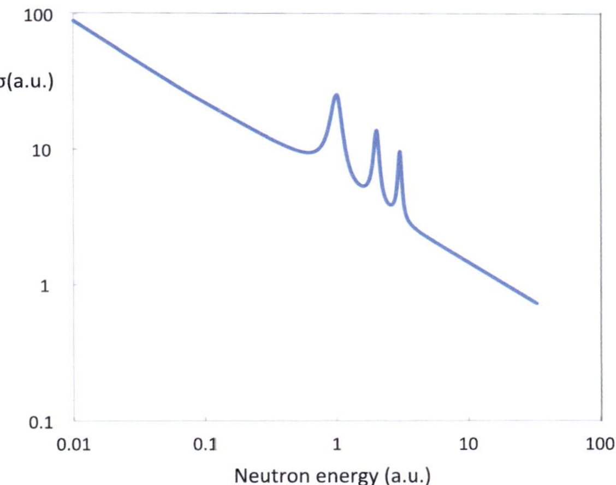

Fig. 8.1 Example of a nuclear cross section for a exoergic reaction. At low energies $\sigma \propto v^{-1}$. Three distinct resonances are present in the intermediate energy range. The cross section attains a constant value at large energies corresponding to the size of the bare nucleus

All the reactions except the last one (again involving the weak force) conserve separately the number of protons and neutrons.

The reaction cross section far from the resonances can be approximated as (see Fig. 8.1)

$$\sigma \approx \pi(R + \frac{1}{k})^2 \tag{8.6}$$

with $k = (2mE)^{1/2}/\hbar$ being the de Broglie wave vector. The explanation for the use of the de Broglie wavelength has been given in Chap. 6. The use of the de Broglie wavelength alone would imply a vanishing cross section as $E \to \infty$. At high energy the particle size reduces to that of the bare nucleus and the cross section attains a finite value. If the reaction involves charged particles the reduction associated with the Gamow factor must be introduced as done in Chap. 6. For neutron induced reactions we have to account for the finite transmission coefficient due to the interaction with the nuclear potential well. Taking for simplicity the value of the transmission coefficient for the potential step calculated in Chap. 6, the cross section has to be modified at low energies by the transmission factor (see Eq. 6.32)

$$\sigma \approx \pi \left(R + \frac{1}{k} \right)^2 \frac{4kk_o}{(k + k_o)^2} \tag{8.7}$$

For very low energies we can expand the expression above for $k << k_o$ yielding

$$\sigma \approx \frac{4\pi}{kk_o} \propto \frac{1}{v} \tag{8.8}$$

that shows that neutron cross sections of exoergic reactions tend to become very large at low energy.

If the incident particle energy is equal to the energy separation of two levels in the compound nucleus we have the phenomenon of resonances with a large increase of the cross section at the resonant energy. Close to the resonance the cross section can be expressed via the *Breit-Wigner* formula

$$\sigma \approx \frac{4\pi}{k^2} \frac{(\frac{\Gamma}{2})^2}{(E - E_{res})^2 + \frac{\Gamma^2}{4}} \tag{8.9}$$

with Γ being the resonance width. This behaviour can be understood by comparing with the cross section for resonant scattering of electromagnetic radiation discussed in Chap. 4 (see Eq. 4.76).

8.2 Interaction Between Reactants

A large number of low-energy reactions take place via the formation of the compound nucleus. The incoming particle is absorbed by the target nucleus to form the compound nucleus that lasts between $10^{-18}s$ and $10^{-16}s$ and then decays. This life is extremely long if compared with the characteristic transit time of a nucleon inside a nucleus (taking a velocity of the nucleon $v \approx c/10$ and a nuclear radius of $10^{-14}m$ the transit time is $\approx 10^{-21}s$).

The interesting feature of the compound nucleus is that it does not "remember" how the nucleus has been formed and its decay process depends uniquely on the available energy. A classical example taken from Ref. [1] are the reactions between a proton and a ^{63}Cu nucleus and between an alpha particle and a ^{60}Ni nucleus. In both cases the compound nucleus ^{64}Zn is formed that, in turn, decays through three channels: a neutron plus ^{63}Zn, two neutrons plus ^{62}Zn and a proton plus a neutron plus ^{62}Cu. If we plot the cross sections for the three channels as a function of the incoming particle energy we observe that the cross sections in each channel are identical for the two reactions. In doing the comparison we have obviously to account for the difference in the available energy of the two reactions that leads to a shift of the maximum. However, the shape of each pair of curves is remarkably similar.

The compound nucleus is not the only mechanism of interaction. The other class of reactions are the *Direct reactions* in which the incoming particle produces the expulsion of fragments from the target nucleus. Below 10 MeV per nucleon the de Broglie wavelength is about $10 fm$ and therefore much larger than the distance between nucleons ($\approx 1.2 fm$). Above 400 MeV per nucleon the de Broglie wavelength is $\leq 1 fm$ and the incoming particle can interact with the individual nucleons. Direct reactions are therefore subdivided depending on the energy of the incoming particle.

- Low energy direct reactions.
  - Stripping. Part of the incoming particle is taken by the target nucleus, e.g. $A(d, p)B$, $A(\alpha, d)B$, $A(\alpha, t)B$
  - Pick-up. The incoming particle captures part of the target nucleons, e.g. $A(p, d)B$, $A(p, t)B$, $A(\alpha, ^6 Li)B$
- High energy direct reactions.
  - Spallation. The incoming particle ($E > 100$ MeV) kicks out a few nucleons leaving the residual target in an excited state that decays through evaporation of protons or neutrons or through fission.

8.3 The Fission Process

The fission process was observed by Hahn and Strassmann in 1938 [2]. They were able to show that by bombarding uranium with neutrons barium was formed. Immediately after the discovery Meitner and Frisch [3] proposed fission as the mechanism leading to the barium production.

To understand the energetics of the fission process we need to go back to the liquid drop model and ask what is the modification of the energy if the nucleus is deformed from a spherical to an ellipsoidal shape at constant nuclear volume. This deformation can be achieved by going from a sphere of radius R to an ellipsoid with major radius $a = R(1 + \epsilon)$ and minor radius $b = R/(1 + \epsilon)^{1/2}$ (Fig. 8.2).

The volume term is unaffected. Similarly, we will assume that the symmetry and pairing term are not changed in the deformation process. We need to focus on the change in the surface and Coulomb terms. The calculation is presented in Appendix A.

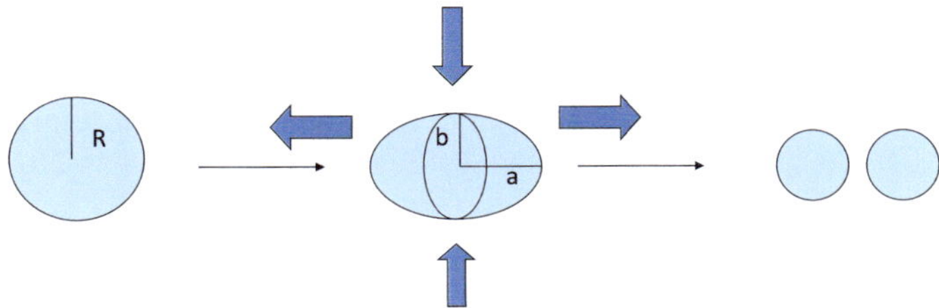

Fig. 8.2 The fission process can be modelled as a continuous deformation of the nucleus from a spherical shape to an ellipsoidal shape at constant volume. Beyond a certain deformation the nucleus breaks up in two parts that recover the spherical shape

For small deformations ($\epsilon << 1$) the surface of the ellipsoid S can be expanded as

$$S = 4\pi R^2 \left(1 + \frac{2\epsilon^2}{5} - \frac{52\epsilon^3}{105} + O(\epsilon^4)\right) \tag{8.10}$$

The surface of the ellipsoid is larger than the surface of the sphere and this reduces the binding energy.

The Coulomb term can be evaluated assuming a uniform charge density inside the ellipsoid

$$W = \frac{\epsilon_o}{2} \int d\mathbf{r} E^2 = \int d r \rho(\mathbf{r}) \Phi(\mathbf{r}) \tag{8.11}$$

with the electrostatic potential given by [4]

$$\Phi(\mathbf{r}) = \pi a b^2 \rho_o \int_0^\infty \frac{dv}{(b^2+v)(a^2+v)^{1/2}} \left[1 - \left(\frac{r^2}{b^2+v} + \frac{z^2}{a^2+v}\right)\right] \tag{8.12}$$

For small deformations the Coulomb energy can be written as

$$W = \frac{3}{5} \frac{Z^2 e^2}{4\pi\epsilon_o R} \left(1 - \frac{\epsilon^2}{5} - \frac{4}{7}\epsilon^3 + O(\epsilon^4)\right) \tag{8.13}$$

In this case the deformation increases the average distance between protons with the result of increasing the binding energy.

The variation of the mass of the nucleus due to the deformation up to order ϵ^3 is therefore given by

$$\delta M(Z, A)c^2 = -\delta E_{binding} = b_2 A^{2/3} \frac{2\epsilon^2}{5} - \frac{b_3 Z^2}{A^{1/3}} \frac{\epsilon^2}{5} - \\ - \frac{4}{105} \left(13 b_2 A^{2/3} + 15 \frac{b_3 Z^2}{A^{1/3}}\right) \epsilon^3 \tag{8.14}$$

Note that the quadratic term has no definite sign whereas the cubic term has always the same sign. At the lowest $O(\epsilon^2)$ order, for nuclei with $Z^2/A > 2b_2/b_3 = 49$, $\delta M(Z, A) < 0$. This means that the deformed state is energetically favoured over the undeformed state (in going from the deformed to the deformed state energy is released). The deformation then grows in time until the nucleus splits in two parts. This phenomenon is the spontaneous fission that can be observed in heavy nuclei. Taking the values of Z corresponding to the stability condition Eq. (7.11), the condition for spontaneous fission is satisfied for

$$A > 190(1 + 0.0076\, A^{2/3})^2 \tag{8.15}$$

i.e. for $A > 350$. However, the simple model used here tends to overestimate the threshold for spontaneous fission that occurs in reality for nuclides of the actinides series such as ^{250}Cm (half life 10^4years) or ^{252}Cf (half life 86 years), i.e. for $Z^2/A >$ $37 - 38$. Uranium has a low probability of spontaneous fission: ^{235}U half life is $\approx 3 \times 10^{17}$years.

For nuclei with $Z^2/A < 49$ work has to be done in order to deform their shape and, once the deforming force is released, the nucleus will start to oscillate around the spherically symmetric state. As the next order term in the deformation (the cubic term in Eq. (8.14)) is always negative, if the deformation is large enough it will progress spontaneously until fission is produced. The situation is depicted in Fig. 8.3.

In order for nuclei with $Z^2/A < 49$ to split enough energy must be provided to bring the nucleus above the maximum of the potential well. The energy needed to bring the nucleus from its ground undeformed state to the top of the potential well is the *activation energy*. It can be estimated from Eq. (8.14) by looking at the stationary points where

$$\frac{\partial Mc^2}{\partial \epsilon} = b_2 A^{2/3} \frac{4\epsilon}{5} - \frac{b_3 Z^2}{A^{1/3}} \frac{2\epsilon}{5} - \frac{12}{105}(13b_2 A^{2/3} + 15\frac{b_3 Z^2}{A^{1/3}})\epsilon^2 = 0 \tag{8.16}$$

There are two points satisfying the equation. One correspond to $\epsilon = 0$ (the bottom of the potential well corresponding to a stable state—a small perturbation tends to bring back the system to the equilibrium point) and the other to $\epsilon = \epsilon_c$ with

$$\epsilon_c = \frac{7}{13} \frac{1 - \frac{b_3 Z^2}{2b_2 A}}{1 + \frac{15 b_3 Z^2}{13 b_2 A}} \tag{8.17}$$

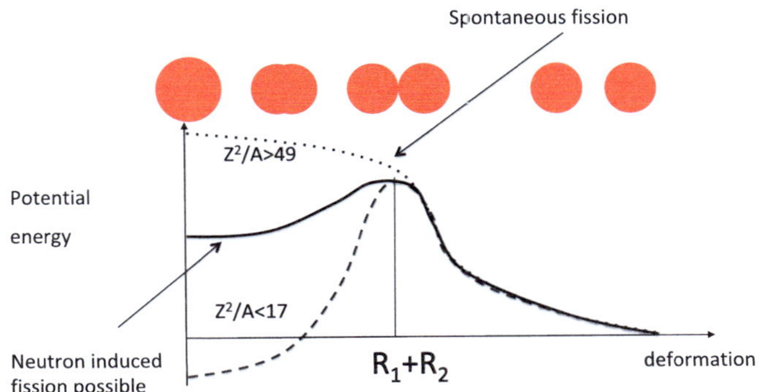

Fig. 8.3 Change in energy as a function of the deformation. Three cases can be distinguished: for $Z^2/A > 49$ (dotted line) the fission process occurs spontaneously. The nucleus is at the top of a potential hill and a small deformation is amplified; for $59 > Z^2/A > 17$ (continuous line) the nucleus is in a potential well and for fission to occur a net activation energy must be provided; for $Z^2/A < 17$ (dashed line) the fission process is energetically unfavourable

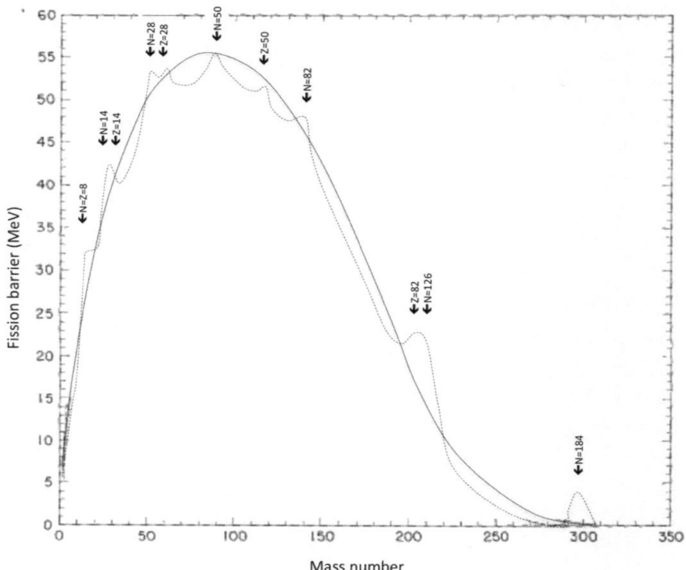

Fig. 8.4 Activation energy for the fission process versus mass number. Reprinted from Ref. [5] with permission from Elsevier. The continuous line is the result using the liquid drop model whereas the dashed line includes the correction associated with the discrete quantum levels of the nucleus

corresponding to an unstable state (a small perturbation tends to bring the system far from the equilibrium point).

Note that these expressions are valid only for $Z^2/A < 49$. For $Z^2/A > 49$ the change in energy is shown in Fig. 8.3 and the equilibrium point $\epsilon = 0$ is now unstable.

The simple expression for the deformation used in this section provides a qualitative correct picture of the fission process. To make quantitative predictions of the critical deformation and activation energy requires using more complex expressions for the deformation. An accurate determination of the activation energy has been made in Ref. [5] and the result is shown in Fig. 8.4.

8.4 The Energetics of the Compound Nucleus Formation in the Neutron Induced Fission Process

How can we provide the nucleus with the activation energy that is necessary for the fission process to occur? The simplest way is to have a neutron absorbed by the nucleus. The difference in the energy between the compound nucleus and the sum of the energies of the original nucleus and the neutron is what is available to produce the desired deformation.

In Table 8.1 the change in the rest energy of the compound nucleus is shown for the formation of the ^{236}U and ^{239}U compound nuclei.

Table 8.1 Energy variation of the compound nuclei

	^{236}U (MeV)	^{239}U (MeV)
$b_1 A$	15.6	15.6
$-b_2 A^{2/3}$	−1.8	−1.8
$-b_3 Z^2/A^{1/3}$	+1.3	+1.3
$-b_4 (Z-N)^2/A$	−9.1	−9.5
$b_5 \delta (Z-N)/A^{1/2}$	+0.78	−0.78
Total	6.8	4.8

Each term of the mass formula is compared. The result is that the change in energy is different in the two cases. For ^{236}U the available energy is 6.5 MeV[1] whereas for the ^{239}U the available energy is 4.8 MeV. This difference has profound consequences. Indeed, by looking at Fig. 8.4 we note that the activation energy for ^{236}U and ^{239}U is almost identical (6.2 MeV and 6.6 MeV, respectively). Thus, the formation of the compound ^{236}U nucleus provides enough energy to activate the process even if the neutron has negligible kinetic energy, whereas in the case of the ^{239}U compound formation in order to have fission we need to provide some extra energy (≈ 1.8 MeV) in the form of the kinetic energy of the neutron. In order to split the ^{235}U nucleus thermal neutrons (i.e. neutrons with the thermal energy characteristic of the reactor core) are enough whereas to split the ^{238}U nucleus fast neutrons are needed. Using ^{235}U it is possible to take advantage of the cross section increases (as $1/v$) at low energy (with typical values in the range of several hundreds barns). On the contrary we expect that the cross section for fission of ^{238}U by low energy neutron will be almost negligible. This is indeed what happens experimentally, as shown in Fig. 8.5.

The origin of the difference between ^{235}U and ^{238}U is in the pairing term, as can be seen by looking at Table 8.1. In the production of ^{236}U an even/odd nucleus is transmuted in an even/even nucleus that is characterised by a larger binding energy and so provides an extra energy gain with respect to the ^{239}U formation in which an even/even nucleus is transmuted into an even/odd (less bound) nucleus. This observation suggests that the even/odd nuclei (as ^{235}U, ^{233}U or ^{239}Pu) can be split by thermal neutrons whereas even/even nuclei (as ^{238}U or ^{232}Th) require fast neutrons.

Nuclides that undergo fission are called *fissionable*. If in addition they are split by thermal neutrons they are called *fissile*. Materials such as ^{232}Th are not fissile. However, after neutron capture they can transmute to a fissile nuclide and for this reason they are called *fertile* nuclides.

---

[1] This is the difference in mass obtained from the experimental measurements, slightly smaller than the value obtained from the liquid drop model in Table 8.1.

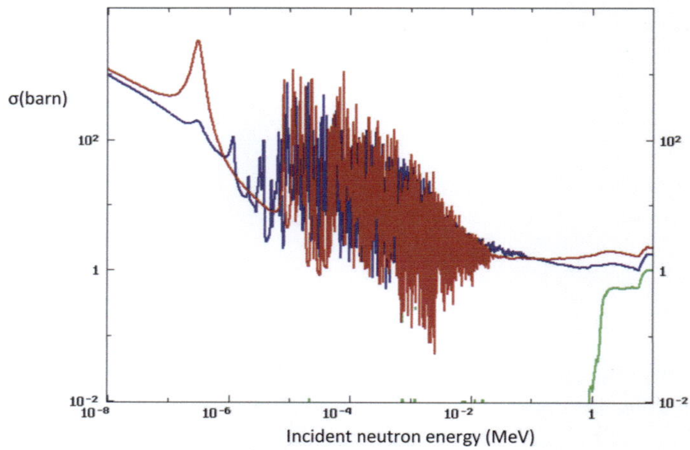

Fig. 8.5 Fission cross section from the ENDF libraries [6] of ^{235}U (blue), ^{239}Pu (red) and ^{238}U (green). Both ^{235}U and ^{239}Pu are fissile nuclei whereas ^{238}U is a fissionable nucleus

8.5 Fission of Uranium Isotopes

Natural uranium is made of two isotopes: ^{238}U (99.3%) with a half-life $4.468 Gy$ and ^{235}U (0.7%) with half-life $704 My$. The next long lasting isotopes are ^{236}U (half life $23 My$), ^{234}U ($245 ky$) and ^{233}U ($159 ky$) and are produced artificially.

The fission process leads to a number of products:

- Two fission fragments with A around 95 and A around 140, respectively, releasing a total energy of about 170 MeV;
- 2-3 neutrons (prompt neutrons) of energy around 2 MeV per neutron;
- The two fragments are typically far from the stability curve and are neutron-rich so that they undergo a series of beta decays to move closer to the stability curve with the production of electrons and (anti) neutrinos releasing an energy of about 20 MeV;
- Gammas are also produced in the process both as prompt gammas (8 MeV) and gammas from the decays of the excited states of the fission fragments (8 MeV).

The total energy released in the fission process is 210 MeV.

The cross section for radiative neutron capture (n, γ) and the fission cross section are shown in Fig. 8.6 for ^{235}U and ^{238}U.

The energy domain between $10 eV$ and $10^4 eV$ is dominated by the effect of the resonances. In the case of ^{235}U the (n, γ) cross section is always smaller than the fission cross section by about an order of magnitude with the fission cross section achieving a value of $10^3 b$ at $E = 10^{-2} eV$. Conversely, the fission cross section of ^{238}U is larger than the (n, γ) cross section only for $E \geq 1$ MeV and drops to values $< 10^{-2} b$ at lower energies whereas the (n, γ) cross section achieves values of the order $10^3 b - 10^4 b$ in the resonance region.

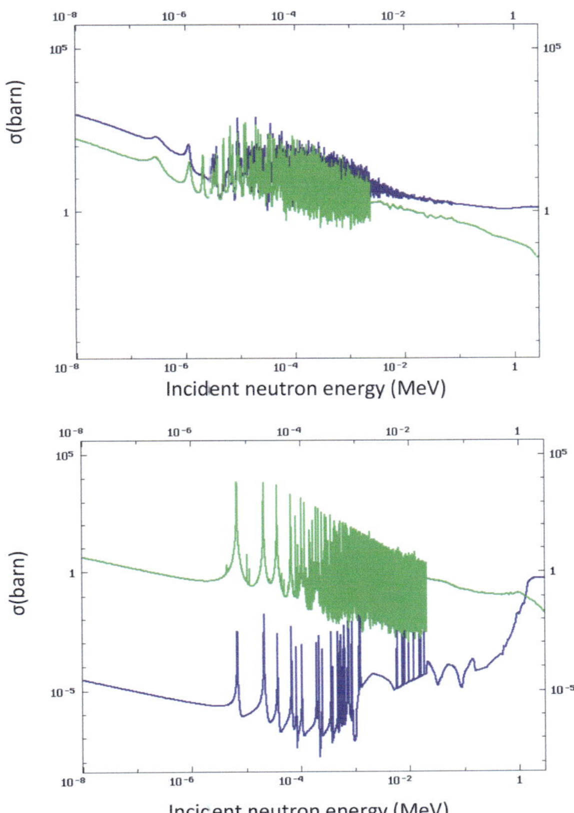

Fig. 8.6 Fission cross section (blue) versus radiative capture (green) from the ENDF libraries [6] for ^{235}U (upper figure) and ^{238}U (lower figure). In the case of ^{235}U the cross section for radiative capture is always smaller than the fission cross section. In the case of ^{238}U this is true only for energies above 1 MeV

Therefore the easiest way to produce energy from nuclear fission is to rely on a small concentration of fissile elements such as ^{235}U and to increase the probability of fission by slowing down the neutrons (produced at energies of about 2 MeV) by collisions with a moderator made of a low-Z material such as hydrogen or carbon. In the next chapter the dynamics of fission reactors will be considered in more detail.

8.6 Thorium Cycle

A possible alternative to the use of uranium is the thorium cycle. Thorium is present on Earth almost as a single isotope ^{232}Th (abundance 99.98%) that is three times more abundant than uranium. Since it is an even-even nuclide it is not fissile but can be used to produce ^{233}U via neutron capture

$$^{232}Th + n \rightarrow ^{233}Th + \gamma \rightarrow ^{233}Pa + e + \bar{\nu}_e$$
$$^{233}Pa \rightarrow ^{233}U + e + \bar{\nu}_e \quad (8.18)$$

The first advantage of thorium is that the amount of transuranic elements produced by burning ^{233}U is much lower than the amount obtained by burning ^{235}U. Indeed, the capture of a neutron by ^{233}U leads with 92% probability to fission and to the production of ^{234}U in the remaining 8%. The fission probability is much higher than that of ^{235}U and ^{239}Pu.

Furthermore, the final result of the $(n, 2n)$ reactions (see Eqs. (8.19–8.21) is ^{232}U that is characterised by decay products (such as ^{208}Tl) that are strong gamma emitters. This makes it difficult manipulating the waste in order to separate fissile products (such as ^{233}U) for nuclear proliferation.

$$^{232}Th + n \rightarrow ^{233}Th + \gamma \rightarrow ^{233}Pa + e + \bar{\nu}_e$$
$$^{233}Pa \rightarrow ^{233}U + e + \bar{\nu}_e$$
$$^{233}U + n \rightarrow ^{232}U + 2n \quad (8.19)$$

$$^{232}Th + n \rightarrow ^{233}Th + \gamma \rightarrow ^{233}Pa + e + \bar{\nu}_e$$
$$^{233}Pa + n \rightarrow ^{232}Pa + 2n$$
$$^{232}Pa \rightarrow ^{232}U + e + \bar{\nu}_e \quad (8.20)$$

$$^{232}Th + n \rightarrow ^{231}Th + 2n$$
$$^{231}Th \rightarrow ^{231}Pa + e + \bar{\nu}_e$$
$$^{231}Pa + n \rightarrow ^{232}Pa + \gamma \rightarrow ^{232}U + e + \bar{\nu}_e \quad (8.21)$$

8.7 Fusion Reactions

Fusion reactions involve light nuclei and are energetically favoured for $A < 60$. The main fusion reactions are listed in Table 8.2.

As discussed in Chap. 6, the fusion cross sections are expressed as follows

$$\sigma(E) = S(E)\frac{e^{-\frac{B_G}{E^{1/2}}}}{E} \quad (8.22)$$

where the term in the exponent is the Gamow factor with $B_G = \pi q_1 q_2/(2\epsilon_o h)$ $(2m_r)^{1/2}$. The factor $S(E)$ is the astrophysical factor that can be found e.g. in [7]. Figure 8.7 shows the cross section as a function of energy for the fusion reactions foreseen for energy production.

Table 8.2 Fusion reactions

Reactants	Products	Branching ratio (%)
$D + T \rightarrow$	${}^4He(3.5\,\mathrm{MeV}) + n(14.1\,\mathrm{MeV})$	
$D + D \rightarrow$	$T(1.01\,\mathrm{MeV}) + p(3.02\,\mathrm{MeV})$	50
	${}^3He(0.82\,\mathrm{MeV}) + n(2.45\,\mathrm{MeV})$	50
$D +^3 He \rightarrow$	${}^4He(3.6\,\mathrm{MeV}) + p(14.7\,\mathrm{MeV})$	
$T + T \rightarrow$	${}^4He + 2n + 11.3\,\mathrm{MeV}$	
${}^3He + T \rightarrow$	${}^4He + p + n + 12.1\,\mathrm{MeV}$	51
	${}^4He(4.8\,\mathrm{MeV}) + D(9.5\,\mathrm{MeV})$	43
	${}^5He(2.4\,\mathrm{MeV}) + p(11.9\,\mathrm{MeV})$	6
$p +^6 Li \rightarrow$	${}^4He(1.7\,\mathrm{MeV}) +^3 He(2.3\,\mathrm{MeV})$	
$p +^7 Li \rightarrow$	$2^4He + 17.3\,\mathrm{MeV}$	20
	${}^7Be + n - 1.6\,\mathrm{MeV}$	80
$D +^6 Li \rightarrow$	$2^4He + 22.4\,\mathrm{MeV}$	
$p +^{11} B \rightarrow$	$3^4He + 8.7\,\mathrm{MeV}$	

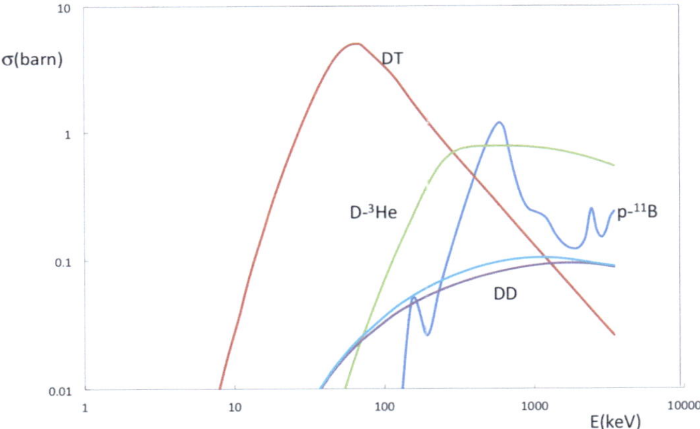

Fig. 8.7 Fusion cross sections. The DT reaction has the largest cross section and its peak is at the lowest energies. The D^3He reaction has a maximum cross section of about $1b$ but requires much larger energies. The DD reaction has a cross section about two orders of magnitude below the DT reaction at low energy. The $p^{11}B$ has a resonance around 700 keV with a cross section above $1b$

The various reactions shown in Fig. 8.7 have pros and cons. The DT reaction is the reaction with the largest cross section and it achieves its maximum at the lowest energy. However it has two main issues:

- Tritium is not available and must be produced inside the reactor;
- The reaction produces high-energy neutrons that activate the reaction chamber and damage the structural materials.

The D^3He reaction requires higher energies but it does not produce neutrons (note however that neutrons are produced by the DD reactions that involve part of the fuel). Unfortunately, ^3He is not available on Earth.

The DD reaction has no problem with the availability of fuel (sea water contains on average $24mg$ of deuterium per litre). It does produce neutrons but at lower energy than the DT reaction leading to a reduced issue with neutron induced damage and activation. The problem with the DD reaction is that its cross section is about two orders of magnitude lower than the DT reaction.

The $p^{11}B$ reaction does not produce neutrons (note however that neutrons can be produced through the $^{11}B(\alpha, n)^{14}N$ reaction). Its cross section has a maximum above $1b$ but requires very high energies.

8.8 Maxwellian Reactivity. The Gamow Peak

Rather than the cross section, it is convenient to consider the Maxwellian reactivity defined in Chap. 3 as the average over the distribution function of the reactants of the product σu (with u the relative velocity)

$$< \sigma u >= \int\int \frac{d\mathbf{v}_a}{(\pi^{1/2} v_{ta})^3} e^{-\frac{v_a^2}{v_{ta}^2}} \frac{d\mathbf{v}_b}{(\pi^{1/2} v_{tb})^3} e^{-\frac{v_b^2}{v_{tb}^2}} \sigma(m_r u^2/2) u \tag{8.23}$$

Since the cross section is typically given as a function of the energy in the centre of mass system, we change variables from $\mathbf{v}_a$ and $\mathbf{v}_b$ to $\mathbf{V}_{cdm}$ and $\mathbf{u}$. Upon assuming that species a and b have Maxwellian distributions with the same temperature it is easy to show (see Chap. 3) that

$$< \sigma u >= \int \frac{d\mathbf{u} e^{-\frac{u^2}{v_{tr}^2}}}{(\pi^{1/2} v_{tr})^3} \sigma(m_r u^2/2) u \tag{8.24}$$

where $v_{tr} = (2T/m_r)^{1/2}$ and m_r the reduced mass. The integrand is the product of two exponential functions, one dominant at low energy and the other at high energy. The exponent has a maximum for

$$\frac{d}{dE}\left(\frac{E}{T} + \frac{B_G}{E^{1/2}}\right) = \frac{1}{T} - \frac{B_G}{2E^{3/2}} = 0 \tag{8.25}$$

i.e. for $E = E_G = (T B_G/2)^{2/3} \equiv m_r u_G^2/2$. The maximum of the product of the exponential functions corresponds to the so-called *Gamow peak*.

To evaluate the integral it is convenient to expand the integrand around $E = E_G$ and keep only the energy dependence of the exponential factor, yielding

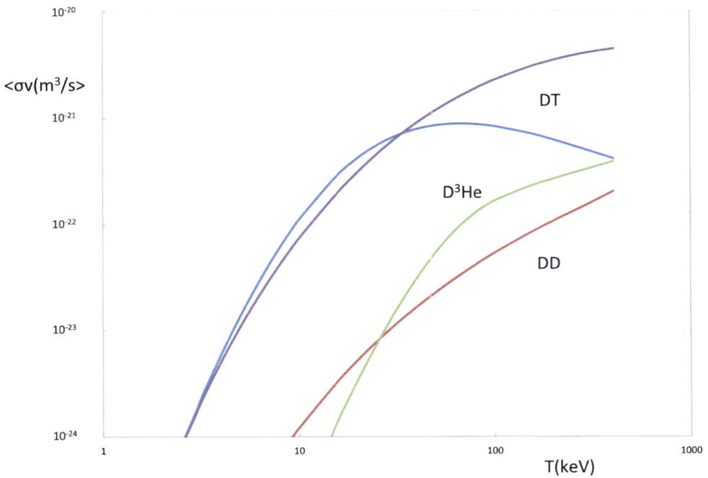

Fig. 8.8 The Maxwellian reactivity for the DT reaction from [8] (blue line) is compared with the simple estimate given in Eq. (8.26) (purple line). The other curves are the Maxwellian reactivities for the D^3He (green) and DD (red) reactions

$$
\begin{aligned}
< \sigma u >= & \frac{4u_G^2}{m_r \pi^{1/2} v_{tr}^3} e^{-\frac{3B_G^{2/3}}{(4T)^{1/3}}} \frac{S(E_G)}{E_G} \int dE e^{-\frac{3B_G(E-E_G)^2}{8E_G^{5/2}}} \\
& \approx \frac{8\,S(E_G)}{m_r^2 v_{tr}^3} e^{-\frac{3B_G^{2/3}}{(4T)^{1/3}}} \left(\frac{8}{3B_G} (\frac{B_G T}{2})^{5/3} \right)^{1/2}
\end{aligned}
\tag{8.26}
$$

As a first approximation we can approximate Eq. (8.26) as $< \sigma u >= C_1 exp(-C_2/T^{1/3})/T^{2/3}$, with C_1 and C_2 depending on the specific fusion reaction. Better approximations have been given in the literature for the various reactions. For the DT, DD and D^3He the best approximation has been given in Ref. [8]. The $p^{11}B$ reaction has been reviewed in Ref. [9].

In Fig. 8.8 the fusion reactivity or the DT reaction from [8] is compared with the simple estimate given in Eq. (8.26).

The number of reactions for unit time and unit volume can be evaluated from the reactivity as follows (see Chap. 3)

$$
\frac{dN_{ab}}{dt} = \frac{n_a n_b < \sigma u >_{ab}}{1 + \delta_{ab}} \tag{8.27}
$$

Note the factor $1 + \delta_{ab}$ that avoids double counting the reaction between particles as shown in Fig. 8.9.

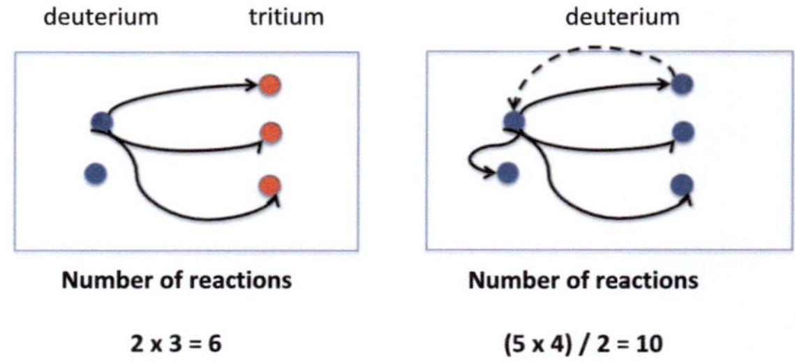

Fig. 8.9 For reactions between particles of the same species the frequency of events is proportional to the square of the density divided by two since the reaction between particle a and particle b is the same as the reaction between particle b and particle a

8.9 The DT Fusion Cycle

While deuterium is stable and extremely abundant, tritium has an half-life of 12.3y and must be produced inside a fusion reactor operated with a DT mixture. To this goal we can use the neutron produced in the DT reaction and a lithium blanket surrounding the reaction chamber. Note that lithium has two stable isotopes ^7Li (92.4%) and ^6Li(7.6%). The relevant reactions are

$$n +^6 Li \rightarrow \alpha + T \tag{8.28}$$

$$n +^7 Li \rightarrow \alpha + T + n \tag{8.29}$$

The cross section of the two reactions is shown in Fig. 8.10. The ^6Li reaction is exoergic and has the characteristic $(1/v)$ behaviour at low energy. The ^7Li reaction is endoergic and occurs only at high energy with a threshold for $E = 2.8\,\text{MeV}$.

8.10 Production of Radio-Nuclides

Artificially produced radio-nuclides have a number of applications in medical treatments. They are produced by irradiating the parent nuclide with neutrons. The evolution of the nuclide is described by the following equation

$$\frac{dN_1}{dt} = -N_n N_1 \sigma v = -\Phi_n \sigma N_1 \tag{8.30}$$

with $\Phi_n \equiv N_n v$ the neutron flux and σ the cross section for the transmutation process. The child nuclide, with decay constant λ_2, evolves according to the following equation

$$\frac{dN_2}{dt} = -\lambda_2 N_2 + \Phi_n \sigma N_1 \tag{8.31}$$

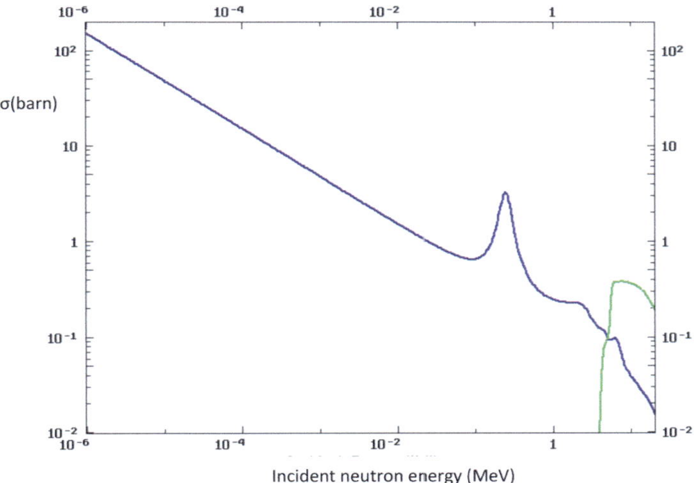

Fig. 8.10 The (n, Li) cross section from the ENDF libraries [6] is shown for the two stable isotopes of Li, namely ^6Li (blue) and ^7Li (green). The $(n,^6 Li)$ reaction is exoergic whereas the $(n,^7 Li)$ reaction is endoergic and has a threshold

The same couple of equations describes the production of fissile elements by neutron irradiation of fertile elements. The solution can be found as discussed in Chap. 7 for the case of the decay chains. Three characteristic time scales appear in these equations: the time scale for parent isotope transmutation $1/(\Phi_n \sigma)$, the child nuclide half life $\ln 2/\lambda_2$ and the duration of the irradiation. Although this set of equations, even in the case of more complex chains, can be solved by simple algebra, it is often convenient to exploit the difference in the three time scales to get a rapid determination of the produced radionuclides (see exercises).

8.11 Suggestions for Further Readings

The discovery of fission and the Manhattan Project is well described in Ref. [10], that reports a number of information about the scientists who had to take a difficult decision at the time of the second World War.

8.12 Exercises

Problem 8.1 Using the Weizsäcker formula evaluate the energy produced in the fission of ^{235}U in two nuclei with $A = 90$ and $A = 142$. Evaluate the energy released in the beta decays of the fission fragments.

Problem 8.2 Evaluate the number of DT fusion reactions per unit volume and unit time in a plasma with $50\%/50\%$ DT composition and temperature $T = 20\,\text{keV}$.

Problem 8.3 Using a file excel plot the Maxwellian reactivity for the DT reaction using the Bosch-Hale fit in the range 1–30 keV.

Problem 8.4 A target made of $1g$ of ^{59}Co is irradiated with a flux of thermal neutrons with intensity $10^{12}cm^{-2}s^{-1}$. In the reaction the radioactive isotope ^{60}Co is produced which has a half life of $5.27y$. Evaluate the activity of ^{60}Co after one year of irradiation assuming a cross section for the transmutation process of $40b$.

Problem 8.5 In a fusion reactor tritium is produced via the $^6Li(n, \alpha)T$ reaction. Evaluate the amount of tritium produced in one year from 10^{21} ^6Li atoms assuming a neutron flux of $5 \times 10^{14}n/(cm^2)s$ and a cross section for the reaction (averaged over the neutron spectrum) of $10b$. How much tritium will be available at the end of the irradiation?

Problem 8.6 Consider a fusion reactor with a mixture of 99% of deuterium and 1% of tritium at the temperature $T = 50\,\text{keV}$. The reaction chamber is surrounded by a blanket with water. Using the ENDF libraries determine the capture probability by hydrogen of the neutrons emitted by the reactor and the probability of oxygen activation via the reaction $^{16}O(n, p)^{16}N$.

References

1. S.N. Ghoshal, An experimental verification of the theory of compound nucleus. Phys. Rev. **80**, 939 (1950)
2. O. Hahn, F. Strassmann, Über den Nachweis und das Verhalten der bei der Bestrahlung des Urans mittels Neutronen entstehenden Erdalkalimetalle. (On the detection and characteristics of the alkaline earth metals formed by irradiation of uranium with neutronsl). Naturwissenschaften **27**(1), 11–15 (1939)
3. L. Meitner, O.R. Frisch, Disintegration of uranium by neutrons: a new type of nuclear reaction. Nature **143**, 239–240 (1939)
4. O.D. Kellogg, *Foundations of Potential Theory* (Dover, New York, 1953)
5. W.D. Myers, W.J. Swiatecki, Nuclear masses and deformations. Nucl. Phys. **81**, 1 (1966)
6. Evaluated Nuclear Data Files (ENDF). https://www-nds.iaea.org/exfor/endf.htm
7. http://www.astro.ulb.ac.be/nacreii/index.html?individualreaction.html
8. H.-S. Bosch, G.M. Hale, Improved formulas for fusion cross-sections and thermal reactivities. Nucl. Fusion **32**, 611 (1992)
9. W.M. Nevins, The thermonuclear fusion rate coefficient for $p -^{11} B$ reactions. Nucl. Fusion **40**, 310 (2000)
10. R. Rhodes, *The Making of the Atomic Bomb* (Simon & Schuster, New York, 1986)

Chapter 9
Fission Reactors

Abstract *In this chapter the concept of criticality is introduced and a description of the main features of a fission power plant is provided.*

This book is mostly aimed at providing the basis for fusion as an energy source. Therefore the detailed description of fission power plants is outside our scope. For a full treatment we refer to classical textbooks such as [1]. A general overview of fission reactors and the associate challenges can be found in Ref. [2]. In this chapter a short overview of the dynamics of fission reactors is presented mostly to provide a comparison with the fusion power plant described later in this book.

In each fission event the number of neutrons produced is between two and three. These neutrons can produce further fission events leading to the production of a *chain reaction* (see Fig. 9.1). Neutron losses reduce the number of neutrons that are available to maintain the chain reaction and a stationary condition can be achieved.

The number of neutrons produced per unit time is given by the following equation

$$\frac{dn}{dt} = (k - 1)\frac{n}{\tau} \tag{9.1}$$

where k is the neutron multiplication coefficient and the characteristic time τ is of the order of 10^{-4} s. In the equation above the factor k accounts for all the production and absorption processes and the factor -1 for the absorption of the neutron starting the chain reaction. A stationary condition is achieved for $k = 1$, known as *criticality* condition.

As we have seen in Chap. 8 the cross section of fissile radionuclides such as ^{235}U is about a factor 1000 larger at thermal energies than at the birth energies of neutrons $\approx$1 MeV thanks to the $1/v$ behaviour of the neutron cross sections. This suggests that, in order to maximise the neutron production, neutrons must be slowed down to thermal energies via collisions with light materials (water or graphite). This is the approach followed in the so called *thermal reactors* and all but two of the reactors in operation at the moment are indeed thermal reactors. As we will see later thermal reactors are easier to operate, require a limited amount of fuel enrichment in fissile elements but have some drawback as they burn only a limited part of the fissionable material and produce long lived radioisotopes by neutron capture (minor actinides)

F. Romanelli, *Physics of Nuclear Energy*, Springer Series in Plasma Science and Technology, https://doi.org/10.1007/978-981-97-9609-0_9

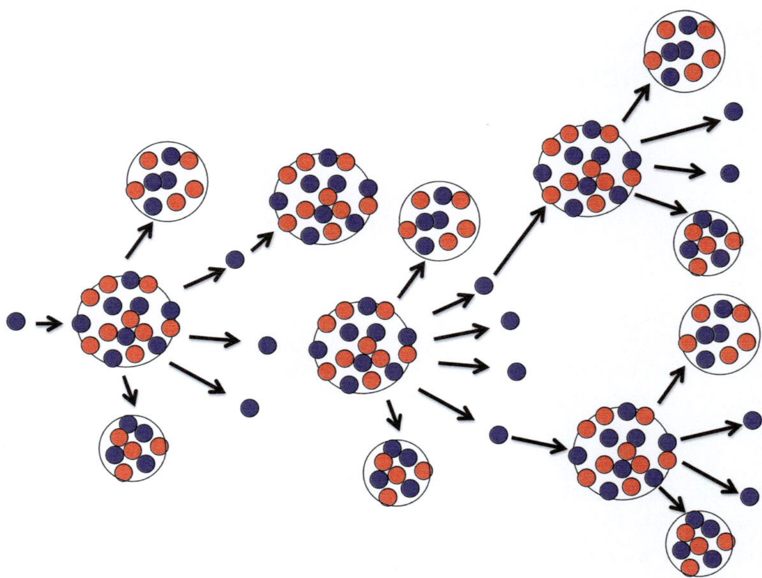

Fig. 9.1 Chain reaction. A neutron absorbed by an uranium nucleus leads to a fission and the emission of 2–3 neutrons. The first generation neutrons can produce new fissions or undergo radiative capture. If on average the number of neutrons in each generation is constant the chain reaction is self sustained. This condition is called criticality

that are difficult to dispose of. The second approach *(fast reactors)* consists in relying entirely on the fissions made by high-energy *(fast)* neutrons ($E > 100\,\text{keV}$). In this case, to achieve criticality the enrichment in fissile material must be larger than in thermal reactor but a more efficient use of the fissionable elements is possible and the production of minor actinides is reduced. In the case of fast reactor any neutron moderation must be avoided and the cooling action must be performed by a high-mass number material (sodium or lead).

Before discussing the criticality condition it is necessary to introduce the concept of *flux per unit energy* that is commonly used to describe the dynamics of nuclear systems. Making reference to Fig. 9.2 let us consider a system of mono energetic neutron beams of intensity $I_j \equiv n_{nj} v$ impinging on a target volume ΔV and density n_T. The neutrons in each beam have the same energy $E = mv^2/2$ (and therefore the same v) but the direction of the velocity is different. The frequency of events in the volume ΔV is simply given by $\sum_j n_T \sigma(E) I_j \Delta V$ with $\sigma(E)$ the interaction cross section. As the beams have the same energy it is possible to take v outside the summation to obtain $n_T \sigma(E) v \Delta V \sum_j n_{nj}$. If instead of a discrete set of beams we have a continuous distribution the summation will be replaced by an integral over the solid angle to obtain

$$n(\mathbf{r}) = \int_{4\pi} d\Omega_v n(\mathbf{r}, \Omega_v) \tag{9.2}$$

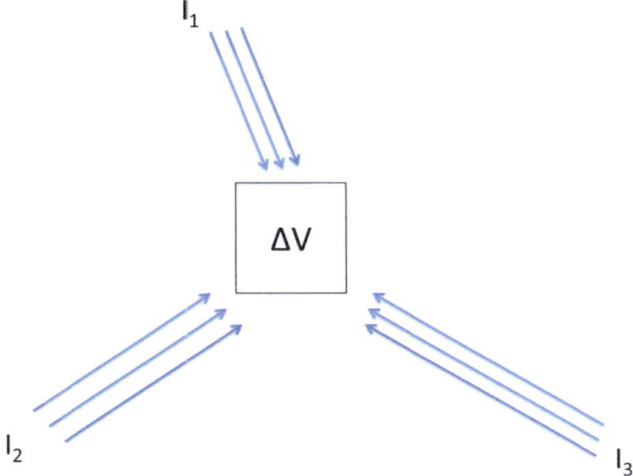

Fig. 9.2 It is customary to describe the dynamics of a nuclear system in terms of the flux per unit energy as illustrated here for a set of three neutron beams with the same energy impinging on the volume ΔV

where Ω_v is the solid angle in velocity space. The quantity $n(\mathbf{r})$ is the total number of neutrons at $\mathbf{r}$ whereas $n(\mathbf{r}, \Omega_v)$ is the *neutron density distribution function*. The differential beam intensity can be defined as

$$dI(\mathbf{r}, \Omega_v) = n(\mathbf{r}, \Omega_v) v d\Omega_v \tag{9.3}$$

which will produce a contribution to the number of events per unit volume given by

$$dF(\mathbf{r}, \Omega_v) = n_T \sigma(E) n(\mathbf{r}, \Omega_v) v d\Omega_v \tag{9.4}$$

with the total number of events per unit volume given by $\int_{4\pi} dF = n_T \sigma(E) n(\mathbf{r}) v$. This approach can be generalised to the case of a continuous energy distribution of neutrons (instead of mono energetic beams) by defining a *neutron density distribution per unit energy* $n(\mathbf{r}, E, \Omega_v)$ such that

$$n(\mathbf{r}) = \int_0^\infty dE \int_{4\pi} d\Omega_v n(\mathbf{r}, E, \Omega_v) \tag{9.5}$$

Note that the neutron density distribution per unit energy is related to the distribution function $f_n(\mathbf{r}, \mathbf{v})$ introduced in Chap. 2 by

$$n(\mathbf{r}, E, \Omega_v) \equiv \frac{4E^{1/2}}{(2m)^{3/2}} f_n(\mathbf{r}, \mathbf{v}) \tag{9.6}$$

If the distribution function is isotropic (i.e. independent of the solid angle) all the integrals in $d\Omega_v$ provide a contribution 4π. In this case we define the neutron distribution per unit energy given by

$$n(\mathbf{r}, E) \equiv 4\pi \frac{4E^{1/2}}{(2m)^{3/2}} f_n(\mathbf{r}, v) \tag{9.7}$$

Similarly, the *flux per unit energy* can be defined as

$$\phi(\mathbf{r}, E) = n(\mathbf{r}, E)v(E) \equiv 8\pi \frac{E}{m^2} f_n(\mathbf{r}, v) \tag{9.8}$$

From the flux per unit energy it is possible to determine all the interaction rates per unit volume through

$$F(\mathbf{r}) = \int_0^\infty n_T(\mathbf{r})\sigma(E)\phi(\mathbf{r}, E)dE \tag{9.9}$$

Therefore the goal of neutronic calculations is to determine $\phi(\mathbf{r}, E)$ for each configuration of interest. This requires sophisticated codes that are divided in two general classes: deterministic codes that solve the Boltzmann equation (see Chap. 11) and probabilistic codes such as the *Monte Carlo N-Particle Transport Code (MCNP)* [3]. Probabilistic codes follow the evolution of a neutron from its creation to the absorption or loss due to leakage, including all the possible reactions with the probability of each reaction accounted for via the reaction cross section.

As a simple illustration of the calculation of the flux per unit energy we consider the evolution of neutrons in a homogeneous moderator made of hydrogen (density n_H) in the absence of absorption processes, a situation that will apply to thermal reactor dynamics. To this aim we go back to Eqs. (3.39), (3.40) and (3.41) for the probability of a neutron to have an energy E_{i+1} after the ith collision with E_i the energy before the collision

$$\frac{dW}{dE_{i+1}} = \frac{(A+1)^2}{4AE_i} \tag{9.10}$$

for $E_i \geq E_{i+1} \geq (1-\alpha)E_i$ and 0 otherwise, with α defined in Eq. (3.38) and $\alpha = 1$ for hydrogen. Following [1] it is possible to find an equation for the interaction rate per unit volume and unit energy $F(E)$ (we neglect the dependence on $\mathbf{r}$ since by assumption the system is homogeneous) by looking at the source and sink of particles in an interval dE around the energy E (Fig. 9.3).

Neutrons are generated at an energy E_o at a rate S. Following a collision there is a probability $dW_1 = dE/E_o$ that they will be scattered in dE. In addition, all the neutrons that already underwent collisions to energies $E \leq E\prime \leq E_o$ will contribute to populate dE. The neutrons that undergo a collision in the interval $dE\prime$ is, by definition of $F(E)$, $F(E\prime)dE\prime$. Therefore neutrons that will populate the interval dE, taking into account the probability to change energy from $E\prime$ to E in a collision, is given by

$$\int_E^{E_o} \frac{F(E\prime)dE\prime dE}{E\prime} \tag{9.11}$$

We have now to balance the sum of these two contributions with the losses due to the collision of particles with energy E that by definition is $F(E)dE$, yielding

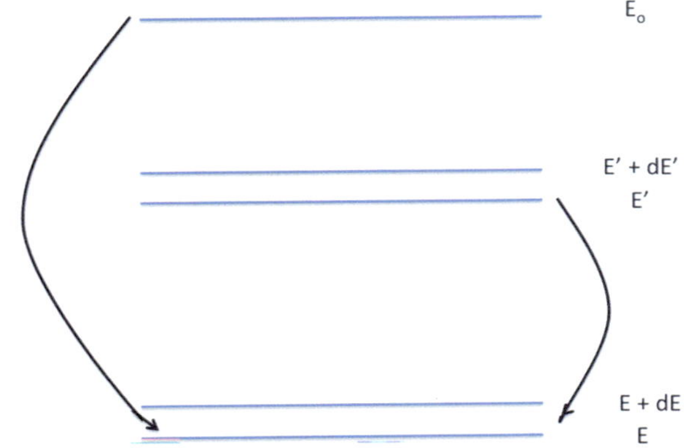

Fig. 9.3 The mechanism of neutron slowing down in a homogeneous moderator made of hydrogen can be illustrated by looking at the balance of sources and sinks in the energy interval dE around E

$$F(E)dE = \frac{SdE}{E_o} + \int_E^{E_o} \frac{F(E\prime)dE\prime dE}{E\prime} \tag{9.12}$$

Upon dividing by dE and differentiating with respect to E we obtain

$$\frac{dF}{dE} = -\frac{F}{E} \tag{9.13}$$

with solution (taking into account that from Eq. (9.12) $F(E_o) = S/E_o$)

$$F(E) = \frac{S}{E} \tag{9.14}$$

Using the definition $F(E) = n_H \sigma_s \phi(E)$, with σ_s the elastic scattering cross section, we finally obtain

$$\phi(E) = \frac{S}{n_H \sigma_s(E) E} \tag{9.15}$$

The cross section for elastic scattering is shown in Fig. 9.4. The flux shown in Eq. (9.15) can be expressed in terms of the *flux for unit lethargy* by noting that from the definition of lethargy $u_L = \ln(E_o/E)$ (see Chap. 1) we have $du_L = dE/E$ which implies that the flux per unit lethargy is almost constant under the assumptions used for this calculation.

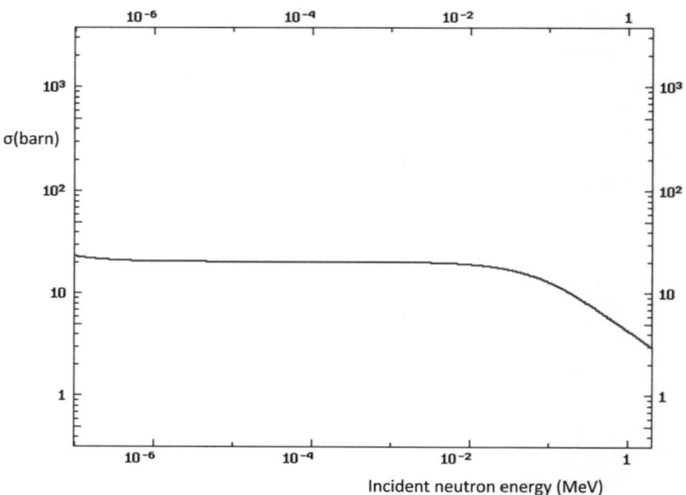

Fig. 9.4 The cross section from the ENDF library [4] for elastic collision between a neutron and a proton is shown as a function of the neutron energy

9.1 Criticality of Thermal Reactors

The neutron spectrum of a thermal reactor is the sum of three components [5] here expressed in terms of the flux per unit energy

- The part of the spectrum in thermal equilibrium with the moderator at temperature T_m can be described by a Maxwellian distribution.

$$\phi_{th}(E) = n_n(r)\frac{8E}{m^2\pi^{1/2}v_{tm}^3}e^{-E/T_m} \tag{9.16}$$

with v_{tm} the neutron thermal velocity and n_n is the neutron density in physical space.
- The fission neutron spectrum that can be approximated as

$$\phi_{fission}(E) \propto \frac{2E^{1/2}}{\pi^{1/2}\epsilon^{3/2}}e^{-E/\epsilon} \tag{9.17}$$

with the characteristic energy ϵ equal to 1.29 MeV for the ^{235}U isotope.
- The intermediate energy spectrum given in Eq. (9.15).

Typical neutron spectra of a thermal and a fast reactor are shown in Fig. 9.5.

The neutron multiplication coefficient k evaluated without taking into account losses through the boundary is denoted as k_∞ and it is usually expressed with the four-factor formula [1]

$$k_\infty = \eta_{rf}\, f_{tu}\epsilon_{fast}\, p_{ra} \tag{9.18}$$

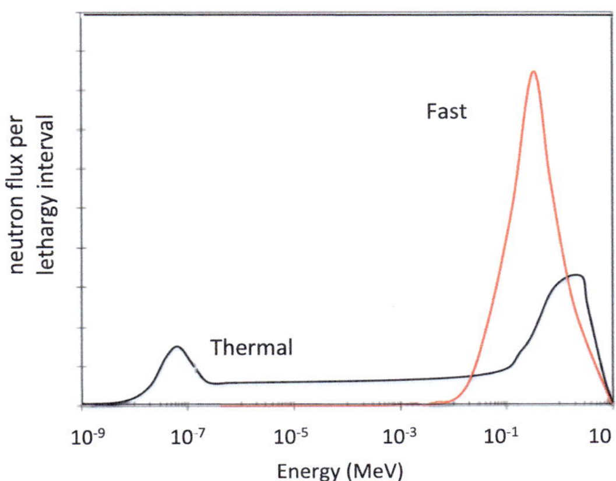

Fig. 9.5 Neutron energy spectrum for a fast and a thermal reactor. In the case of a fast reactor the amount of neutrons below 20 keV is negligible. Note that this is the higher boundary of the region of resonant radiative capture (see Fig. 8.6)

with the various terms explained below and the expressions for the various factors evaluated [1] under the assumption of a homogeneous reactor core.

The *neutron reproduction factor* η_{rf} represents the ratio between the number of neutrons created in the fission process and the number of neutrons absorbed by the fuel

$$\eta_{rf} = \nu \frac{\sigma_f(U)}{\sigma_f(U) + \sigma_r(U)} \tag{9.19}$$

with $\nu = 2 - 3$ the average number of neutrons per fission, σ_f the fission cross section and σ_r the cross section for all the other absorption processes by the fuel. In the following we will denote the total absorption cross section with $\sigma_a \equiv \sigma_f + \sigma_r$. A value for the cross sections averaged over the neutron energy spectrum must be taken here. If the fuel is made of several isotopes, the expression above has to be weighted also over the fuel composition. For example, in the case of a mixture of ^{235}U and ^{238}U, if f_{235} is the fraction of ^{235}U ($f_{235} = 0.7\%$ for natural uranium) we have

$$\eta_{rf} = \frac{f_{235}\nu_{235}\sigma_f(^{235}U) + (1 - f_{235})\nu_{238}\sigma_f(^{238}U)}{f_{235}\sigma_a(^{235}U) + (1 - f_{235})\sigma_a(^{235}U)} \tag{9.20}$$

To achieve values of η_{rf} well above unity is necessary because, as we will see, of the four factors, f_{tu} and p_{ra} are always below one and ϵ_{fast} is typically slightly above one. Figure 9.6 shows the dependence of η_{rf} on f_{235} in the case of a thermal and of a fast neutron spectrum using the following representative parameters.

Thermal spectrum

$$\nu_{235} = 2.4,\ \nu_{238} = 0,\ \sigma_f(^{235}U) = 580b,\ \sigma_a(^{235}U) = 680b,$$
$$\sigma_f(^{238}U) = 0,\ \sigma_a(^{238}U) = 2.7b; \tag{9.21}$$

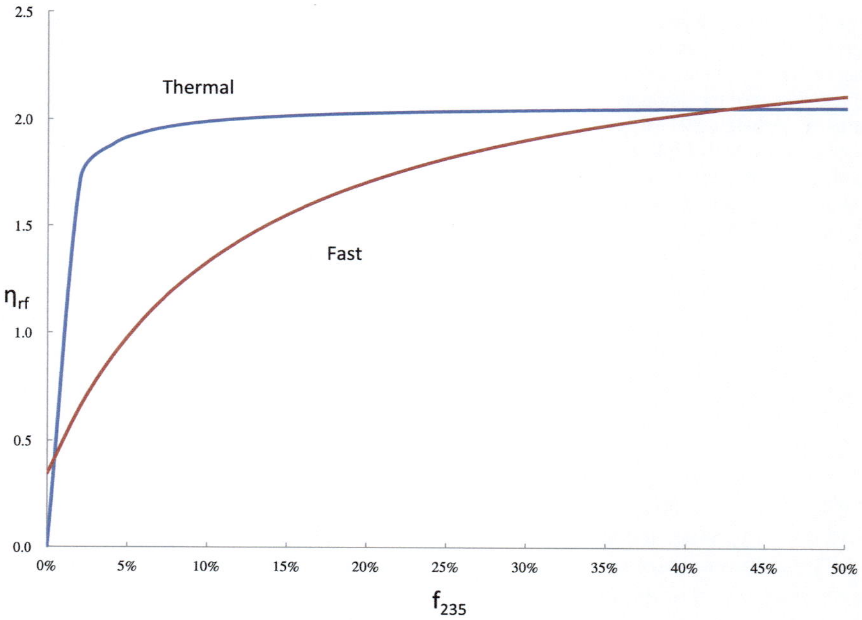

Fig. 9.6 Neutron reproduction factor as a function of the enrichment for a thermal and a fast reactor

Fast spectrum

$$\nu_{235} = 2.5, \nu_{238} = 2.6, \sigma_f(^{235}U) = 1.25b, \sigma_a(^{235}U) = 1.4b,$$
$$\sigma_f(^{238}U) = 0.02b, \sigma_a(^{238}U) = 0.15b. \quad (9.22)$$

In both cases we need to increase the fraction of ^{235}U above the natural composition (enrichment quantified by the enrichment factor $R = f_{235}/(1 - f_{235})$) in order to achieve $\eta_{rf} > 1$. However in the case of thermal spectrum a fraction of 5% is enough to achieve the maximum value whereas in the case of a fast reactor we need to enrich around 20%.

The *thermal utilization factor* f_{tu} is defined as the ratio of the neutrons absorbed by the fuel divided by the neutrons absorbed by all the materials (fuel included)

$$f_{tu} = \frac{n_U \sigma_a(U)}{n_U \sigma_a(U) + \sum_C n_C \sigma_a(C)} \quad (9.23)$$

with n_U (n_C) the total number of fuel (materials other than fuel) atoms per unit volume[1] By definition $f_{tu} < 1$. When hydrogen is used as moderator the most important

[1] Since the core is assumed homogeneous n_U and n_C are also the total number of atoms divided by the reactor core volume.

process is the neutron radiative capture to form deuterium. The cross section for neutron capture in hydrogen and deuterium is shown in Fig. 9.7. Since for deuterium the absorption cross section is about three orders of magnitude lower than for hydrogen, deuterium is used as moderator in heavy-water reactors.

In the case of a thermal reactor with an enrichment of 5% the capture cross section is $\sigma_a(H) \approx 0.5b^2$ to be compared with an absorption cross section of about $\sigma_a(U) \approx 5\% \times 680b + 95\% \times 2.7b = 36b$. In the core of a reactor the amount of fuel is about $100t$ whereas the amount of water is about $180t$ (of which $20t$ are made of hydrogen atoms). The ratio $n_U/n_H = 0.5n_U/n_W$ is therefore $n_U/n_H \approx (100/20)(1/238) = 2\%$ yielding

$$f_{tu} \approx 2\% \times 36b/(2\% \times 36b + 0.5b) = 0.78 \tag{9.24}$$

Increasing the fuel/moderator ratio increases the value of f_{tu}.

The *fast fission factor* ϵ_{fast} is the ratio between the number of fission due to both thermal and fast neutrons and the number of thermal fissions.

$$\epsilon_{fast} \equiv \frac{\int_0^\infty \phi(E) n_U \sigma_U(E) \nu(E) dE}{\int_{th} \phi(E) n_U \sigma_U(E) \nu(E) dE} \tag{9.25}$$

with integral in the denominator carried out over thermal energies only. A fit of the results for the case of a reactor fuelled with uranium and moderated with water is [1]

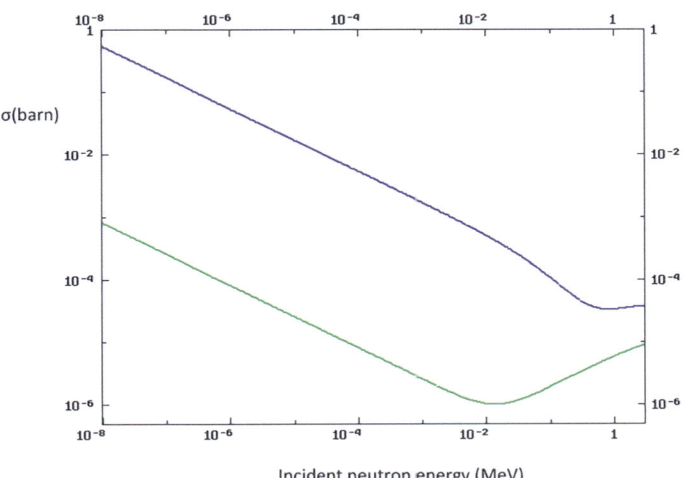

Fig. 9.7 Neutron capture cross section for H (blue) and D (green) from the ENDF library [4]. The low capture cross section of deuterium allows the use of natural uranium in CANDU reactors

[2] Other absorption mechanisms such as neutron absorption on oxygen are neglected here.

$$\epsilon_{fast} \approx \frac{1 + 0.690\frac{n_U}{n_W}}{1 + 0.563\frac{n_U}{n_W}} \tag{9.26}$$

The fast fission factor is typically between 1 and 1.2.

The *resonance escape probability* p_{ra} is the probability for a neutron to avoid the radiative capture by ^{238}U. The evaluation of p_{ra} involves the integration over the resonance region and can be performed under a number of assumptions. A simple expression for the factor p_{ra} is given in Ref. [1]

$$p_{ra} \approx \exp\left(- \frac{2.73}{\bar{\xi}} \left(\frac{n_U}{n_W \sigma_{sW}} \right)^{0.514} \right) \tag{9.27}$$

with $\bar{\xi}$ the average change in lethargy per collision and σ_s the cross section for elastic scattering with the moderator. The dependence of the neutron multiplication coefficient on n_U/n_W is shown in Fig. 9.8.

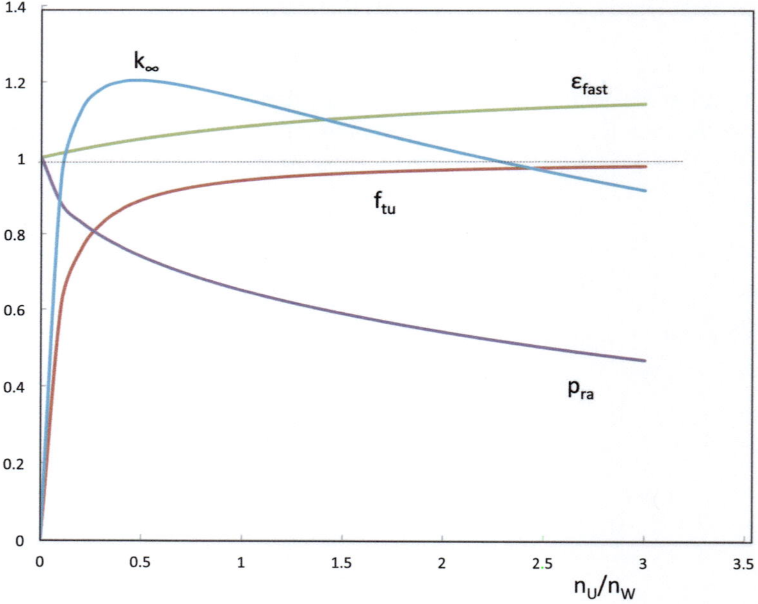

Fig. 9.8 Neutron multiplication factor vs. n_U/n_W for $\eta_{rf} = 1.744$ and $\sigma_s = 44.8b$ and a 2% enrichment

9.2 Fission Reactor Layout

The main elements of a fission reactor are shown in Fig. 9.9. In general, we can distinguish among the elements listed below.

- Fuel assembly. Fuel is made by the fissile elements in the form of solid compounds. In the case of uranium the fuel is in the form of pellets of uranium oxide UO_2 contained in metallic cylindrical rods. Several fuel rods form a *fuel assembly*.
- Moderator. The moderator slows the high-energy neutrons down to thermal energies for which the fission cross section is higher. The moderator is made of a low-A element in order to maximise the energy transfer in collisions. The maximum energy transfer occurs for collisions with protons, therefore light water (H_2O) is a widely used material. The drawback is the high cross section for radiative capture of the neutron. For this reason, heavy water (D_2O) is sometimes used. Alternatively, graphite is employed as moderator.
- Coolant. The heat generated in the nuclear processes must be removed by a suitable fluid. This can be water, carbon dioxide or liquid metals, the latter being required in fast reactor to avoid any moderation effect by the coolant.
- Absorber. The reactor control requires the possibility of reducing the neutron population by inserting a neutron absorbing material located inside *control rods*.

9.3 Reactor Kinetics

Equation 9.1 for $k > 1$ implies an exponential increase in the neutron population. Since the time constant τ is of the order of 10^{-4} s, even for k slightly above 1 it would be difficult to actively control the reactor evolution. Fortunately, the neutron emission related to the fission process takes place on different time scales. Most of the

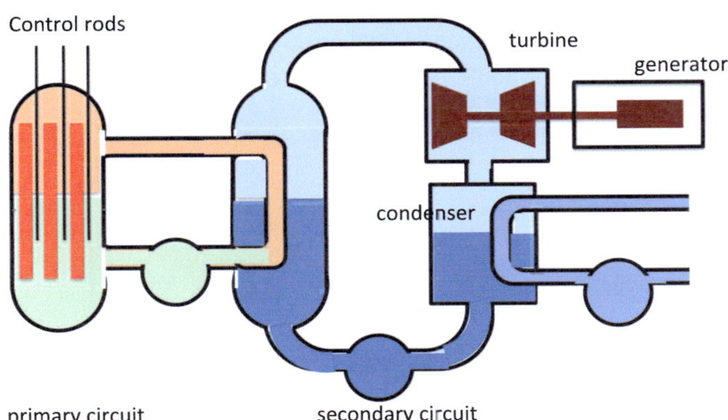

Fig. 9.9 Layout of a pressurised water reactor plant

neutrons are emitted immediately (within 10^{-16} s) and are called *prompt* neutrons. However, a fraction of about 1% of the neutrons emitted per fission event (1.6% in the case of ^{235}U -since on average each fission event produces 2.5 neutrons this corresponds to 0.64% delayed neutrons per neutron) is emitted on a much longer time scale (tens of seconds) due to the neutron emission by a *delayed neutron precursor* generated by the decay of the fission fragments. Examples of this processes are shown below

$$
\begin{aligned}
&^{93}Rb \rightarrow ^{93}Sr + e + \bar{\nu}_e \\
&\qquad ^{93}Sr \rightarrow ^{93}Y + e + \bar{\nu}_e \\
&\qquad\qquad ^{93}Sr \rightarrow ^{92}Sr + n(1.4\%)
\end{aligned}
\tag{9.28}
$$

$$
\begin{aligned}
&^{141}Cs \rightarrow ^{141}Ba + e + \bar{\nu}_e \\
&\qquad ^{141}Ba \rightarrow ^{141}La + e + \bar{\nu}_e \\
&\qquad\qquad ^{141}Ba \rightarrow ^{140}Ba + n(0.03\%)
\end{aligned}
\tag{9.29}
$$

The *delayed* neutrons generated by the precursors allow the active control of a reactor. Indeed, if the reactor is subcritical for prompt neutrons only but supercritical when delayed neutrons are included, the evolution of the neutron population takes place on the time scale of the delayed neutrons.

Assume a fraction β_{del} of delayed neutrons per generated neutron ($\beta_{del} = 0.64\%$ in the case of ^{235}U). It is convenient to introduce a quantity called *reactivity* given by

$$\rho \equiv \frac{k-1}{k\beta_{del}} \tag{9.30}$$

In order to understand how delayed neutrons helps in reactor control it is convenient to consider a simplified model with two equations for the evolution of the number of total neutrons n and of precursor nuclei n_P

$$\frac{dn}{dt} = \frac{(1-\beta_{del})k - 1}{\tau} n + \lambda n_P \tag{9.31}$$

$$\frac{dn_P}{dt} = \frac{\beta_{del} k}{\tau} n - \lambda n_P \tag{9.32}$$

with λ the decay constant of the precursor. In order to determine the time scale of evolution we can look for solutions of the form $n = n(0)e^{\gamma t}$ and $n_P = n_P(0)e^{\gamma t}$. After substitution the following equation is fond that must be satisfied by γ to avoid the trivial solution $n(0) = n_P(0) = 0$

$$\gamma^2 + \gamma[\lambda + \frac{\beta_{del}k}{\tau}(1-\rho)] - \lambda\frac{\beta_{del}k}{\tau}\rho = 0 \tag{9.33}$$

A simple inspection of this quadratic equation (see Fig. 9.10) shows that for $\rho < 0$ ($k < 1$) the two solutions correspond to $\gamma < 0$ and a neutron number decreasing with time, consistently with the fact that the reactor is subcritical. For $\rho > 0$ one of the solutions becomes positive and corresponds to an exponentially *increasing* number of neutrons but on a much longer time scale than τ. Indeed taking e.g. $\rho = 1$, Eq. (9.33) yields $\gamma \approx \lambda$ for $\lambda \gg \beta_{del}/\tau$ and $\gamma \approx (\lambda\beta_{del}/\tau)^{1/2}$ in the opposite limit $\lambda \ll \beta_{del}/\tau$. For $\rho \gg 1$ the solutions converge to the value $\gamma \rightarrow (k-1)/\tau$ as in the case in which the delayed neutron dynamics is neglected.

For operating a fission reactor it is necessary to know how the reactivity changes with changing parameters. The *temperature coefficient* α is defined as $\alpha \equiv d\rho/dT$ and, in order to have stable operation, the condition $\alpha < 0$ must hold in such a way that the reaction to an increase of fission power (and temperature) is a reduction of the reactivity and ultimately of the neutron production rate until equilibrium conditions are recovered. A negative temperature coefficient can be due to several effects:

- An increase in temperature reduces the density of the moderator and as a consequence the neutron production.
- An increase in temperature increases also the thermal neutron energy and this reduces the fission cross section $\sigma \propto 1/v$. In the case of ^{239}Pu however the effect

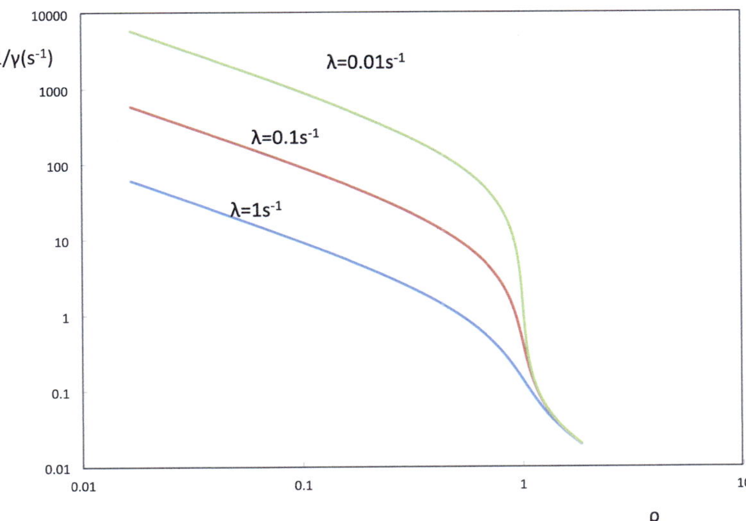

Fig. 9.10 The time scale of evolution of a power plant is shown as a function of the reactivity $\rho \equiv (k-1)/(\beta_{del}k)$ for different values of the decay constant λ of the precursor. For $\rho < 1$ the time scale is related to λ

is an increase of the cross section due to the low energy resonance visible e.g. in Fig. 8.5.

- An increase in temperature widens the resonances for the absorption by ^{238}U again decreasing the reactivity.

9.4 Classification of Fission Reactors

The present fleet of fission reactors is shown in Table 9.1

The vast majority of the existing reactors are *pressurised water reactors (PWR)* with light-water acting both as a moderator and as a coolant. The fuel assembly is immersed in liquid water inside a steel pressure vessel that acts as primary containment. In the primary cooling circuit a pressure of 15.5 MPa maintains water in the liquid phase at a temperature between 270 and 320 °C. In the secondary circuit steam is generated that drives a turbine. In this way the secondary circuit is not contaminated. The temperature coefficient is negative ensuring stable operation.

The *boiling water reactor (BWR)* differs from the PWR as there is a single circuit and water reaches the boiling temperature in the core. The design is simpler than that of a PWR but this is balanced by the more difficult maintenance since the turbine can be contaminated.

The *pressurised heavy water reactor (PHWR)* uses heavy-water as moderator and for this reason can use natural uranium. The design of the reactor allows for core refuelling without interrupting the reactor operation.

The *advanced gas cooled reactor (AGR)* uses CO_2 as coolant that can reach 650 °C temperature and therefore 41% conversion efficiency.

The *light water graphite moderated reactor (LWGR)* is a Russian design (RBMK) initially developed for plutonium production. Water can boil at low pressure in the core. The main issue with this design is that the temperature coefficient is positive because increase of core temperature reduces the neutron absorption but has no effect on the moderation.

9.5 Fuel Utilisation

Thermal reactors do not have an optimal utilisation of the fissionable elements. Taking as an example a light-water reactor with fresh fuel made of $33g$ of ^{235}U and $967g$ of ^{238}U, after three years of operation the composition will be modified as follows: $8g$ of ^{235}U, $4.6g$ of ^{236}U, $943g$ of ^{238}U, $8.9g$ of Pu isotopes (of which about 2/3 fissile), $35g$ of various fission products (^{90}Sr, ^{137}Cs, ^{93}Zr, ^{99}Tc, ^{85}Kr) and 0.65 g of minor actinides. Therefore the vast majority of what is left after three years, with the exception of the fission fragments and the minor actinides, is material that could be reused.

Table 9.1 Existing nuclear reactors [6]. LEU = Low Enriched Uranium. MOX = Mixed Oxide fuel

	PWR (Pressurized Water Reactor)	BWR (Boiling Water Reactor)	PHWR (Pressurized Heavy Water Reactor)	LWGR (Light-Water Cooled Graphite Moderated Reactor)	AGR (Advanced Gas Cooled Reactor)	(LM)FBR (Fast Breeder Reactor)	HTGR (High-temperature gas-cooled reactor)
Fuel	LEU, MOX	LEU, MOX	Natural Uranium, LEU	LEU with Thorium	Natural Uranium	Plutonium mixed with natural or depleted Uranium	
Fissile production	Converter	Converter	Converter	Converter	Converter	Breeder	Converter
Moderator	Light-water	Light-water	Heavy-water	Graphite	Graphite	None	Graphite
Thermal or Fast	Thermal	Thermal	Thermal	Thermal	Thermal	Fast	Thermal
Coolant	Light-water	Light-water	Heavy-water	Light-water	Carbon dioxide	Liquid metal	Helium
Number in operation	301	42	46	11	8	2	1
Under construction at the end of 2022	49	2	3			4	
GWe	292.8	60.8	24.3	7.4	4.7	1.4	0.2
Main countries	USA, France, Japan, Russia, China, South Korea	USA, Sweden	Canada, India	Russia	UK	Russia	China

In most of the present reactors, based on the once-through cycle, what is left after three years is considered waste and disposed of. With this approach the estimated amount of uranium resources would not be enough to make fission a long term alternative to fossil fuels. As mentioned above, it is necessary to pursue the route of fast reactors that are able to breed fuel by transmuting fertile nuclide into fissile nuclides in excess of what is burnt in the reactor. The spent fuel already produced could be used to produce additional energy. This requires reprocessing the fuel with the separation of nuclear fuel from true waste (fission fragments and minor actinides).

9.6 Suggestions for Further Readings

A good overview of the present status of fission research is given in [2]. A classical text for the fission reactor dynamics is Ref. [1].

References

1. J. Lamarsh, *Introduction to Nuclear Reactor Theory* (Addison-Wesley, Reading, 1972)
2. E. De Sanctis, S. Monti, M. Ripani, *Energy from Nuclear Fission - An Introduction* (Springer, 2016)
3. X.-5 Monte Carlo Team x, MCNP–A General N-Particle Transport Code, Version 5. Volume I: Overview and Theory, LA-UR-03-1987 (Los Alamos National Laboratory, USA, 2005)
4. Evaluated Nuclear Data Files (ENDF). https://www-nds.iaea.org/exfor/endf.htm
5. IAEA, Neutron Fluence Measurements, IAEA Technical Report Series n. 107, IAEA, Vienna (1970). https://inis.iaea.org/collection/NCLCollectionStore/_Public/34/065/34065175.pdf
6. IAEA, Nuclear Power Reactors in the World, Reference Data Series No. 2 2023 Edition, https://www-pub.iaea.org/MTCD/Publications/PDF/RDS-2-43_web.pdf

Chapter 10
The Penetration of Radiation Through Matter

Abstract *The interaction of radiation with matter is discussed in the case of massive charged particles, electrons and gammas.*

10.1 Introduction

The characteristics of the interaction between radiation and matter depend on the specific nature of radiation:

- In the case of heavy charged particles (alpha particles, protons, etc.) the incoming beam slow down due to the excitation and ionisation processes of the target atoms. All the beam particles are stopped at the same distance (called *range*) inside the target.
- In the case of electromagnetic radiation the beam intensity is exponentially attenuated mainly due to photoelectric effect, Compton scattering and pair production.
- Electrons have a more complicated behaviour. Due to their low mass they are subject to high acceleration and lose energy via Bremsstrahlung. Their trajectories are irregular and a range is not well defined.

The following quantities are usually defined

- Stopping power. The average value of the energy lost per unit length.
- Specific ionization. Number of ion pairs produced per unit length.
- Straggling. Fluctuation in the range due to the fluctuation in the energy loss.

10.2 Coulomb Collisions

In Chap. 3 we have derived the expression of the velocity in the laboratory frame after a collision (Eq. (3.30)). For target at rest Eq. (3.30) becomes

F. Romanelli, *Physics of Nuclear Energy*, Springer Series in Plasma Science and Technology, https://doi.org/10.1007/978-981-97-9609-0_10

$$\mathbf{v}_{1f} = \mathbf{v}_{1i}\left(\frac{m_r}{m_2} + \frac{m_r}{m_1}\cos\chi\right) + \frac{m_r}{m_1}v_{1i}\sin\chi\mathbf{n} \tag{10.1}$$

with particle 1 being a beam particle and particle 2 a target particle. The deflection angle χ in the centre of mass frame has to be determined by looking at the specific interaction. In the case of Coulomb interaction the deflection angle was determined by Rutherford [1]

$$\tan\left(\frac{\chi}{2}\right) = \frac{q_1q_2}{4\pi\epsilon_o m_r u^2 b} \equiv \frac{b_o}{2b} \tag{10.2}$$

A simple derivation can be obtained in the limit of large impact parameter b. In this case the deflection angle is small and can be determined using perturbation theory. The change in the velocity is obtained by integrating the acceleration along the unperturbed trajectory given by $x = ut$ and $y = b$ (see Fig. 10.1).

The acceleration in the centre of mass frame is $\mathbf{a} = q_1q_2/(4\pi\epsilon_o m_r r^2)(\mathbf{r}/r)$ and the change in the particle velocity along y is

$$\Delta v_y = \int_{-\infty}^{t} dt\prime \frac{q_1q_2}{4\pi\epsilon_o m_r}\frac{y(t\prime)}{r(t\prime)^3} \approx \frac{q_1q_2}{4\pi\epsilon_o m_r}\int_{-\infty}^{t} dt\prime \frac{b}{(u^2t^2+b^2)^{3/2}} = \frac{q_1q_2}{2\pi\epsilon_o m_r ub} \tag{10.3}$$

From Eq. (3.29) the change in velocity is related to the deflection angle by $\Delta v_y = u\sin\chi$. Therefore, for small deflection angles we obtain

$$\chi \approx \frac{q_1q_2}{2\pi\epsilon_o m_r u^2 b} = \frac{b_o}{b} \tag{10.4}$$

in agreement with the $\chi \to 0$ limit of the Rutherford formula.

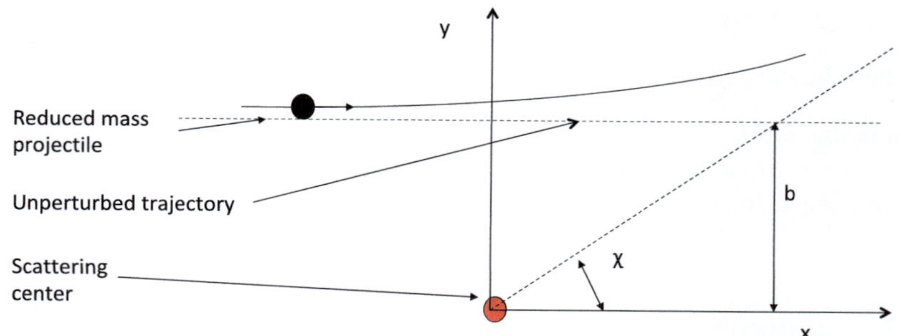

Fig. 10.1 For small deflection angles (large impact parameters) the deflection angle can be determined by integrating the acceleration due to the Coulomb force along the unperturbed trajectory (dashed line)

The long-range effect of the Coulomb interaction has important implications in the dynamics of charged particles. Since plasmas are dominated by the Coulomb force, the considerations presented in this chapter applies equally to plasmas.

In Chap. 3 it as been explained how to estimate the cross section on the basis of the maximum impact parameter b_M for which there is an appreciable deflection angle χ_{min}. In the case of Coulomb collisions the deflection angle monotonically decreases with b and the definition of χ_{min} is somehow arbitrary. By choosing $\chi_{min} \ll 1$ from the Rutherford formula we have

$$b_M = \frac{b_o}{\chi_{min}} \tag{10.5}$$

and the cross section for a deflection at χ_{min} $\sigma_{\chi_{min}} = \pi b_M^2 = \pi \frac{b_o^2}{\chi_{min}^2}$. The collision frequency for a χ_{min} deflection will be

$$\nu_{\chi_{min}} = n_2 u \pi \frac{b_o^2}{\chi_{min}^2} \tag{10.6}$$

However, a charged particle entering the target will feel the effect of the Coulomb interaction with all the charged particles in the target. All these interactions will produce a cumulative effect due to multiple collisions. What is the deflection angle under the effect of multiple collisions?

To answer this question we consider the direction of the incoming beam particle as the generatrix of a sheaf of planes corresponding to different angles ϕ with a fixed direction in space (see Fig. 10.2).

To evaluate the deflection angle due to multiple collisions we first select the plane defined by the unperturbed trajectory of the incoming particle (the generatrix of the sheaf) and the position of the target particle at fixed impact parameter b on that plane

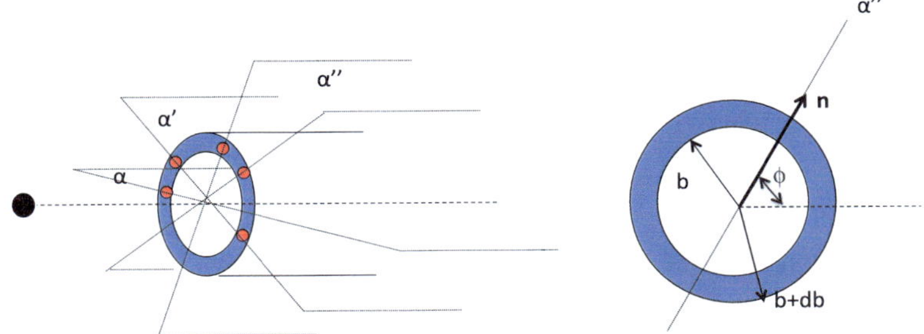

Fig. 10.2 Collision geometry. The average value of a certain quantity due to the effect of multiple collisions can be evaluated by looking first at the deflection angle in the plane determined by the unperturbed trajectory (dashed line) of the beam particle (black) and the target particle (red) at fixed b. Then the quantity under examination is integrated over all the possible angles ϕ and finally the result is integrated over all the impact parameters b

and then sum the effect of all the planes (at fixed b) by integrating over the angle ϕ. Finally, we integrate over all the possible values of b. The beam particle will undergo collisions that will change its velocity but since the target is assumed homogeneous, on each plane the particle will be deflected at positive χ and negative χ with the same probability. However, the average of χ^2 will be different from zero and will increase as the particle move inside the target. Its value as a function of time can be obtained by summing the square of the deflection angle over the number of target particles encountered by the beam particle moving with velocity u

$$< \chi^2 >= n_2 ut \int_0^\infty 2\pi b db \chi^2 = n_2 ut \int_0^\infty 2\pi b db \left(\frac{b_o}{b}\right)^2 \tag{10.7}$$

where we have assumed the expression for χ valid for large b. The results looks at first sight contradictory. Indeed the integral over the impact parameter is logarithmically divergent for both small and large values of b. If we introduce a lower cut-off b_{min} and an upper cut-off b_{max} to the integral over b we can overcome the problem of the singularity. The resulting value will be weakly dependent on both b_{min} and b_{max}. But how to choose the values of the two cut-offs? The divergence for small b values is a consequence of our approximation for the deflection. On the contrary, the divergence for large b is the consequence of the long-range character of the Coulomb interaction and therefore must be cured with specific physical arguments that depend, in general, on the characteristics of the target material. However, independently of the specific choice for b_{min} and b_{max} the effect of multiple collisions will reduce the time needed to deflect the beam particle at a given angle. We can estimate the time needed to produce an average deflection of χ_{min} via multiple collisions as

$$\frac{1}{t_{\chi_{min,mult}}} = n_2 u 2\pi \frac{b_o^2}{\chi_{min}^2} \ln \frac{b_{max}}{b_{min}} = 2 \ln \frac{b_{max}}{b_{min}} \nu_{\chi_{min}} \tag{10.8}$$

The numerical factor $(2\ln(b_{max}/b_{min})$ is $\gg 1$ (since $b_{min} \ll b_{max}$). Thus, we can conclude that the time needed to deflect on average a particle via multiple collisions is approximately a factor $2\ln(b_{max}/b_{min})$ shorter than the characteristic time to deflect by the same angle a particle via single beam-target collisions. This result characterises the processes due to Coulomb interaction.

10.3 Slowing Down of Heavy Charged Particles

A heavy charged particle moving through a target is slowed down by the collisions with the target material. The change in energy ΔE_1 in a single collision has been evaluated in Eq. (3.32). To obtain the change due to multiple collisions we have to perform the same summation as done to evaluate the average of the square of the deflection angle

$$dE_1 = n_2 u_i dt \int d\phi b db \Delta E_1 =$$
$$= n_2 u_i dt \int d\phi b db[-m_r(1-\cos\chi)\mathbf{u}_i \cdot \mathbf{V}_{com} + m_r u_i \sin\chi \mathbf{n}\cdot\mathbf{V}_{com}] \qquad (10.9)$$

Looking at Fig. 10.2, it is possible to see that the average over ϕ of the term proportional to $\mathbf{n} \cdot \mathbf{V}_{com}$ exactly vanishes since for any angle ϕ there is a contribution equal in magnitude and opposite in sign associated with $\phi + \pi$. The contribution proportional to $\mathbf{u}_i \cdot \mathbf{V}_{com}$ is independent of ϕ and can be evaluated straightforwardly, yielding

$$\frac{dE_1}{dt} = n_2 u_i \int 2\pi b db[-m_r(1-\cos\chi)\mathbf{u}_i \cdot \mathbf{V}_{com}] \qquad (10.10)$$

Since the effect of multiple small-angle collisions is larger than the effect of single large-angle collisions, the deflection angle χ can be approximated by its expression for large impact parameters. Once again we have a logarithmically divergent integral that we can treat as done in the previous section

$$\frac{dE_1}{dt} = n_2 u_i \pi b_o^2 \ln\left(\frac{b_{max}}{b_{min}}\right)(-m_r \mathbf{u}_i \cdot \mathbf{V}_{com}) \qquad (10.11)$$

Let us assume at this point the case of a target at rest. In this case $u_i = v_{1i}$ and $V_{cdm} = m_r/m_2 v_{1i}$. Upon using the definition of b_o, we have

$$\frac{dE_1}{dt} = -\frac{q_1^2 q_2^2}{4\pi\epsilon_o^2} \ln\left(\frac{b_{max}}{b_{min}}\right) \frac{n_2}{m_2 v_1} \qquad (10.12)$$

If the target is made of more than one species (for example nuclei and electrons) the change in energy will have to be summed over all the species. *Therefore Eq. (10.12) shows that the largest contribution to the change in energy is due to the species with the lowest mass.* This result could at a first sight appear to contradict what we discussed in Chap. 1 where we have seen that the best neutron moderator must have a mass close to the mass of the particle to be slowed down. What we find now is that energy is lost preferentially to electrons.

The reason for this seeming discrepancy is the role of small angle deflections. Indeed, the change in energy in a single collision as given by Eq. (3.32) is the same for any kind of elastic collision described by a central potential. For a target at rest Eq. (3.32) yields

$$\Delta E_1 = -2\frac{m_r^2}{m_1 m_2}(1-\cos\chi)E_1 \qquad (10.13)$$

that indeed shows that the maximum change is obtained for $\chi = \pi$ and $m_1 = m_2$. However, in the case of Coulomb collisions what really matters are the collisions at low deflection angle. For these collisions the change in energy is given by

$$\Delta E_1 = -\frac{m_r^2}{m_1 m_2}\left(\frac{b_o}{b}\right)^2 E_1 = -\frac{m_r^2}{m_1 m_2}\frac{q_1^2 q_2^2}{2\pi\epsilon_o}\frac{1}{(m_r v_1^2 b)^2}E_1 \approx \frac{q_1^2 q_2^2}{2\pi\epsilon_o}\frac{1}{2m_2 v_1^2 b^2} \tag{10.14}$$

that once integrated over the impact parameter yields Eq. (10.12).

Finally, we need an estimate for the cut off values b_{max} and b_{min}. This depends on the characteristic of the target. For interaction with a solid target the assumption of interaction with free electrons will remain valid until the typical revolution time of the electron around the nucleus ($\approx 1/\nu$ with ν the revolution frequency) is longer than the duration of a collision (that we can estimate, including relativistic corrections, as $b/\gamma v$). This set a limit on the maximum impact parameter $b_{max} = \gamma v/\nu$ above which the duration is so long that the electron has the time of making multiple revolutions around the nucleus and therefore can no longer be considered as a free particle. We have already noted that the logarithmic divergence at low b is a consequence of our approximation. For $b \leq b_o$ the deflection angle does not diverge but it attains a finite value ($\chi \to \pi$for $b \to 0$). Thus, we might simply assume $b_{min} = b_o$ to evaluate the integral (see exercise). As we will see this is the recipe for b_{min} in the case of a plasma as target material.[1] For targets other than plasmas a second estimate can be made based on the uncertainty principle yielding a value for b_{min} given by the electron de Broglie wavelength

$$b_{min} = \frac{\hbar}{m_e \gamma v} \tag{10.15}$$

as an electron cannot be localised at distances below the de Broglie wavelength. Clearly we will have to take for b_{min} the maximum between b_o and $\hbar/(m_e\gamma v)$.Upon inserting into Eq. (10.12) the expression for b_{max} and b_{min}, the Bohr formula [2, 3] is obtained

$$\frac{dE_1}{dt} = -\frac{Z_1^2 e^4}{4\pi\epsilon_o^2}\ln\left(\frac{m_e\gamma_1^2 v_1^2}{\hbar\nu}\right)\frac{n_e}{m_e v_1} \tag{10.16}$$

with the average ionisation energy $\hbar\nu$ usually approximated by the expression [4, 6]

[1] This value of b_{min} is often obtained on the basis of the following argument. The maximum energy that can be exchanged between the charged particle and the electron is given by Eq. (10.13) with $\chi = \pi$. If this energy exchange must be obtained in a single large angle collision the corresponding impact parameter from Eq. (10.14) would be $b = b_o/2$. This clearly contradicts the small angle assumption that requires $b >> b_o$. Therefore we will simply assume that te maximum energy exchange argument leads to $b_{min} = b_o$.

$$\hbar\nu = 9.1 Z_2 \left(1 + \frac{1.9}{Z_2^{2/3}}\right) eV \tag{10.17}$$

valid for $Z_2 \geq 4$. The stopping power can be obtained from Eq. (10.16) with the substitution $dx = v_1 dt$

$$\frac{dE_1}{dx} = -\frac{Z_1^2 e^4}{4\pi\epsilon_o^2} \ln\left(\frac{m_e \gamma_1^2 v_1^2}{\hbar\nu}\right) \frac{n_e}{m_e v_1^2} \tag{10.18}$$

A more precise calculation of the stopping power is due to Bethe [7]

$$\frac{dE_1}{dx} = -\frac{Z_1^2 e^4}{4\pi\epsilon_o^2} \left[\ln\left(\frac{2m_e \gamma_1^2 v_1^2}{\hbar\nu}\right) - \beta^2\right] \frac{n_e}{m_e v_1^2} \tag{10.19}$$

The Bethe–Bohr formulas can be used to evaluate the range of an heavy charged particle in a medium. Equation 10.18 can be integrated by separation of variables between the target boundary ($x = 0$) and the range R where $E_1 = 0$

$$R_1 = \frac{4\pi\epsilon_o^2}{Z_1^2 e^4} \frac{m_e}{n_e} \frac{m_1}{2} \int_0^{v_{1a}^2} dv_1^2 \frac{v_1^2}{\ln \frac{m_e \gamma_1^2 v_1^2}{\hbar\nu}} \tag{10.20}$$

with v_{1a} the initial particle velocity. Note that the integrand is a function of v_1 only and the expression for the range can be cast in the form

$$R_a = \frac{A_a}{Z_a^2} F\left(\frac{E_a}{A_a}\right) \tag{10.21}$$

Therefore, knowing the expression of F for a species it is possible to determine the range of another heavy charged particles. For example, the range of a proton in air is approximately given by $R_p(m) = (E_p(\text{MeV})/9.3)^{1.8}$. To evaluate the range of a ^3He nucleus at 10 MeV we have first to evaluate the equivalent energy of a proton $E_{pequiv} = 10\,\text{MeV}/3 = 3.3\,\text{MeV}$ and use Eq. (10.21), yielding

$$R_{^3He}(10\,\text{MeV}) = \frac{3}{4}(3.3)^{1.8} = 11.6\,\text{cm} \tag{10.22}$$

To evaluate the range in a different medium we have to correct by the density of the target material. If the target has a mass density ρ, the number of atoms per unit volume will be $n_{atoms} \approx \rho/(A_2 m_{AMU})$ taking into account that stable nuclei have $Z_2 \approx A_2/2$, the electron density will approximately scale as the mass density of the target and an estimate for the range in the material can be easily obtained. The empirical expression $\rho_2 R_2 = \rho_{air} R_{air} (A_2/15)^{1/2}$ [5] provides an estimate that reproduces the experimental data within 15%.

10.4 Interaction Between Beta Radiation and Matter

The energy losses of beta radiation passing through matter are due to ionisation and Bremsstrahlung (see next section). The ionisation losses (that dominate at low energies) can be treated similarly to what has been done for heavy particles taking care in using the proper approximations for this case. The general expression has been derived in Ref. [7]

$$\frac{dE_1}{dx} = -\frac{e^4}{4\pi\epsilon_o^2}\left[\ln\left(\frac{m_e v_1^2}{2\hbar\nu}\right) - \frac{1}{2}\ln 2 + \frac{1}{2}\right]\frac{n_2}{m_e v_{1i}^2} \tag{10.23}$$

This expression closely follows Eq. (10.19) derived for heavy charged particles. Electrons however differ considerably as they do not follow straight lines but are rather scattered in all the directions in such a way that the actual path between two points may be considerably longer than their distance. As a consequence, it is impossible to define a range for electrons. However, in the case of electrons generated in a beta decay an empirical relation can be written for the maximum range as a function of the maximum energy of the electrons from the beta decay [6]

$$\begin{aligned} R(\text{g/cm}^2_{Al}) &= 0.542E(\text{MeV}) - 0.133 \quad 0.8 < E(\text{MeV}) < 3 \\ &= 0.407E(\text{MeV})^{1.38} \quad 0.15 < E(\text{MeV}) < 0.8 \end{aligned} \tag{10.24}$$

Note that the range here is expressed in g/cm^2. In order to have the range in cm the above expression must be divided by the Aluminium density in g/cm^3.

10.5 Bremsstrahlung

Electrons accelerated by the nuclei emit radiation according to the Larmor expression Eq. (4.57) valid in the non-relativistic limit. In order to determine the total power lost by the plasma we first determine the energy emitted in a single collision and we then sum over all the electrons and ions.

The acceleration of an electron in the electrostatic field of an ion of charge Z is given by

$$a = \frac{Ze^2}{4\pi\epsilon_o m_e r^2} \tag{10.25}$$

with $r(t)$ the distance between the electron and the ion. Upon introducing Eq. (10.25) in the Larmor formula we obtain

$$P = \frac{Z^2 e^6}{96\pi^3\epsilon_o^3 c^3 m_e^2 r^4} \tag{10.26}$$

In order to have a rough estimate of the energy radiated in a collision we will consider large impact parameter collisions in which the particle trajectory is only slightly deflected by replacing the distance r with $(b^2 + v_e^2 t^2)^{1/2}$. The integral can be done analytically and yields

$$E_{rad} = \frac{\pi}{2} \frac{Z^2 e^6}{96\pi^3 \epsilon_o^3 c^3 m_e^2 v_e b^3} \tag{10.27}$$

However, the largest contribution is expected from the small impact parameter collisions. The energy lost per unit length is obtained by summing over all the interactions with the target nuclei

$$\frac{dE}{dx} = n_2 \int 2\pi E_{rad} b db \tag{10.28}$$

The integral is divergent for $b \rightarrow 0$ and it is necessary to introduce a lower cut-off. The origin of the divergence is the fact that quantum-mechanical effects become important at small impact parameter. A simple way to evaluate the cut off is to impose the Heisenberg principle $\Delta x \Delta p \approx \hbar$. If we identify Δx with b_{min} and Δp with $m_e v_e$ we obtain

$$b_{min} = \frac{\hbar}{m_e v_e} \tag{10.29}$$

The expression for the stopping power becomes

$$\frac{dE}{dx} = \frac{\pi}{2} \frac{Z^2 e^6}{96\pi^3 \epsilon_o^3 c^3 m_e^2 v_e} 2\pi n_2 \frac{2\pi m_e v_e}{h} = \frac{\pi}{2} n_2 Z^2 m_e c^2 \frac{4\pi}{3} \alpha r_{cl}^2 \tag{10.30}$$

with $\alpha \equiv e^2/(2\epsilon_o hc) \approx 1/137$ the fine structure constant and r_{cl} the classical electron radius.

Equation (10.30) is valid in the non relativistic limit. In the relativistic limit the stopping power is modified as follows (see [8] Chap.15)

$$\frac{dE}{dx} = \frac{16}{3} n_2 Z^2 \alpha r_{cl}^2 \ln\left(\frac{233}{Z^{1/3}}\right) E \tag{10.31}$$

that produce an exponential decay over a characteristic length that varies between a few cm for $low - Z$ solid material to a few mm for $high - Z$ solid materials.

10.6 Interaction Between Gamma Rays and Matter

The attenuation of gamma radiation passing through matter is due mainly to three effects:

- photo electric effect
- Compton effect
- pair production

The photo electric effect is the dominant effect at low gamma energy. It has been introduced in Chap. 5 where we have shown that in order to take place the photon energy must be larger than the extraction work of the electrons. For electrons belonging to the outermost energy levels the photoelectric effect requires low-energy photon. As the photon energy increases the effect can take place with the electrons in the inner levels down to the K-shell electrons that requires the highest energy photons. Therefore the cross section exhibits discontinuities corresponding to the the change in the energy level involved in the process. Depending upon the boundary involved these discontinuities are labelled as K-edge, L-edge and so on. Between neighbouring edges the cross section exhibits a monotonically decreasing behaviour with increasing energy. For example, the cross section for photon energies E_γ above the K-edge and below the electron rest energy is given by [9]

$$\sigma_{ph} = \frac{16\pi}{3} 2^{1/2} \alpha^4 r_{cl}^2 Z^5 \left(\frac{m_e c^2}{E_\gamma} \right)^{3.5} \tag{10.32}$$

The Compton effect is dominant at intermediate energies. The cross section is given by the Klein–Nishina formula [10]

$$\sigma_{Compton} = Z 2\pi r_{cl}^2 \left(\frac{1+k}{k^2} \left(\frac{2+2k}{1+2k} - \frac{\ln(1+2k)}{k} \right) + \frac{\ln(1+2k)}{2k} - \frac{1+3k}{(1+2k)^2} \right) \tag{10.33}$$

with $k \equiv E_\gamma/(m_e c^2)$.

The pair production becomes important for photon energies above 1.022 MeV corresponding to the rest energy of the electron-positron pair. In order to conserve simultaneously energy and momentum this effect can occur only in the presence of the electric field of he nucleus. The cross section for pair production is given by [11]

$$\sigma_{pair} = Z^2 \alpha r_{cl}^2 \frac{2\pi}{3} \left(\frac{k-2}{k} \right)^3 \left(1 + \frac{\rho}{2} + \frac{23\rho^2}{40} + \frac{11\rho^3}{60} + \frac{29\rho^4}{960} \right) \tag{10.34}$$

with

$$\rho \equiv \frac{2k-4}{2+k+2(2k)^{1/2}} \tag{10.35}$$

valid for $k \leq 4$ (Fig. 10.3).

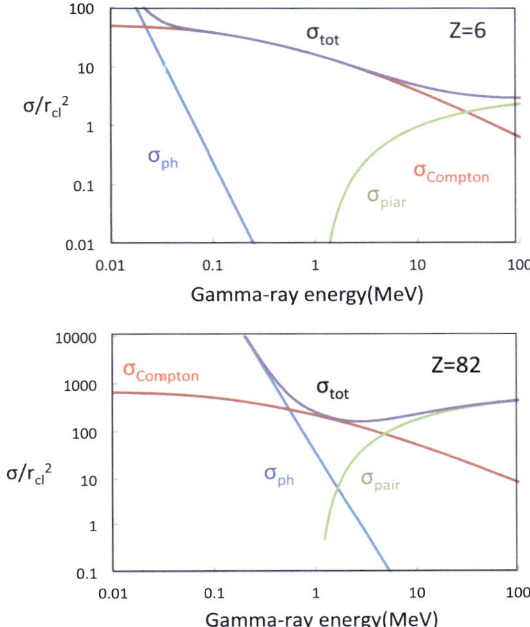

Fig. 10.3 The cross section for the main interactions between gamma rays and matter is shown for carbon and lead. The cross section is normalised to the square of the classical electron radius. The photoelectric effect dominates at low energies, the Compton effect at intermediate energies and the pair production at high energies

10.7 Suggestions for Further Readings

In this chapter we have mostly followed the approach of Ref. [6]. The theoretical analysis of the interaction of charged particles with matter and of radiation is fully addressed in Ref. [8].

10.8 Exercises

Problem 10.1 Show that the using the Rutherford expression for the deflection angle the logarithmic divergence at small impact parameters disappear.

Problem 10.2 Taking $\chi_{min} = \pi/3$ evaluate the electron-electron and electron-ion collision frequency.

Problem 10.3 Starting from the expression of the range of protons in air $R_p(m) = (E_p(\text{MeV})/9.3)^{1.8}$ determine the range of ^3He particles with energy 1 MeV and 10 MeV in Al.

References

1. E. Rutherford, The scattering of and particles and the structure of the atom. Philos. Mag. **21**, 669–688 (1911)
2. N. Bohr, On the theory of the decrease of velocity of moving electrified particles on passing through matter. Philos. Mag. **25**, 10 (1913)
3. N. Bohr, On the decrease of velocity of swiftly moving electrified particles in passing through matter, London, Edinburgh. Dublin Philos. Mag. J. Sci. **30**, 581 (1915)
4. F. Bloch, Zur Bremsung rasch bewegter Teilchen beim Durchgang durch Materie. Ann. Phys. (Leipzig) **408**, 285 (1933)
5. W.H. Bragg, R. Kleeman, Philosophical Magazine **10**, 358 (1905)
6. E. Segre', *Nuclei and Particles* (Benjamin/Cummings Pub. Co., Menlo Park (CA), 1982)
7. H.A. Bethe, J. Ashkin, Passage of Radiations through Matter, in *Experimental Nuclear Physics*, ed. by E. Segre, vol. 1 (Wiley, New York, N.Y. 1953)
8. J.D. Jackson, *Classical Electrodynamics*, 3rd edn. (Wiley, New York, 2021)
9. C.M. Davisson, Interaction of gamma-radiation with matter, in *Alpha-, Beta- and Gamma-ray Spectroscopy*, ed. by K. Siegbahn, vol. 1 (North-Holland Publishing Company, Amsterdam, 1965)
10. O. Klein, Y. Nishina, Über die Streuung von Strahlung durch freie Elektronen nach der neuen relativistischen Quantendynamik von Dirac. Z. Phys. **52**(11–12), 853 and 869 (1929)
11. L.C. Maximon, Simple analytic expressions for the total Born approximation cross section for pair production in a Coulomb field. J. Res. Nat. Bur. Stand. **72B** (Math. Sci.) (1), 79–88 (1968)

Chapter 11
Introduction to Plasmas

Abstract *This chapter provides a simple introduction to the basic concepts of plasmas.*

11.1 How Plasmas are Formed

Plasma is the fourth state of matter. The transformation from the solid to the liquid and the gas state is associated with an increase in the average energy (temperature) of the constituents that progressively overcomes the binding energies of atoms and molecules. In all these states however the electrons remain bound to the nuclei. If the system is further heated electrons move to excited states and eventually escape the Coulomb force of the nucleus. The final state is a superposition of a gas of positive ions and a gas of negative electrons. This state is called *plasma*.

If the gas is at thermodynamic equilibrium, the population of each ionisation state can be determined on the basis of statistical considerations such as those discussed in Chap. 2 and described by the *Saha equation* [1]

$$\frac{n_{i+1}n_e}{n_i} = \left(\frac{2\pi m_e T}{h^2}\right)^{3/2} \frac{2g_{i+1}}{g_i} e^{-E_o/T} \tag{11.1}$$

with n_i the density of ions in the ith ionisation state, n_e the electron density, T the temperature, E_o the ionisation energy of the ith ion state and g_i the degeneracy of the ground state of the ith ion species. The fraction $f_i \equiv n_i/n_o$ of charged particles depends on the gas temperature and density. In the case of hydrogen ($E_o = 13.6\,\mathrm{eV}$), $n_o = n - n_1$ is the density of H neutrals, n_1 is the density of the hydrogen ions and $n_e = n_1$ for the conservation of charge. The Saha equation reduces to a quadratic algebraic equation for $f_1 = n_1/n$

$$\frac{f_1^2}{1 - f_1} = \frac{1}{n}\left(\frac{2\pi m_e T}{h^2}\right)^{3/2} e^{-E_o/T} \tag{11.2}$$

where we have taken $2g_{i+1}/g_i = 1$ because there is no degeneracy for the bare proton but there are 2 states corresponding to the ground level of neutral hydrogen.

F. Romanelli, *Physics of Nuclear Energy*, Springer Series in Plasma Science and Technology, https://doi.org/10.1007/978-981-97-9609-0_11

For $n = 10^{20}\,\mathrm{m}^{-3}$, a typical value for fusion plasmas, the charged particle fraction starts to be significant for temperatures of the order of 7000 K. As the temperature is further increased the ionisation degree increases until, above 10000 K the gas is fully ionised (Fig. 11.1).

The ionisation process is described by specific cross sections depending on how ionisation is produced. The most important process for the plasmas considered in this book is the electron impact ionisation. An electron with energy larger than the ionisation energy of an atom can kick-off an electron from a bound state. In the case of hydrogen the ionisation energy of the ground state is 13.6 eV. The cross section has a maximum around 50 eV and then decreases with energy. Figure 11.2 shows an example of ionisation cross section taken from the ALADDIN library maintained by IAEA [2]. As for the other cross sections, in the presence of a distribution function in velocity the process is determined by an average of the cross section over the distribution function to determine the *ionisation rate coefficients* defined as $< \sigma_{ionisation} v >$ and analogous to the reactivity already introduced in Chap. 3. Simple fit of the ionisation rate coefficients can be found in literature (see e.g. [3]) and are shown in Fig. 11.2 for hydrogen.

Different methods can be used to produce plasmas [4]. The most common method is the application of a DC electric field. The naturally occurring free charges (e.g. due to the effect of cosmic rays) can be accelerated and via collisions they produce other

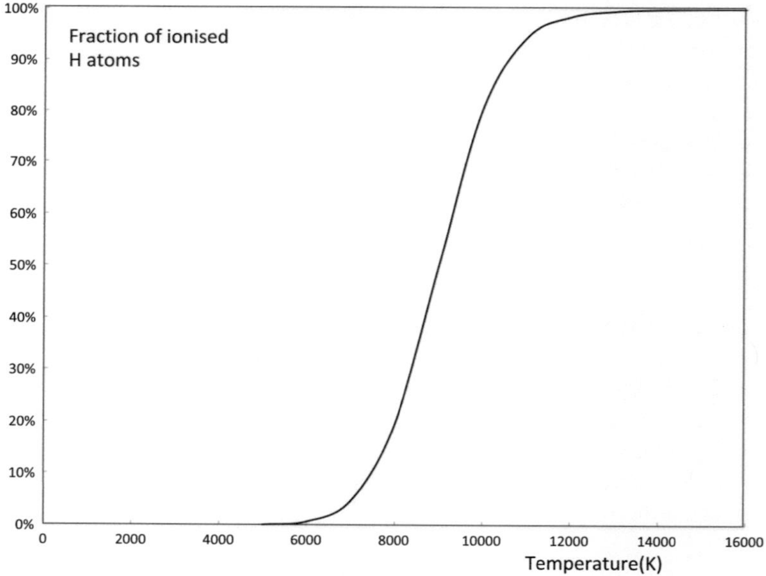

Fig. 11.1 The fraction of ionised particles depends on the temperature and the density of the gas and can be evaluated from the Saha equation. In the case of a hydrogen gas with density $n = 10^{20}\,\mathrm{m}^{-3}$ the fraction is shown in the figure

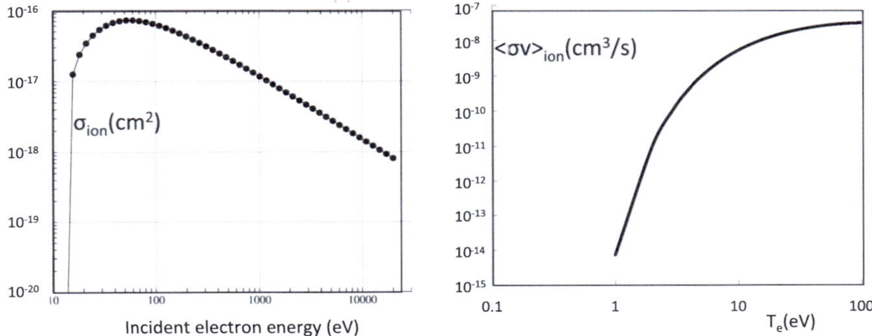

Fig. 11.2 Cross section for the ionisation of hydrogen as a function of energy from the ALADDIN library [2] and ionisation rate for hydrogen vs. electron temperature from [3]

free electrons starting in this way an avalanche of charged particles that is eventually balanced by some loss mechanism to achieve a steady state.

The simplest case of a DC discharge can be produced by two electrodes placed at a distance d inside a tube filled by a low pressure gas. Different regimes can be observed as shown in the voltage-current curve shown in Fig. 11.3. If a potential difference $V = Ed$ (with E the electric field) is applied, the free charges are accelerated towards the electrodes and a small current is established. This current increases with the applied electric field but it saturates at a value equal to the rate of production of free charges corresponding to the collection of all the naturally occurring charges.

Further increasing the electric field, a condition can be achieved in which an electron is accelerated at an energy sufficient to ionise an atom and create a new electron. In this regime, called *first Townsend ionisation regime*, an electron avalanche is produced and the current grows exponentially with the distance d between the electrodes

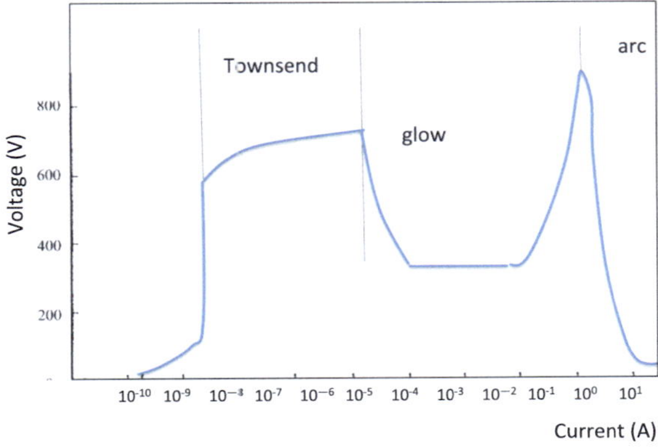

Fig. 11.3 Voltage current curve illustrating the various regimes of a discharge

$$I = I(0)e^{\alpha_T d} \tag{11.3}$$

with α_T the *first Townsend coefficient* and $I(0)$ the current at the cathode. To achieve this condition the electrons must gain enough energy between two collisions, i.e. $eE\lambda_{mfp} \geq E_{ionisation}$, with λ_{mfp} the mean free path, and since $\lambda_{mfp} \propto (1/p)$ we expect that the first Townsend coefficient can be expressed in terms of the gas pressure p and the electric field E. The coefficient α_T can indeed be reproduced by the empirical law

$$\alpha_T = Ape^{-Bp/E} \tag{11.4}$$

with A and B depending on the gas composition. However this regime still relies on a primary source of electrons that is limited and therefore it cannot be self sustained.

The amount of electrons created in the ionisation process is $\Delta I/e = (I(0)/e)(e^{\alpha_T d} - 1)$ and an equal amount of ions is produced. If the electric field is further increased the positive ions are accelerated at sufficiently high energy to produce secondary electron emission at the cathode *(second Townsend ionisation regime)*. The amount of secondary electrons is $\gamma_T \Delta I/e$ with γ_T the *second Townsend coefficient*. If the number of secondary electrons is equal to $I(0)/e$ the discharge can be self sustained. This condition can be expressed as

$$\alpha_T d = \ln\left(1 + \frac{1}{\gamma_T}\right) \tag{11.5}$$

The domain of the Townsend regime is shown in Fig. 11.3. It is called also the *dark discharge regime* since there is little light emission by the gas. Above a threshold in the electric field *breakdown* occurs. The breakdown voltage is determined by the *Paschen law* [5] that can be obtained by combining Eqs. (11.4) and (11.5)

$$V_{breakdown} = E_{breakdown} d = \frac{Bpd}{\ln(Apd) - \ln(\ln(1 + \frac{1}{\gamma_T})} \tag{11.6}$$

The Paschen law is shown in Fig. 11.4.

When breakdown is achieved a *glow discharge* is established that corresponds to the plateau in Fig. 11.3. The atoms that are excited via collisions with electrons, ions or other atoms relax to the ground energy level with emission of radiation. The voltage in this regime is decreased and the current increased. If the electric field is further increased there is a transition of the discharge to the *arc regime* in which electrodes becomes sufficiently hot that electrons can be produced via thermionic and field emission.

For the situations of interest in magnetic fusion confinement, a large DC electric field is produced in toroidal devices that in turn drives a plasma current. The electric field is produced in the presence of an externally imposed magnetic field and the charged particles tend to follow the magnetic field lines. In this case the distance d

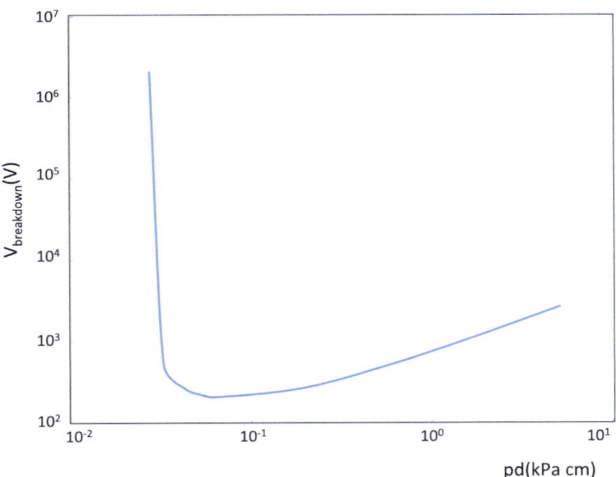

Fig. 11.4 Paschen law. The breakdown voltage for air is shown as a function of the product of the gas pressure and the distance between electrodes

between electrodes is replaced by the *connection length* L_c i.e. the distance between material surfaces measured along the magnetic field line. The current dissipates energy through Joule heating and heats the plasma up to temperatures of the order of 1 keV (one million of degrees).

11.2 How Fusion Plasmas are Confined

As we have seen in Chap. 8 fusion plasmas for energy production require temperatures in the range of 10 keV (or even larger if instead of the DT reaction other fusion reactions are employed). At these temperatures plasmas must be confined avoiding any interaction with the materials of the reaction chamber. There are three main methods to confine them:

1. Magnetic confinement
2. Inertial confinement
3. Electrostatic confinement

The *magnetic confinement* of high temperature plasmas is based on two simple recipes that exploit the charged particle nature of the plasma particles. The first recipe, common to all the magnetic confinement concepts, consists in the production of intense magnetic fields. Charged particles move along the magnetic field in helical trajectories and for magnetic fields of the order of $5T$ are confined in the direction perpendicular to the field within distances of the order of 1–10 mm from magnetic field lines. However particles are almost free to move along the magnetic field lines. The most popular recipe[1] to confine them entirely consists in bending the magnetic

[1] There are magnetic confinement concepts other than toroidal confinement such as mirror devices [6] and Z-pinches [7]. We will not consider in these book these approaches although there is still significant effort for the us of Z-pinches for high-power short-pulse X-rays production or mirrors for space propulsion.

field in such a way that particles following a field line never enter in contact with the material walls (Fig. 11.5). This second recipe is responsible for the characteristic shape of magnetic confinement fusion experiments, the *toroidal* shape.

To make a proper classification of the various toroidal magnetic confinement approaches require a preliminary understanding of plasma equilibria in magnetic field that we will discuss later in this book. For the time being we mention the main lines of research of toroidal confinement represented by *tokamaks* (including the variant called *spherical tokamak*), *stellarators* and *reversed field pinches* [8]. Pulsed plasmas with toroidal geometry are produced in *spheromaks* [9] or *Field Reversed Configurations* [10] but they have at the moment mostly a research interest and will not be considered here. In tokamaks the magnetic field is generated by external coils and by currents flowing in the plasma whereas in the stellarator case it is generated entirely by external coils. Tokamaks at the moment have achieved the better performance in term of generated fusion power with the production of about $60MJ$ of fusion energy on the European experiment JET [11]. However, the requirement of a toroidal current flowing in the plasma makes tokamaks prone to off-normal events called disruptions that lead to a loss of confinement. Stellarators have not yet achieved the tokamak fusion performance but provide inherent avoidance of disruptions.

In this book we will focus on tokamaks. The largest magnetic confinement fusion experiment is the *International Thermonuclear Experimental Reactor (ITER)* [12]. ITER is being presently built in France as part of an international collaborations. For illustration of the main phenomena in magnetic confinement fusion we will often use the ITER parameters: major radius of the torus 6.2 m, minor radius 2 m, magnetic field 5.3 T, plasma current 15 MA, density $n = 10^{20}\,\mathrm{m}^{-3}$ and central temperature 20 keV.

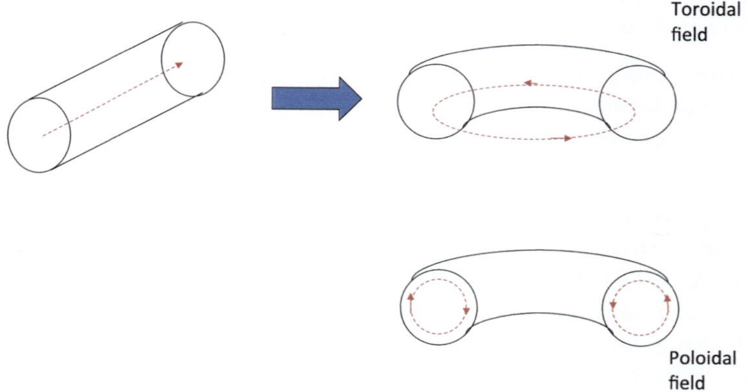

Fig. 11.5 Magnetic confinement is obtained by producing intense magnetic fields. Charged particles move in helical orbits along the magnetic field lines. To confine the plasma also in the direction along the magnetic field, magnetic field lines are bent in such a way that they wrap around toroidal surfaces

A typical cross section of a tokamak fusion power plant is shown in Fig. 11.6.

Inertial confinement is obtained by compressing a spherical solid target [14]. The target is irradiated with electromagnetic waves or energetic particles. The surface layer is ablated and its particles are ejected in the radial direction. This produces a radial compression and eventually the conditions to achieve fusion Fig. 11.7. This scheme can be implemented either with *direct drive* or *indirect drive*. In the former case the laser beams are directed on the surface of the spherical target. This approach requires an excellent uniformity of the irradiation otherwise instabilities are generated that prevent compression. In the second approach the laser beams are directed inside a gold cylinder (*Holraum*) where the spherical target is located. The laser beam light is converted in X-rays and a high degree of uniformity is obtained. The most advanced inertial confinement experiment is the *National Ignition Facility* in the US that, using the indirect drive approach, has achieved for the first time the generation of fusion energy in excess of the energy injected in the reaction chamber [15].

In the simplest example of electrostatic confinement particles are confined in an electrostatic potential well produced by two concentric wire cages (Fig. 11.8). The inner cage is negatively charged at a high potential difference that accelerate the

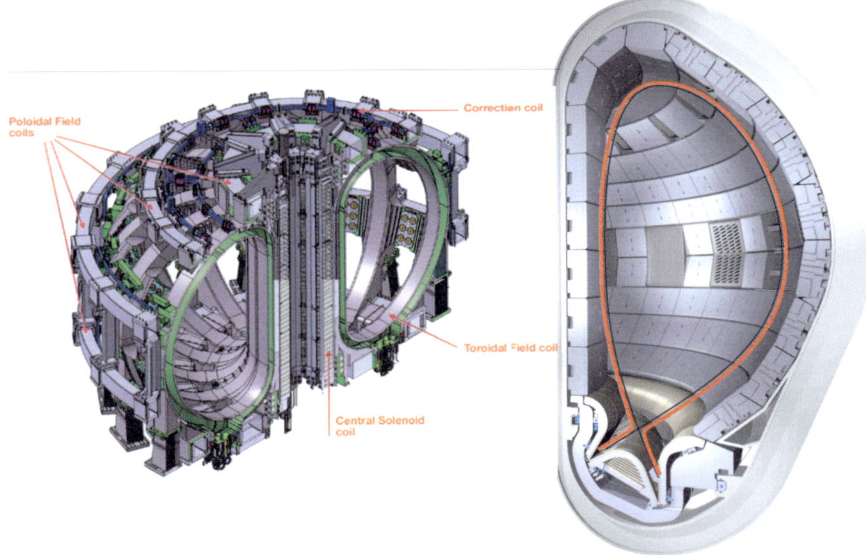

Fig. 11.6 Tokamak fusion power plant. The plasma has a characteristic toroidal shape and is confined by the magnetic field produced by the external coils (left- from Ref. [13]) and by the current flowing in the plasma itself. The reaction chamber (right) is a vessel in which a high degree vacuum is produced. Then gas is injected and a DC electric field ionises the gas producing a plasma. The fusion reactions produce neutrons that are collected by the blanket placed between the plasma and the vacuum vessel. At the bottom of the vacuum vessel target plates are placed where the heat produced by alpha particles is deposited and extracted. Courtesy of Dr. C. Sborchia and of the ITER Organisation

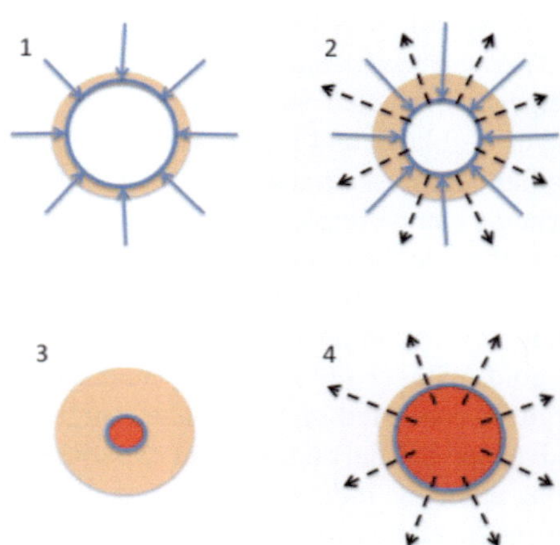

Fig. 11.7 Inertial confinement: The shell target (blue circle) is irradiated with the driver beams (blue arrows) and a spherical cloud of ablated plasma is formed (1). The shell implodes as a consequence of ablation (2). A hot spot of compressed fuel is formed where fusion reactions take place (3). The fuel is burned and an explosion occurs (4)

ions towards the centre. Ions that do not intercept the cathode can collide within the spherical volume bounded by the negative electrode and produce fusion reactions. Also in this case several concepts have been developed but at the moment none of them is at a promising level for net fusion power generation.

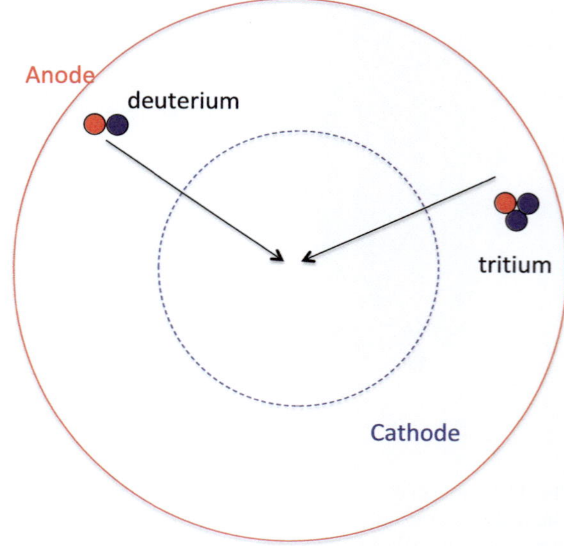

Fig. 11.8 In the simplest version electrostatic confinement is realised with two concentric electrodes. The cathode attracts the ions that are accelerated at the cathode potential and if they are not intercepted by the cathode can penetrate in the inner volume and have some finite probability of fusing

11.3 Kinetic and Fluid Descriptions of Plasmas

Plasma phenomena can be properly described only through a kinetic approach by solving the Boltzmann equation for the evolution of the distribution function under the effect of collisions among different species and of the external electric and magnetic fields

$$\frac{\partial f_j}{\partial t} + \mathbf{v} \cdot \nabla f_j + \frac{q_j}{m_j}(\mathbf{E} + \mathbf{v} \times \mathbf{B}) \cdot \nabla_v f_j = \sum_i C(f_j, f_i) \tag{11.7}$$

where $f_j(t, \mathbf{r}, \mathbf{v})$ is the distribution function of the jth species and $C(f_j, f_i)$ is the collision operator between the ith and the jth species.

Recalling that the distribution function $f(\mathbf{x}, \mathbf{v})$ represents the number of particles per unit volume in phase-space, Eq. (11.7) shows that the distribution function evolves in time due to the following processes (Fig. 11.9):

- particles in the elementary space volume enter from adjacent volumes and leave the volume due to the free streaming with velocity $\mathbf{v}$;
- particles similarly leave/enter the velocity volume due to the acceleration associated with the forces acting on them;
- particles are scattered in/out by collisions with the other particles.

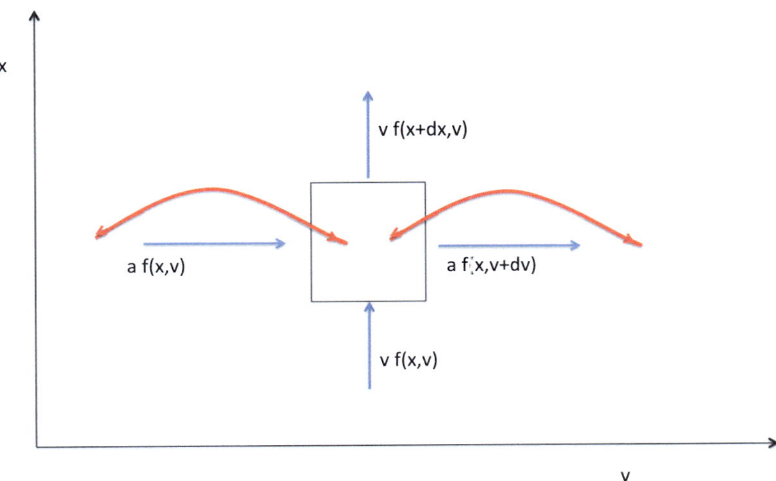

Fig. 11.9 The Boltzmann equation can be illustrated by looking at the particle balance in an elementary volume of phase space. The change per unit time is due to the acceleration a of the particle due to the electric and Lorentz force and to the free streaming of particles with velocity v. In addition collisions kick.out and in particles from different velocity domains. The illustration is for a one dimensional situation

Here we focus on the last term. Its derivation is one of the main contributions of Boltzmann and relies on the assumption that in a collision the velocities of the particles are uncorrelated. The assumption, known as *molecular chaos assumption*, allows writing the collision operator in terms of the individual distribution functions of the colliding species. The derivation can be done for a generic process described by a cross section $\sigma(\mathbf{v}_1, \mathbf{v}_2, \mathbf{v}\prime_1, \mathbf{v}\prime_2)$ where $(\mathbf{v}_1, \mathbf{v}_2)$ are the initial velocities and $(\mathbf{v}\prime_1, \mathbf{v}\prime_2)$ the final velocities. We assume that the process is reversible which implies

$$\sigma(\mathbf{v}_1, \mathbf{v}_2, \mathbf{v}\prime_1, \mathbf{v}\prime_2) = \sigma(\mathbf{v}\prime_1, \mathbf{v}\prime_2, \mathbf{v}_1, \mathbf{v}_2) \tag{11.8}$$

Assume that particles 1 are the impinging particles and particle 2 the target particles. The rate of *decrease* $(\partial f_1^{(-)}/\partial t)d^3v_1$ of particles 1 in a volume of phase space d^3v_1 due to the collisions occurring in the volume with particles 2 is obtained by multiplying the incident flux $uf_1(\mathbf{v}_1)d^3v_1$ by the cross section times the target particle density $f_2(\mathbf{v}_2)d^3v_2$ and integrating over all the possible final velocity states and all the velocities of particle 2 [16], with $u = |\mathbf{v}_1 - \mathbf{v}_2|$

$$\left(\frac{\partial f_1^{(-)}}{\partial t}\right)_2 d^3v_1 = -\int (uf(\mathbf{v}_1)d^3v_1)\sigma(\mathbf{v}_1, \mathbf{v}_2, \mathbf{v}\prime_1, \mathbf{v}\prime_2)(f(\mathbf{v}_2)d^3v_2)d^3v\prime_1 d^3v\prime_2 \tag{11.9}$$

Similarly, the rate of *increase* $(\partial f_1^{(+)}/\partial t)d^3v_1$ of particles in the same volume due to the scattering of particles with initial velocities $\mathbf{v}\prime_1$ $\mathbf{v}\prime_2$ is given by

$$\left(\frac{\partial f_1^{(+)}}{\partial t}\right)_2 d^3v_1 = +\int (u\prime f(\mathbf{v}\prime_1)d^3v\prime_1)\sigma(\mathbf{v}\prime_1, \mathbf{v}\prime_2, \mathbf{v}_1, \mathbf{v}_2)(f(\mathbf{v}\prime_2)d^3v\prime_2)d^3v_1 d^3v_2 \tag{11.10}$$

with $u\prime = |\mathbf{v}\prime_1 - \mathbf{v}\prime_2|$. In the case of elastic collisions we have shown in Chap. 3 that $u = u\prime$. In this case the collision operator can be written, using Eq. (11.8) as follows

$$C(f_1, f_2) = \left(\frac{\partial f_1^{(-)}}{\partial t}\right)_2 + \left(\frac{\partial f_1^{(-)}}{\partial t}\right)_2 = \int d^3v_2 \int d^3v\prime_1 \int d^3v\prime_2 u\sigma(\mathbf{v}_1, \mathbf{v}_2, \mathbf{v}\prime_1, \mathbf{v}\prime_2)\times$$
$$\times[f_1(\mathbf{v}\prime_1)f_2(\mathbf{v}\prime_2) - f_1(\mathbf{v}_1)f_2(\mathbf{v}_2)] \tag{11.11}$$

This is the classical expression of the Boltzmann collision operator. It can be shown that the operator conserves the number of particles, momentum and energy. The expression for $C(f)$ can be simplified in different ways but we will not enter in the details here referring to standard textbooks of plasma physics such as [17, 18].

Equation (11.7) is in general an integro-differential equation that is difficult to solve. Therefore the problem is often reduced to that of solving a set of equation for the velocity moments of the distribution function

$$n_j(t, \mathbf{r}) = \int d^3v f_j(t, \mathbf{r}, \mathbf{v}) \quad density \tag{11.12}$$

$$\mathbf{V}_j(t, \mathbf{r}) = \frac{1}{n_j} \int d^3v \mathbf{v} f_j(t, \mathbf{r}, \mathbf{v}) \quad velocity \tag{11.13}$$

$$\bar{\bar{p}}_j(t, \mathbf{r}) = p_j \bar{\bar{I}} + \bar{\bar{\pi}}_j \equiv \\ \equiv m_j \int d^3v (\mathbf{v} - \mathbf{V}_j)(\mathbf{v} - \mathbf{V}_j) f_j(t, \mathbf{r}, \mathbf{v}) \quad pressure\, tensor \tag{11.14}$$

with $p_j = Tr(\bar{\bar{p}}_j)/3$ the scalar pressure, $\bar{\bar{\pi}}_j$ the viscosity tensor and $T_j = p_j/n_j$ the temperature;

$$\mathbf{q}_j(t, \mathbf{r}) = \frac{m_j}{2} \int d^3v (\mathbf{v} - \mathbf{V}_j)(\mathbf{v} - \mathbf{V}_j)^2 f_j(t, \mathbf{r}, \mathbf{v}) \quad heat\ flux \tag{11.15}$$

The equations for the moments can be obtained from the Boltzmann equation by multiplying by $\mathbf{v}^n$ and integrating over the velocity space. Upon introducing the *convective derivative*

$$\frac{d}{dt} \equiv \frac{\partial}{\partial t} + \mathbf{V}_j \cdot \nabla \tag{11.16}$$

the fluid equations take the following form

$$\frac{dn_j}{dt} + n_j \nabla \cdot \mathbf{V}_j = 0 \quad continuity\, equation \tag{11.17}$$

$$n_j m_j \frac{d\mathbf{V}_j}{dt} + \nabla p_j + \nabla \cdot \bar{\bar{\pi}}_j - q_j n_j (\mathbf{E} + \mathbf{V}_j \times \mathbf{B}) = \mathbf{F}_j \quad momentum \quad equation \tag{11.18}$$

$$\frac{3}{2}\frac{dp_j}{dt} + \frac{5}{2} p_j \nabla \cdot \mathbf{V}_j + \bar{\bar{\pi}}_j : \nabla \mathbf{V}_j + \nabla \cdot \mathbf{q}_j = W_j \quad energy \quad equation \tag{11.19}$$

where in the r.h.s. appear the velocity moment of the collision operator corresponding to the friction force $\mathbf{F}_j$ and the energy exchange W_j experienced by the jth species due to collisions with the other species.

$$\mathbf{F}_j \equiv \sum_i \int d^3v_1 \mathbf{v}_1 C(f_i, f_j) \tag{11.20}$$

and

$$W_j \equiv \sum_i \int d^3v_1 \frac{m_1 v_1^2}{2} C(f_i, f_j) \tag{11.21}$$

The equation for the nth moment involves the moment of order $n+1$. In order to obtain a closed system of equations we need to express the highest order moment in terms of the lower moments. Such a closure assumption typically requires solving the Boltzmann equation under some approximation. Apart from the closure assumption, the fluid equations are *exact* since the infinite hierarchy of fluid equations are equivalent to the Boltzmann equation. All the physical assumptions are therefore included in the closure. Note that this is true for *any* value of collisionality, not, as sometime is stated, for regimes in which the mean free path is shorter than the macroscopic space scales. Fluid equations are valid also in the long mean free path regime provided an appropriate closure assumption (consistent with the value of collisionality) is made.

11.4 Charge Neutrality

The plasmas considered in this book are neutral. Charge neutrality holds not only at macroscopic scales but also at microscopic scales above the *Debye length* scale (see later). If the total number of negative and positive particles must be the same for arbitrarily small volumes the following condition must be satisfied

$$n_e = \sum Z_i n_i \tag{11.22}$$

with $n_e(n_i)$ the electron and ion particle densities and Z_i the electric charge of the ith ion species. Equation (11.22) is the charge neutrality condition.

11.5 The Plasma Parameter

Fusion plasmas can be described as the superposition of various ideal gases. In this chapter we will present a simplified description of the main time and space scales that characterise the system. The confinement scale R_{conf} is defined by the macroscopic size of the confined plasma (the Sun radius or the minor/major radius of toroidal confinement systems—see Fig. 11.10). The shortest space scale is the average inter particle distance that can be estimated as $r_{ipd} = n^{-1/3}$ with n the plasma density. Laboratory plasmas have a confinement scale ranging from $1m$ to $10m$ and an average inter particle distance of the order $10^{-6} - 10^{-7}m$ for magnetically confined plasmas and $10^{-9}m$ for inertially confined plasmas.

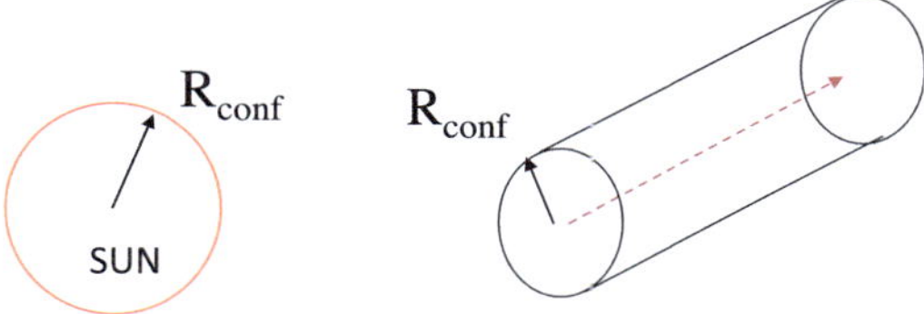

Fig. 11.10 Plasmas in equilibrium conditions are characterised by a macroscopic scale of the order of the system dimension: the Sun radius or the radius of the confinement system

As plasmas are made of charged particles, the natural way for classifying plasmas is by comparing the average Coulomb interaction energy between neighbouring particles (evaluated using the average inter particle distance r_{ipd}) with the average particle kinetic energy T

$$\frac{E_{Coulomb,ave}}{E_{kin,ave}} \approx \frac{e^2/(4\pi\epsilon_o r_{ipd})}{T} = (4\pi\Lambda^2)^{-1/3} \tag{11.23}$$

where we have introduced the *plasma parameter* $\Lambda \propto T^{3/2}/n^{1/2}$. Plasmas can be divided in two classes

- *weakly coupled plasmas* for $\Lambda \gg 1$: the kinetic energy dominates over the Coulomb energy, plasmas are hot and rarefied, a situation typical of ionosphere, space plasmas and fusion plasmas;
- *strongly coupled plasmas* for $\Lambda \ll 1$: the Coulomb energy dominates over the kinetic energy, plasmas are cold and dense, a situation typical of the atmosphere of white dwarf stars, neutron stars, laser produced plasmas and high pressure arc discharges

The plasma parameter can also be expressed as

$$\Lambda = 4\pi n\lambda_D^3 = 4\pi\left(\frac{\lambda_D}{r_{ipd}}\right)^3 \tag{11.24}$$

with λ_D the *Debye length* (see next paragraph).

$$\lambda_D = \left(\frac{\epsilon_o T}{ne^2}\right)^{1/2} \tag{11.25}$$

Therefore, the plasma parameter corresponds (apart from a factor three) to the number of particles in a sphere with radius λ_D. Fusion plasmas are characterised by a very large values of Λ of the order 10^8 (i.e. $\lambda_D >> r_{ipd}$).

11.6 The Physical Meaning of the Debye Length

The Debye length is directly related to the local violation of the charge neutrality condition. To understand how it enters the plasma dynamics let us assume that in a homogenous plasma made of positive and negative charges satisfying charge neutrality at all scales a point charge $-q$ is inserted. The position of the point charge will be taken as the origin of the coordinate system. The plasma will respond by decreasing the density of negative charges around $r = 0$ since they feel the Coulomb repulsion. Similarly, positive charges will be attracted. However, since ions are more massive than electrons, they will not respond to the perturbation. Thus, locally charge neutrality will be violated and an electrostatic perturbation characterised by a scalar potential Φ will be produced. To determine the structure of the potential we need to evaluate the variation of the density of electric charges. To this aim, assuming the ions at rest, only the effect on the electron density has to be evaluated. Thanks to the large electron mobility we can expect that the electrons will react to the perturbation by falling in the local minima of the potential energy $-e\Phi$. The distribution function will be a Maxwell–Boltzmann distribution of the form

$$f(\mathbf{r}, \mathbf{v}) = \frac{n_o}{\pi^{3/2} v_{th}^3} e^{-\frac{v^2}{v_{th}^2} + \frac{e\Phi}{T}} \tag{11.26}$$

with n_o the unperturbed (uniform) electron and ion density density. The effect of the electrostatic potential is to produce a modulation of the density that, for the case of a sinusoidal perturbation, would be as shown in Fig. 11.11.

In the case under examination we need to determine self consistently the scalar potential from the electrostatic induction equation

$$\nabla^2 \Phi = -\frac{\rho}{\epsilon_o} \tag{11.27}$$

with the charge density ρ being the sum of the electron and ion charge densities and the density associated with the point charge $-q$.

$$\rho = -q\delta(\mathbf{r}) + e(n_i - n_e) = -q\delta(\mathbf{r}) + e(n_o - n_o e^{\frac{e\Phi}{T}}) \tag{11.28}$$

where we have substituted for the electron density the expression obtained by integrating over the velocity space the distribution function given in Eq. (11.26). For $e\Phi/T \ll 1$ the exponential in Eq. (11.28) can be expanded for small arguments. The lowest order electron and ion contributions cancel (the ions in this model play the role of a neutralising background) and the electrostatic induction equation takes the following form

$$\nabla^2 \Phi - \frac{\Phi}{\lambda_D^2} = \frac{q}{\epsilon_o}\delta(\mathbf{r}) \tag{11.29}$$

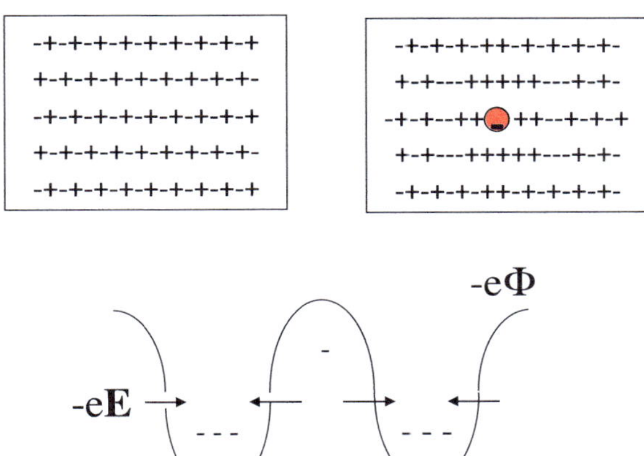

Fig. 11.11 A plasma with a uniform density of negative and positive charges (satisfying charge neutrality at all scales) is perturbed with the insertion of a negative point charge (in red). The plasma reacts by modifying the charge density close to the red charge. The resulting local charge imbalance produces an electrostatic potential Φ. Negative charges in the electrostatic potential (exemplified by a sinusoidal perturbation) falls in the minima of the potential energy whereas the positive charges, due to their large mass, remains at rest. The resulting density perturbation can be evaluated from Eq. (11.26)

The solution is a spherically symmetric potential $\Phi(r)$ satisfying the equation

$$\frac{1}{r^2}\frac{d}{dr}\left(r^2\frac{d\Phi}{dr}\right) - \frac{\Phi}{\lambda_D^2} = \frac{q}{\epsilon_o}\delta(\mathbf{r}) \tag{11.30}$$

with solution

$$\Phi = \frac{q}{4\pi\epsilon_o r}e^{-\frac{r}{\lambda_D}} \tag{11.31}$$

For $r << \lambda_D$ the electrostatic potential has the familiar behaviour as $1/r$. However, for $r >> \lambda_D$ the electrostatic potential vanishes exponentially. After a few Debye lengths the effect of the point charge is no longer appreciated because the plasma shields the charge. Therefore on space scales larger than the Debye length we can assume that charge neutrality holds.

Finally, it is interesting to note that for plasmas where the various species have similar temperatures (as we will see this is a characteristic of fusion plasmas) and densities (a consequence of the charge neutrality condition in the case of the main ion species) the Debye length is the same for electrons and ions.

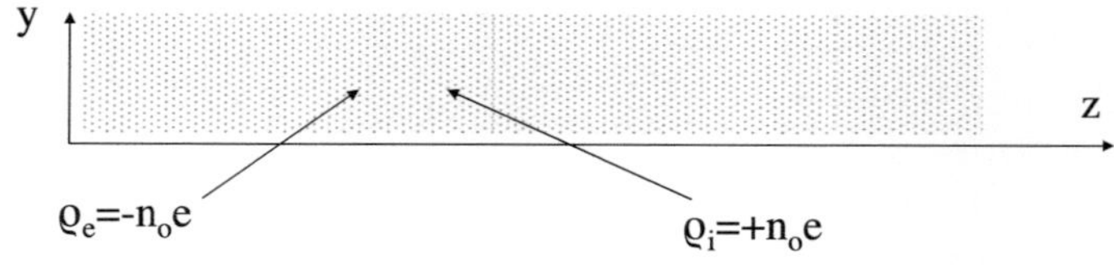

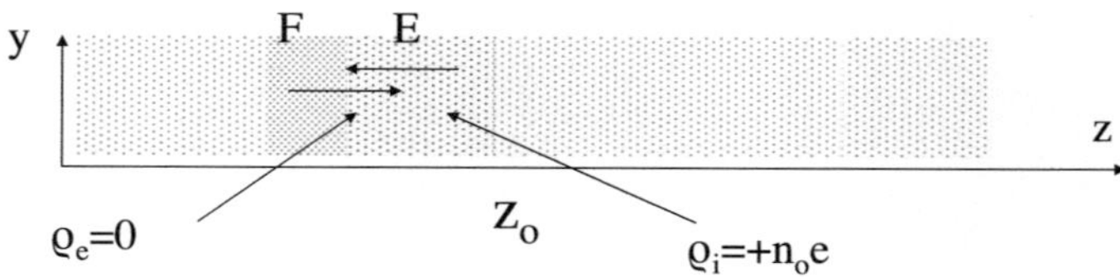

Fig. 11.12 Plasma oscillations. The ions are assumed at rest due to their inertia and act as a neutralising background. A compression of the electrons produces an electrostatic perturbation that acts as a restoring force

11.7 Plasma Oscillations

Homogeneous plasmas can be subject to *electrostatic plasma oscillations* as shown in Fig. 11.12. In the equilibrium state the density of electrons and ions are equal ($n_e = n_i = n_o$). As in the previous paragraph the ions will be assumed at rest due to their inertia. Now suppose to compress the electron density as shown in Fig. 11.12. This compression will produce a charge imbalance and an electric field $\mathbf{E} = -\nabla\Phi$ that will act to restore charge neutrality. If the amplitude of the perturbation is small we can linearise the continuity and momentum equation around the equilibrium solution $n_e = n_o$. Furthermore, we will assume that both electrons and ions are cold fluids ($p_e = p_i = 0$). The electron momentum and continuity equations then become

$$n_o m_e \frac{\partial \mathbf{V}_e}{\partial t} = e n_o \nabla \Phi \tag{11.32}$$

$$\frac{\partial n_e}{\partial t} + n_o \nabla \cdot \mathbf{V}_e = 0 \tag{11.33}$$

whereas the electric induction equation takes the form

$$\nabla^2 \Phi = -\frac{e(n_o - n_e)}{\epsilon_o} \tag{11.34}$$

Looking for a solution of the form $\Phi \propto e^{-i\omega t + ikz}$, the dispersion relation is straightforwardly obtained

$$\omega^2 = \omega_{pe}^2 \tag{11.35}$$

with $\omega_{pe} = (n_e e^2/(\epsilon_o m_e))^{1/2}$ the electron plasma frequency introduced in Chap. 4. It is important to note that Eqs. (11.35) and (4.47) describe two different kinds of oscillation. Equation (11.35) describes a non propagating electrostatic wave whereas Eq. (4.47) describes an electromagnetic transverse wave. The electrostatic wave becomes a propagating wave if the assumption of cold fluid is removed. Assuming for the sake of simplicity that the electron temperature is not affected by the perturbation, the electron momentum equation becomes

$$n_o m_e \frac{\partial \mathbf{V}_e}{\partial t} = e n_o \nabla \Phi - T_e \nabla n_e \tag{11.36}$$

yielding the modified dispersion relation

$$\omega^2 = \omega_{pe}^2 + v_{te}^2 k^2 \tag{11.37}$$

It is possible to define a plasma frequency $\omega_{pj} = (n_j Z_j^2 e^2/(\epsilon_o m_j))^{1/2}$ for each species in the plasma. It is worth noting that the Debye length can be expressed as

$$\lambda_{Dj} = \frac{v_{thj}}{\omega_{pj}} \tag{11.38}$$

11.8 Magnetised Plasmas

The presence of a magnetic field introduces further space and time scales given by the Larmor radius and the cyclotron frequency. In this case both the space and time scales are very different for electrons and ions.

In the plane perpendicular to the magnetic field particles move in circular orbits of radius $\rho_{Larmor,j} \equiv v_{\perp j}/\Omega_j$ with $\Omega_j \equiv q_j B/m_j$ the cyclotron frequency. Magnetised plasmas correspond to values of ρ_L much smaller than the macroscopic scale R_{conf}.

The ratio between the Debye length and the Larmor radius can be expressed as

$$\frac{\lambda_{Dj}}{\rho_L} = \frac{\Omega_j}{\omega_{pj}} \tag{11.39}$$

For parameters typical of thermonuclear plasmas this ratio is $<< 1$ for ions and ≈ 1 for electrons.

11.9 Suggestions for Further Readings

There are several books that can be used for an introduction to plasmas. The classical books of Chen [17] and Miyamoto [18] are recommended. For inertial confinement Ref. [14] is recommended.

11.10 Exercises

Problem 11.1 Evaluate the minimum breakdown voltage for hydrogen ($A = 510\,\mathrm{m^{-1}Torr^{-1}}$, $B = 1.25 \times 10^4\,\mathrm{Vm^{-1}Torr^{-1}}$) and helium ($A = 300\,\mathrm{m^{-1}Torr^{-1}}$, $B = 3.4 \times 10^4\,\mathrm{Vm^{-1}Torr^{-1}}$).

Problem 11.2 Show that Eq. (11.31) is the solution of Eq. (11.30).

Problem 11.3 Order the relevant plasma space and time scales for parameters typical of a magnetically confined fusion plasma $n = 10^{20}\,\mathrm{m^{-3}}$, $T = 10\,\mathrm{keV}$, $B = 5\,\mathrm{T}$.

References

1. M.N. Saha, LIII. Ionization in the solar chromosphere. Philosoph. Maga. Series 6. **40**(238), 472–488 (1920)
2. https://www-amdis.iaea.org/ALADDIN/
3. C.S. Voronov, A practical fit formula for ionization rate coefficients of atoms and ions by electron impact: Z=1-28. Atom. Data Nucl. Data Tables **65**, 1–35 (1997)
4. H. Conrads, M. Schmidt, Plasma Sources. Sci. Technol. **9**, 441 (2000)
5. F. Paschen, Über die zum Funkenübergang in Luft, Wasserstoff und Kohlensäure bei verschiedenen Drucken erforderliche Potentialdifferenz. Annalen der Physik. **273**(5), 69–96 (1889)
6. A.V. Burdakiv, Modern magnetic mirrors and their fusion prospects. Plasma Phys. Control. Fusion **52**, 124026 (2010)
7. M.G. Heines, A review of the dense Z-pinch. Plasma Phys. Control. Fusion **53**, 093001 (2011)
8. L. Marrelli et al., The Reversed Field Pinch. Nucl. Fusion **61**, 023001–1 (1921)
9. T.R. Jarboe, Review of spheromak research. Plasma Phys. Controlled Fusion **36**, 945 (1994)
10. L.C. Steinhauer, Review of field-reversed configurations. Phys. Plasmas **18**, 070501 (2011)
11. C. Maggi et al., JET special issue. Nucl. Fusion **63**, 112001 (2023)
12. www.iter.org
13. C. Sborchia et al., The ITER magnet systems: progress on construction. Nucl. Fusion **54**, 013006 (2014)
14. S. Atzeni, J. Meyer-Ter-Vehn, *The Physics of Inertial Fusion* (Oxford University Press, 2004)
15. A.B. Zylstra et al., Nature **601**, 542 (2022)
16. F. Reif, *Fundamentals of Statistical and Thermal Physics* (McGraw-Hill, 1965)
17. F. Chen, *Introduction to Plasma Physics*, 3rd edn. (Springer, 2016)
18. K. Miyamoto, *Plasma Physics for Controlled Fusion*, 2nd edn. (Springer, 2016)

Chapter 12
Collisions in Plasmas

Abstract *This chapter provides a discussion of the collisional processes in thermonuclear plasmas. Using the formalism developed in Chap.* 10 *the slowing down of energetic ions is discussed. The notion of plasma resistivity is introduced and a discussion of runaway electrons is presented.*

12.1 Coulomb Logarithm in Plasmas

The analysis of Coulomb collisions in plasmas closely follows the approach presented in Chap. 10. The main difference is the choice of the upper and lower cut-off in the impact parameter. As already discussed, the divergence at low impact parameters is an artefact due to the use of approximation of small deflection angles that is not valid for $b \to 0$. Therefore a reasonable assumption for the lower cut-off is $b_{min} = b_o$. The divergence at large impact parameters is instead a consequence of the long range of the Coulomb force. However, it has been shown in Chap. 11 that at distances above λ_D any charge fluctuation is screened by the reaction of the plasma. Therefore the Debye length is the natural upper cut off. The ratio b_{max}/b_{min} can be written as

$$\frac{b_{max}}{b_{min}} = \frac{\lambda_D}{b_o} = \left(\frac{\epsilon_o T}{ne^2}\right)^{1/2} \frac{2\pi\epsilon_o m_r u^2}{e^2} = \Lambda \frac{m_r u^2}{2T} \approx \Lambda \tag{12.1}$$

Note that this quantity is in principle a function of the relative velocity u, however, since it appears as the argument of a logarithm, such a dependence is usually neglected. Typical values are in the range $\ln \Lambda \approx 15 - 20$.

12.2 Slowing Down of Energetic Ions in Plasmas

Fusion plasmas have several species of energetic ions: the 3.5 MeV fusion generated alpha particles, the energetic ions produced in other fusion reactions and the ions produced by the injection of neutral beams in the range of $1 MeV$ to heat the plasma.

F. Romanelli, *Physics of Nuclear Energy*, Springer Series in Plasma Science and Technology, https://doi.org/10.1007/978-981-97-9609-0_12

In order to determine how these species transfer their energy to the thermal population it is necessary to average the stopping power calculated in Eq. (10.11) over the distribution function in velocity space of the thermal plasma particles that we have generically labelled as species 2. A Maxwell distribution function will be assumed with temperature T_2. The average over the distribution function will be denotes with $< \ldots >$. The resulting expression is

$$< \frac{dE_1}{dt} > \approx -\frac{q_1^2 q_2^2}{4\pi\epsilon_o^2} \ln \Lambda n_2 \int f(v_2) d^3 v_2 \frac{\mathbf{V}_{\mathrm{com}} \cdot \mathbf{u}_i}{m_r u_i^3} \tag{12.2}$$

with

$$f(v_2) = \frac{1}{(\pi^{1/2} v_{t2})^3} e^{-v_2^2/v_{t2}^2} \quad v_{t2} = \left(\frac{2T_2}{m_2} \right)^{1/2} \tag{12.3}$$

Noting that

$$\frac{\mathbf{V}_{\mathrm{com}} \cdot \mathbf{u}_i}{m_r u_i^3} = \frac{\mathbf{v}_1 \cdot \mathbf{u}_i}{m_r u_i^3} - \frac{1}{m_1 u_i} \tag{12.4}$$

It is possible to write Eq. (12.2) in the following form

$$< \frac{dE_1}{dt} > \approx -\frac{q_1^2 q_2^2}{4\pi\epsilon_o^2} \ln \Lambda n_2 \left(\frac{\mathbf{v}_1 \cdot \mathbf{e}_v}{m_r} - \frac{\psi_v}{m_1} \right) \tag{12.5}$$

with

$$\mathbf{e}_v \equiv \int f(v_2) d^3 v_2 \frac{(\mathbf{v}_1 - \mathbf{v}_2)}{|\mathbf{v}_1 - \mathbf{v}_2|^3} \tag{12.6}$$

$$\psi_v \equiv \int f(v_2) d^3 v_2 \frac{1}{|\mathbf{v}_1 - \mathbf{v}_2|} \tag{12.7}$$

These integrals are analogous to those entering in the evaluation of the electrostatic field and potential of a charge distribution with spherical symmetry. Their explicit evaluation is presented in the Appendix B. Upon substituting Eqs. (B.8), (B.11) and (B.13) into Eq. (12.5) the final expression for the slowing down of the particle 1 from a a background of particles 2 is obtained

$$< \frac{dE_1}{dt} > = -\frac{q_1^2 q_2^2}{4\pi\epsilon_o^2 v_1 m_2} \ln \Lambda n_2 \left[\Phi\left(\frac{v_1}{v_{t2}} \right) - \left(1 + \frac{m_2}{m_1} \right) \frac{2 v_1}{\pi^{1/2} v_{t2}} e^{-\frac{v_1^2}{v_{t2}^2}} \right] \tag{12.8}$$

with $\Phi(x)$ the error function. This expression is totally general.

A further simplification, valid for the case in which particle 1 is an energetic ion, can be obtained by considering the two limits $v_1 \gg v_{t2}$ (that applies to the slowing down of alpha particles by thermal ions) and $v_1 \ll v_{t2}$ (that applies to the slowing down of alpha particles by thermal electrons). The limiting forms of the error function are

$$\Phi(x) \approx 1 - e^{-x^2}/(\pi^{1/2}x) \tag{12.9}$$

for $x \gg 1$ and

$$\Phi(x) \approx (2/\pi^{1/2})(x - x^3/3) \tag{12.10}$$

for $x \ll 1$. Therefore the final expression for the slowing down of an energetic ions in a plasma can be expressed as follows

$$< \frac{dE_1}{dt} >= -\frac{Z_1^2 e^4}{4\pi\epsilon_o^2 v_1} \ln \Lambda \left[\frac{n_i Z_i^2}{m_i} + \frac{4}{3\pi^{1/2}} \frac{n_e}{m_e} \left(\frac{m_e E_1}{m_1 T_e} \right)^{3/2} \right] \tag{12.11}$$

Equation 12.11 shows that above a critical energy the dominant mechanism for slowing down is provided by the collisions with electrons. The critical energy can be obtained from Eq. 12.11 by equating the ion and electron contributions

$$E_{crit} = \frac{m_1}{m_e} \left(\frac{3\pi^{1/2}}{4} \frac{m_e}{m_i} \frac{n_i Z_i^2}{n_e} \right)^{2/3} T_e \tag{12.12}$$

For typical thermonuclear plasma conditions the critical energy is about 400 keV. Therefore, the fusion generated alpha particles transfer almost 90% of their energy to the electrons before being slowed down by ions. This result has important implications on the regimes of operation. As the alpha particles mainly heat the electrons, in order for the electrons to heat the ions the electron temperature must be (marginally) larger than the ion temperature. As we will see in the next chapter, radiation losses increase with increasing electron temperature. Furthermore, the heat transport due to plasma turbulence is typically larger for electrons than for ions. Thus, the requirement on the electron temperature makes the achievement of reactor conditions more difficult (or even impossible for some reaction!).

The time scale for slowing down *(the slowing down time τ_{SD})* can be determined from Eq. 12.11 that can be integrated by separation of variable

$$\tau_{SD} \equiv - \int_0^{E_{1o}} \frac{dE_1}{< \frac{dE_1}{dt} >} =$$

$$= \frac{2}{3} \frac{4\pi\epsilon_o^2}{Z_1^2 e^4 \ln \Lambda} \left(\frac{2}{m_1} \right)^{1/2} \frac{3\pi^{1/2}}{4} \frac{m_e}{n_e} \left(\frac{T_e m_1}{m_e} \right)^{3/2} \ln \left[1 + \frac{m_i}{n_i Z_i^2} \frac{n_e}{m_e} \frac{4}{3\pi^{1/2}} \left(\frac{m_e E_{1o}}{m_1 T_e} \right)^{3/2} \right] \tag{12.13}$$

In practical units the slowing down time can be expressed as

$$\tau_{SD} = 12ms \frac{A_1}{Z_1^2} \frac{T(\text{keV})^{3/2}}{n_e(10^{20}\,\text{m}^{-3})} \tag{12.14}$$

For typical plasma conditions of ITER the slowing down time is of the order of 1 s.

Equation 12.8 can also be used to determine the energy transfer between different species. In this case we need to average Eq. (12.8) over a Maxwell distribution for species 1 taking the temperatures of the two species distinct. As shown in Appendix B the final expression is

$$P_{12} = -\frac{q_1^2 q_2^2}{4\pi\epsilon_o^2} \ln \Lambda \frac{n_1 n_2}{m_1 m_2} \frac{4(T_1 - T_2)}{\pi^{1/2}(v_{t1}^2 + v_{t2}^2)^{3/2}} \equiv \frac{n_1(T_1 - T_2)}{\tau_{12}^e} \tag{12.15}$$

To be noted that the characteristic relaxation time τ_{12}^e is much shorter for particles with similar mass

$$\frac{\tau_{ee}}{\tau_{ij}^e} \approx \left(\frac{m_e}{m_p}\right)^{1/2} \frac{(A_i A_j)^{1/2}}{(A_i + A_j)^{3/2}} \tag{12.16}$$

$$\frac{\tau_{ee}}{\tau_{ei}^e} \approx \frac{m_e}{A_i m_p} \tag{12.17}$$

with A_i the mass number of the ith ion species and

$$\tau_{ee} = \frac{3}{4(2\pi)^{1/2}} \frac{m_e^{1/2}(4\pi\epsilon_o)^2 T_e^{3/2}}{n_e e^4 \ln(\Lambda)} \approx 3.44 \times 10^{11} \frac{T_e(\text{eV})^{3/2}}{n_e(\text{m}^{-3}) \ln \Lambda} s \tag{12.18}$$

Note that τ_{ee} is the typical time between two electron-electron collisions. For ITER parameters $T_e = 10$ keV and $n_e = 10^{20}\ \text{m}^{-3}$ we have $\tau_{ee} \approx 0.17$ ms. For the same parameters, the ion-ion temperature equilibration time is of the order $\tau_{ij}^e \approx 10$ ms and the electron ion equilibration time $\tau_{ei}^e \approx 600$ ms.Therefore in most of the situations considered in this book it is reasonable to assume that all the ion species reach the same temperature but this could be not necessarily true for the electron species. However, as we will see in Chap. 14 (and briefly mentioned above), reactor conditions also require $T_e \approx T_i$.

12.3 Resistivity and Runaway Electrons

Plasma confinement in tokamaks requires a current flowing in the plasma. The simplest way to generate the plasma current is via magnetic induction to create an electric field E in the toroidal direction that accelerates the electrons. Electric fields can be created also during off-normal events (disruptions) and their magnitude can be very large.

In the absence of collisions, electrons would be accelerated indefinitely by the electric field. Collisions with the background plasma counteract the electron acceleration and allow a steady state to be achieved in which the force associated to the electric field is balanced by the friction force between electrons and plasma.

Following the approach used for the alpha particle slowing down, we consider first the momentum loss of a test particle of mass m_1 colliding with a plasma described by a Maxwell distribution for both ions and electrons. It should be noted that the distribution function may not be necessarily a Maxwell distribution (and indeed in the case of the electrons the electric field may cause a significant departure from a Maxwell distribution). Nevertheless we consider first this case to determine the relevant collision rates. The change in momentum in a single collision is given by Eq. (3.33). Upon applying the same argument used to determine the alpha particle slowing down by a background of particles of species 2, the friction force *in the non-relativistic limit* can be expressed as

$$
\begin{aligned}
< \frac{d\mathbf{p}_1}{dt} > &\approx -\frac{q_1^2 q_2^2}{4\pi\epsilon_o^2 m_r} \ln \Lambda n_2 \mathbf{e}_v = \\
&= -\frac{q_1^2 q_2^2}{4\pi\epsilon_o^2 m_r} \ln \Lambda n_2 \frac{\mathbf{v}_1}{v_1^3} \int_0^{v_1} f(v_2) 4\pi v_2^2 dv_2 \qquad (12.19)
\end{aligned}
$$

with $\mathbf{e}_v$ evaluated in Appendix B (see Eq. (B.8)). Equation (12.19) shows that the friction force increases linearly for small test particle velocities ($v_1 \ll v_{t2}$) whereas in the opposite limit it decreases as $(1/v_1^2)$.

If instead of a test particle we have a distribution of particles of species 1 we can determine the average friction force. Clearly, the average over a Maxwell distribution of Eq. (12.19) is identically zero. In order to have a net effect the distribution function must have a non zero average velocity. The exact expression of the distribution function in this case would require to solve the Boltzmann equation with an appropriate form of the collision operator and a force that produces a net velocity such as an electric field (see e.g. Ref. [1]). For the sake of illustration let us assume that the distribution function is a shifted Maxwell distribution $f_1 \propto \exp[-(\mathbf{v}_1 - \mathbf{U}_1)^2)/v_{t1}^2]$ and expand for small values of U_1/v_{t1}. The expression of the average friction force becomes

$$
\mathbf{F}_{12} \equiv \int d^3 v_1 \frac{e^{-v_1^2/v_{t1}^2}}{\pi^{3/2} v_{t1}^3} \frac{2\mathbf{v}_1 \cdot \mathbf{U}_1}{v_{t1}^2} < \frac{d\mathbf{p}_1}{dt} > \approx n_1 m_1 < \nu_{12} > \mathbf{U}_1 \qquad (12.20)
$$

with

$$
< \nu_{12} > = \frac{Z_2^2 n_2}{n_e \tau_{ee}} \left(\frac{m_r m_e}{m_1^2} \right)^{1/2} \qquad (12.21)
$$

and m_r the reduced mass. With the expressions above we can determine the current density generated by an electric field. The generation of a net plasma current involves only the electron-ion collisions because the momentum exchange for collisions of particles of the same species vanishes. The balance between the electric force and the friction force yields

$$< \nu_{ei} > m_e U_e = eE \tag{12.22}$$

A net current density $j = n_e e U_e = \sigma E$ is established with the conductivity σ given by

$$\frac{1}{\sigma_{Spitzer}} \equiv \eta_{Spitzer} = \\ = \frac{m_e}{n_e e^2 \tau_{ee}} \frac{Z_2^2 n_2}{n_e} \approx 6.6 \times 10^{-8} \frac{n_i Z_i^2}{n_e} T(\text{keV})^{-3/2} (\text{ohm-m}) \tag{12.23}$$

and $\eta_{Spitzer}$ the Spitzer resistivity where we have neglected $O(m_e/m_p)$ corrections. Note that for $T \approx 1$ keV the plasma resistivity is of the same order of the copper resistivity at room temperature $\eta_{Cu} \approx 1.7 \times 10^{-8}$ ohm-m.

If more than one ion species is present, the ion charge in Eq. (12.23) must be replaced with the same of the single ion charge weighted on the ion distribution. We define the *effective charge* Z_{eff} as

$$Z_{eff} \equiv \frac{\sum Z_i^2 n_i}{n_e} \tag{12.24}$$

and replace $n_i Z_i^2 / n_e$ with Z_{eff} in Eq. (12.23).[1] Due to the charge neutrality condition Eq. (11.22), the effective charge is independent of the density. Thus, the resistivity turns out to be independent of the plasma density since the ion density dependence of the collision frequency is balanced by the dependence of the current density on the electron density. The resistivity decreases with increasing electron temperature as a consequence of the decrease of the Coulomb cross section with increasing energy. Note that this behaviour is opposite to what happens to the resistivity in metals. The decrease of the resistivity with increasing temperature makes it easier to maintain the plasma current at reactor relevant temperatures. However, in case of off-normal events (disruptions) very large electric fields can be created. In typical situations the disruptions is initiated by a rapid cooling down of the plasma (thermal quench), followed by the current decay on a much longer time scale. During the rapid cooling down the electric field becomes very large and this can produce a large number of *runaway electrons*.

The origin of runaways can be understood on the basis of Eq. (12.19). The friction force acting on electron with velocity $v_e \gg v_{te} \gg v_{ti}$ is due to both electron-electron and electron-ion collisions and can be written as follows

$$< \frac{d\mathbf{p}_1}{dt} > = -\frac{2 n_e e^4}{4\pi \epsilon_o^2 m_e} \ln \Lambda \left(1 + \frac{Z_{eff}}{2}\right) \frac{\mathbf{v}_e}{v_e^3} \tag{12.25}$$

[1] The effective charge is the charge weighted on the *charge* concentration $Z_i n_i / n_e$ that is the correct weight as by charge neutrality $\sum Z_i n_i / n_e = 1$.

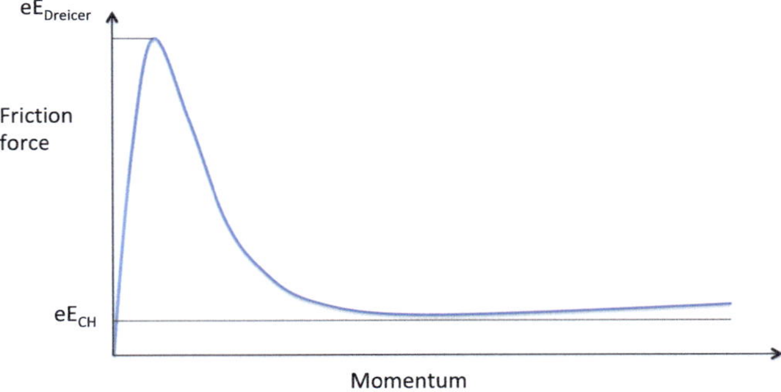

Fig. 12.1 Friction force vs. electron momentum

As a result, above a critical velocity an electron is accelerated indefinitely. This is the mechanism for runaway production. The critical velocity can be obtained by balancing the electric and the friction forces, yielding

$$v_{ecrit} = \left(\frac{2n_e e^3 \ln \Lambda \left(1 + \frac{Z_{eff}}{2} \right)}{4\pi \epsilon_o^2 m_e E} \right)^{1/2} \tag{12.26}$$

If the electric field is not too large the runaway mechanism involves only the electrons in the Maxwell distribution tails. However, if the critical velocity becomes close to the electron thermal velocity a substantial modification of the electron distribution function can be expected. The electric field at which $v_{crit} = v_{te}$ is the Dreicer field [2] $eE_{Dreicer} \equiv m_e v_{the} \nu_{ei}$. Typical fusion plasma conditions correspond to $E \ll E_{Dreicer}$ therefore to small deviations from a Boltzmann distribution and to a low rate of runaway production.

The monotonic decrease of the friction force with the relative velocity is correct only for non relativistic velocities. As relativistic corrections start to be important the friction force attains a minimum [3] as shown in Fig. 12.1. This means that a minimum electric field E_c has to be achieved in order to have runaway generation.

$$E_c = \frac{n_e e^3 \ln \Lambda}{4\pi \epsilon_o^2 m_e c^2} \tag{12.27}$$

For densities of the order $n_e \approx 10^{20}\,\mathrm{m}^{-3}$ the critical electric field is $\approx 7.5 \times 10^{-2}$ V/m and is much larger than the electric field needed to sustain current in stationary conditions (in ITER for example $E \approx 2 \times 10^{-3}$ V/m). However, during transient events the critical field may be exceeded and runaways electrons can be generated.

If runaway electrons can be produced, the critical question becomes the rate at which the runaway population is growing. The runaway production rate depends on the rate at which electrons with velocity above the critical velocity for runaway production are generated. Several mechanisms can play a role. The Dreicer mechanism described above consists in the collisional diffusion through the runaway boundary. Collisions continuously scatter electrons beyond the point in velocity space at which they are continuously accelerated. If the electric field is much smaller than the Dreicer field this mechanism yield a production rate that is exponentially small. A proper evaluation of the runaway production rate S can be found in Ref. [1]. For the case $Z_{eff} = 1$ S is given by the following expression

$$S = n_e \nu_{ee} \left(\frac{E}{E_{Dreicer}} \right)^{-3/8} e^{-\frac{E_{Dreicer}}{4E} - \left(\frac{2E_{Dreicer}}{E} \right)^{1/2}} \tag{12.28}$$

A second mechanism is the *hot-tail generation* [4] that takes place when the cooling time is much faster than the time for thermal equilibration of the high energy electrons. Other mechanisms can play a role in a fusion power plant such as the electron generation by tritium decay or the Compton scattering with the photons produced by gamma decay of the radioactive materials of the wall [5].

However there is a mechanism of runaway multiplication that is expected to dominate the runaway production at high plasma current: the effect of knock-on collisions between energetic electrons and bulk electrons [6]. If the energy gained by the runaway electron is sufficiently large, its collision with bulk electrons will produce a new runaway ending up with two runaway electrons.

The mechanism of the amplification can be understood with a simple model. As shown in the derivation of Eq. (10.3), in the collision between an ultra relativistic electron and a background electron the change in momentum along the direction orthogonal to the unperturbed trajectory is

$$\Delta p_y = \frac{e^2}{2\pi \epsilon_o c b} \tag{12.29}$$

where the relative velocity has been approximated as $u \approx c$. If this variation is larger than the critical velocity for runaway generation as given in Eq. (12.26), the secondary electron will runaway. This condition can be written as

$$b \leq b_{sr} = \frac{e^2}{2\pi \epsilon_o c m_e v_{ecrit}} = \frac{e^2}{2\pi \epsilon_o c m_e} \left(\frac{4\pi \epsilon_o^2 m_e E}{2 n_e e^3 \ln \Lambda \left(1 + \frac{Z_{eff}}{2} \right)} \right)^{1/2} \tag{12.30}$$

The rate of production of secondary runaway electrons can be written as

$$\frac{\partial n_{sr}}{\partial t} = n_{sr} n_e (\pi b_{sr}^2 c) = n_{sr} \frac{eE}{2 m_e c \ln \Lambda \left(1 + \frac{Z_{eff}}{2} \right)} \tag{12.31}$$

As it will be discussed in more detail later, the electric field can be related with the change in the plasma current I_p as follows

$$2\pi R_o E = L\frac{dI_p}{dt} \tag{12.32}$$

with $L \approx \mu_o R_o$ the plasma self inductance and R_o the major radius of the torus. Therefore, Eq. 12.31 can be integrated by separation of variables yielding

$$n_r = n_r(0)e^{\frac{I_p}{I_A \ln \Lambda}} \tag{12.33}$$

where $I_A = 4\pi m_e c/(\mu_o e) \approx 17\,\text{kA}$ is the Alfvén current. Thus the amount of secondary runaway generation depends on the plasma current. Taking $\ln \Lambda \approx 20$ we expect a moderate effect in machines with plasma current of the order 1 MA but a large effect in machines such as ITER with plasma current larger by an order of magnitude.

12.4 Suggestions for Further Readings

The slowing down of massive charged particles in plasmas is a well established subject and can be found in classical textbook such as Refs. [7, 8]. The issue of runaway electron has been the subject of many investigations over the years [9]. The subject has attracted more interest after the discovery of the secondary runaway electrons generation mechanism in connection with its impact on the operation of large fusion experiments such as ITER. Recent reviews on this subject are Refs. [10, 11].

12.5 Exercises

Problem 12.1 Determine the slowing down time of an alpha particle for parameters typical of a magnetically confined fusion plasma $n = 10^{20}\,\text{m}^{-3}$, $T = 10\,\text{keV}$.

References

1. M.D. Kruskal, I.B. Bernstein, Runaway electrons in and ideal Lorentz plasma. Phys. Fluids **7**, 407 (1964)
2. H. Dreicer, Electron and ion runaway in a fully ionised gas. Phys. Rev. **115**, 238 (1959)
3. J.W. Connor, R.J. Hastie, Relativistic limitations on runaway electrons. Nucl. Fusion **15**, 415 (1975)
4. H. Smith, Runaway electron generation in a cooling plasma. Phys. Plasmas **12**, 122505 (2005)

5. J.R. Martin-Solis, Formation and termination of runaway beams in ITER disruptions. Nucl. Fusion **57**, 066025 (2017)
6. M.N. Rosenbluth, S.V. Putvinski, Theory for avalanche of runaway electrons in tokamaks. Nucl. Fusion **37**, 1355 (1997)
7. F. Chen, *Introduction to Plasma Physics*, 3rd edn. (Springer, 2016)
8. K. Miyamoto, *Plasma Physics for Controlled Fusion*, 2nd edn. (Springer, 2016)
9. H. Knoepfel, D.A. Spong, Runaway electrons in toroidal discharges. Nucl. Fusion **19**, 785 (1979)
10. A.H. Boozer, Runaways electrons and ITER. Nucl. Fusion **57**, 056018 (2017)
11. B.N. Breizman, P. Aleynikov, E.M. Hollmann, M. Lehnen, Physics of runaways electrons in tokamaks. Nucl. Fusion **59**, 083001 (2019)

Chapter 13
Toroidal Magnetic Confinement

Abstract *Plasmas are confined by magnetic fields generated by a system of coils and by the plasma itself. In this chapter a simple introduction to the basic concepts of toroidal equilibrium is given. The expression of the magnetic field generated by the main components (wires, coils, solenoids) is derived. The large aspect ratio limit is introduced.*

13.1 Magnetostatics

In this chapter a set of stationary currents is considered as an approximation to plasma magnetic confinement. In practical situations the currents are time dependent but the evolution is sufficiently slow that the displacement current can be neglected.

The magnetic field generated by the current density $\mathbf{j}$ satisfies Ampére's equation

$$\nabla \times \mathbf{B} = \mu_o \mathbf{j} \tag{13.1}$$

that can be inverted to give

$$\mathbf{A}(\mathbf{r}) = \frac{\mu_o}{4\pi} \int d^3 r\prime \frac{\mathbf{j}(\mathbf{r}\prime)}{|\mathbf{r} - \mathbf{r}\prime|} \tag{13.2}$$

with $\mathbf{B} = \nabla \times \mathbf{A}$ and $\mathbf{A}$ being the vector potential. Equation (13.2) can be used to determine the vector potential and, in turn, the magnetic field from the current density distribution. The explicit expression for the magnetic field is

$$\mathbf{B}(\mathbf{r}) = \frac{\mu_o}{4\pi} \int d^3 r\prime \frac{(\mathbf{r} - \mathbf{r}\prime) \times \mathbf{j}(\mathbf{r}\prime)}{|\mathbf{r} - \mathbf{r}\prime|^3} \tag{13.3}$$

However, rather than using Eq. (13.3), it is usually simpler to take advantage of the symmetry of the problem and use Ampére's law in integral form

$$\mu_o \int dS \mathbf{n} \cdot \mathbf{j} = \oint \mathbf{dl} \cdot \mathbf{B} \tag{13.4}$$

F. Romanelli, *Physics of Nuclear Energy*, Springer Series in Plasma Science and Technology, https://doi.org/10.1007/978-981-97-9609-0_13

where use has been made of the Stokes theorem. In evaluating the integral on the r.h.s. we must remember that the normal **n** to the surface must be taken according to the rule shown in Fig. 4.3.

The following solutions have to be noted.

13.1.1 Field Produced by a Wire

Taking the wire direction to coincide with the Z axis, the field must be symmetric with respect to rotations around the Z axis ($\partial/\partial\phi = 0$) and with respect to translations along Z ($\partial/\partial Z = 0$) (see Fig. 13.1). Let us assume that the wire cross section is a circle of radius a. Then, the magnetic field lines must be circles with the magnetic field intensity being only a function of the distance R from the center of the wire. The explicit dependence can be determined from Eq. (13.4) taking as the integration path a circle of radius R

$$\mu_o \int dS\mathbf{n}\cdot\mathbf{j} = \mu_o \int_0^R 2\pi R\prime dR\prime j_Z = \mu_o I(R) = \oint \mathbf{dl}\cdot\mathbf{B} = 2\pi R B_\phi \qquad (13.5)$$

or

$$B_\phi = \frac{\mu_o I(R)}{2\pi R} \qquad (13.6)$$

with $I(R)$ the plasma current flowing through a circle of radius R. The vector potential for this problem can be obtained from $B_\phi = -\partial A_Z/\partial R$, yielding $A_Z = \mu_o I/(2\pi)\ln(R/R_o)$ with R_o being an arbitrary constant.

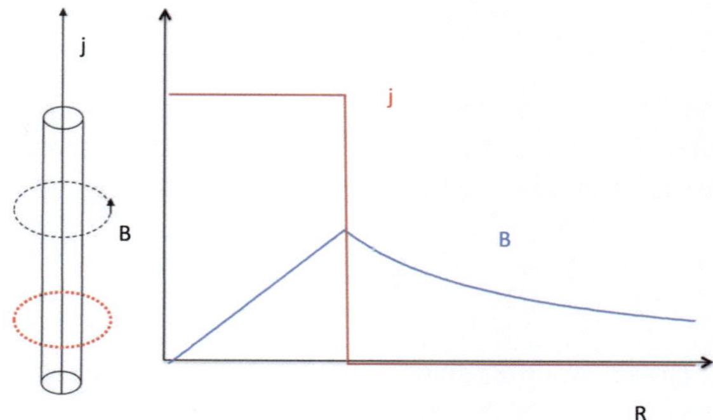

Fig. 13.1 Magnetic field of a single wire with circular cross section. Symmetry considerations impose the direction of the magnetic field along θ. The dependence on R inside the wire depends on the current distribution and is illustrated here for the case of a constant current density

13.1.2 Field Produced by an Infinite Cylindrical Solenoid

In this case it is convenient to make the Z axis to coincide with the solenoid axis (Fig. 13.2). The current density flows in the θ direction and the magnetic field is directed along Z. Symmetry considerations imply again $(\partial/\partial\phi = 0)$ and $(\partial/\partial Z = 0)$. To apply Ampére's law in integral form we chose a rectangular path with one vertical leg inside the solenoid and the other leg outside.

Outside the solenoid the field vanishes, therefore the only contribution comes from the vertical path inside the solenoid. Let us assume that, given the outer radius R_{out} and the inner radius R_{in} of the solenoid, the current density j_ϕ is independent of R. Then, Eq. (13.4) yields

$$\mu_o \int dS\mathbf{n} \cdot \mathbf{j} = \mu_o(R_{out} - R_{in})Lj_\phi = \oint \mathbf{dl} \cdot \mathbf{B} = LB_Z \tag{13.7}$$

or $B_Z = \mu_o(R_{out} - R_{in})j_\phi$. The result is independent of L as it should be since the length L is arbitrary.

The solution is a constant magnetic field inside the solenoid and no magnetic field outside. Between R_{in} and R_{out} the magnetic field can be evaluated in the same way taking the inner vertical path at an arbitrary radius R between R_{in} and R_{out}. We have simply to replace R_{in} with R, yielding

$$B_Z = \mu_o(R_{out} - R)j_\phi \tag{13.8}$$

The field matches the inner value at $R = R_{in}$ and vanishes at $R = R_{out}$.

The fact that the magnetic field of an infinite solenoid vanishes for $R \geq R_{out}$ can be easily demonstrated by first performing the integral in Eq. (13.3) along z with the z-axis taken along the solenoid axis. We assume here cylindrical coordinates $(\mathbf{r}_\perp, Z)$.

$$\int d^3r\prime \frac{(\mathbf{r} - \mathbf{r}\prime) \times \mathbf{j}(\mathbf{r}\prime)}{|\mathbf{r} - \mathbf{r}\prime|^3} = \int d^2r_\perp\prime dz\prime \frac{(\mathbf{r}_\perp - \mathbf{r}_\perp\prime) \times \mathbf{j}(\mathbf{r}\prime) + (Z - Z\prime)\hat{\mathbf{Z}} \times \mathbf{j}(\mathbf{r}\prime)}{[|\mathbf{r}_\perp - \mathbf{r}_\perp\prime|^2 + (Z - Z\prime)^2]^{3/2}} \tag{13.9}$$

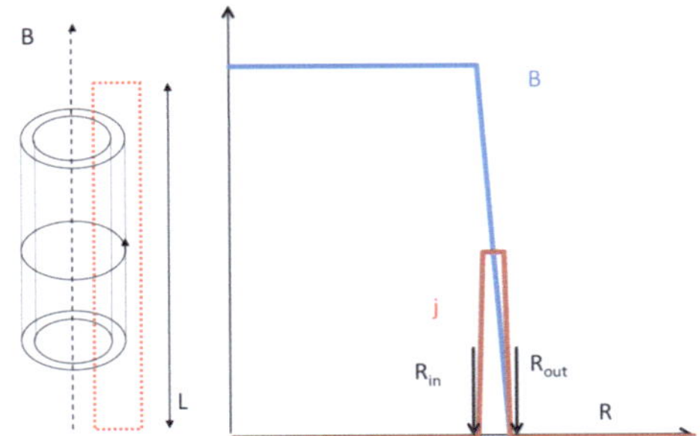

Fig. 13.2 Magnetic field of an infinite cylindrical solenoid with circular cross section. In this case symmetry considerations impose the direction of the magnetic field along z

the second term in the numerator provides a vanishing contribution for an infinite solenoid since it is an odd function in $(Z - Z\prime)$. As to the first term the integral in $Z\prime$ can be performed analytically yielding

$$\int d^3 r\prime \frac{(\mathbf{r} - \mathbf{r}\prime) \times \mathbf{j}(\mathbf{r}\prime)}{|\mathbf{r} - \mathbf{r}\prime|^3} = \int d^2 r_{\perp}\prime \frac{(\mathbf{r}_{\perp} - \mathbf{r}_{\perp}\prime) \times \mathbf{j}(\mathbf{r}\prime)}{|\mathbf{r}_{\perp} - \mathbf{r}_{\perp}\prime|^2} \tag{13.10}$$

Looking at Fig. 13.3 it is possible to show that the integral vanishes for a point external to the solenoid. Indeed, evaluating the contribution to Eq. (13.10) from two elements on the conductor corresponding to the same angle $d\theta$, the contribution to the integral is equal in magnitude and opposite in sign. Therefore the magnetic field outside the solenoid must vanish. Repeating the calculation for a point inside the solenoid, the contributions from the two elements now add together. Note that this argument is independent of the shape of the solenoid cross section.

13.1.3 Field Produced by a Toroidal Solenoid

The toroidal solenoid can be treated similarly to the cylindrical solenoid. Let us consider a cross section of the magnet in the plane $Z = 0$ and assume that the solenoid current flows in the annular regions $R_{MI} \leq R \leq R_{Me}$ and $R'_{Mi} \leq R \leq R'_{Me}$. We take the integration path made of two circles, one inside and the other outside the torus and connected by two straight paths (Fig. 13.4).

The only contribution comes from the path inside the solenoid, yielding

$$\mu_o \int dS\mathbf{n} \cdot \mathbf{j} = \mu_o \pi (R^2 - R^2_{Mi}) j_M = \oint \mathbf{dl} \cdot \mathbf{B} = 2\pi R B_\phi for R_{Me} \geq R \geq R_{Mi}$$

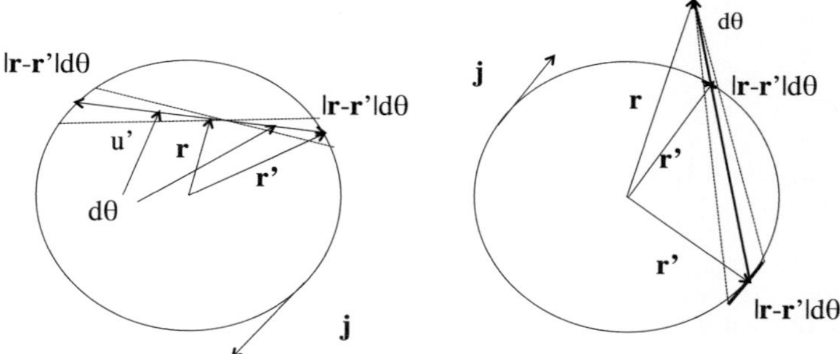

Fig. 13.3 Magnetic field of an infinite cylindrical solenoid. Outside the solenoid the magnetic field is zero because the contributions of the two conductor elements corresponding to the same angle $d\theta$ are equal in magnitude and opposite in sign. Inside the solenoid the two contributions add together

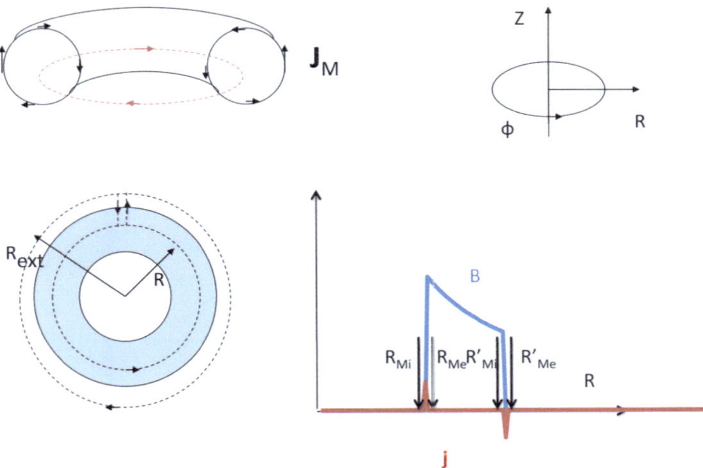

Fig. 13.4 Magnetic field of a toroidal solenoid. Choosing first the equatorial plane, symmetry considerations impose that the magnetic field is in the ϕ direction. The integration path is chosen accordingly and only the inner path provides a finite contribution

$$\mu_o \int dS\mathbf{n} \cdot \mathbf{j} = \mu_o \pi (R_{Me}^2 - R_{Mi}^2) j_M = \oint \mathbf{dl} \cdot \mathbf{B} = 2\pi R B_\phi for R'_{Mi} \geq R \geq R_{Me} \tag{13.11}$$

where we have assumed a constant current density j_M inside the conductor. Thus, inside the solenoid $B_\phi = \mu_o I_M/(2\pi R)$, with $I_M = \pi(R_{Me}^2 - R_{Mi}^2) j_M$. The maximum field is obtained for $R = R_{Me}$. It is convenient to normalise the field to the value B_o at $R = R_o$, yielding

$$B_\phi = B_o \frac{R_o}{R} \tag{13.12}$$

The choice of R_o is arbitrary but we will typically choose R_o to correspond to the middle of the solenoid.

Taking an arbitrary Z and repeating the calculation, the same result is obtained. When performing the integral $\int dS\mathbf{n} \cdot \mathbf{j}$ at a generic Z, upon denoting with χ the angle between the current in the magnet and the vertical direction, the scalar product $\mathbf{n} \cdot \mathbf{j} = j_M \cos\chi$ whereas $dS = dS_o/\cos\chi$ with dS_o the surface element taken in the cross section normal to j. Thus, the toroidal field is independent of Z (see Fig. 13.5).

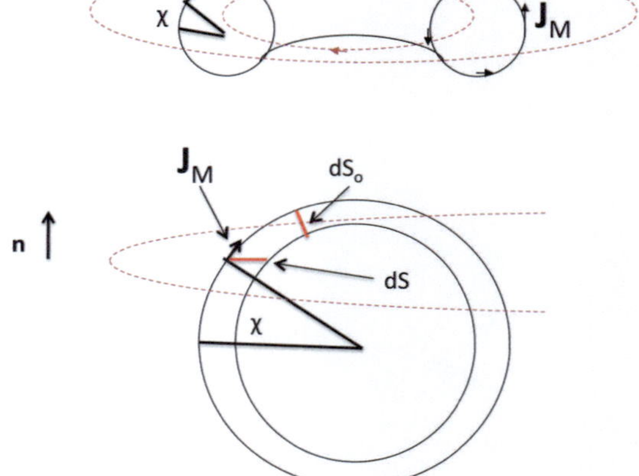

Fig. 13.5 Magnetic field of a toroidal solenoid. In this case the integration path is on a plane different from the equatorial plane. The result is identical to the obtained on the equatorial plane since in the integration the surface element dS is related to the angle χ by $dS = dS_o/\cos\chi$ but the normal component of the current density is given by $\mathbf{n}\cdot\mathbf{j} = j_M \cos\chi$

13.1.4 Field Produced by a Circular Coil

The final relevant case is the field produced by a single coil of radius R_o. Here, symmetry considerations do not help too much. We know from Eq. (13.4) that the vector potential is along the ϕ directions and that $(\partial/\partial\phi = 0)$.

The expression of the ϕ component of the vector potential comes directly from Eq. (13.2).

$$\begin{aligned} A_\phi(\mathbf{r}) &= \frac{\mu_o}{4\pi}\int d^3r\prime\frac{\hat{\boldsymbol{\phi}}\cdot\mathbf{j}(\mathbf{r}\prime)}{|\mathbf{r}-\mathbf{r}\prime|} = \frac{\mu_o}{4\pi}\int dR\prime d\phi\prime dZ\prime\frac{\hat{\boldsymbol{\phi}}\cdot\mathbf{j}(\mathbf{r}\prime)}{|\mathbf{r}-\mathbf{r}\prime|} = \\ &= \frac{\mu_o I}{4\pi}R_o\int d\phi\prime\frac{\cos(\phi-\phi\prime)}{[R_o^2+R^2-2R_oR\cos(\phi-\phi\prime)+(Z-Z_o)^2]^{1/2}} \end{aligned} \tag{13.13}$$

where we have assumed for the current density the form $j_\phi(\mathbf{r}) = I\delta(R-R_o)\delta(Z-Z_o)$ and $\delta(x)$ is the Dirac delta. The solution can be expressed as

$$A_\phi(\mathbf{r}) = \frac{\mu_o I}{\pi k}(\frac{R_o}{R})^{1/2}[(1-\frac{k^2}{2})K(k^2)-E(k^2)] \tag{13.14}$$

with $k^2 = 4R_oR/[(R_o+R)^2+(Z-Z_o)^2]$ and $K(k^2)$ and $E(k^2)$ being complete elliptic integrals (see Appendix C)

$$K(k^2) = \int_0^{\pi/2} d\theta(1-k^2\sin^2\theta)^{-1/2} \quad E(k^2) = \int_0^{\pi/2} d\theta(1-k^2\sin^2\theta)^{1/2} \tag{13.15}$$

It is instructive to look at two limits. Evaluating the vector potential close to the coil $r \equiv [(R - R_o)^2 + (Z - Z_o)^2]^{1/2} \to 0$, i.e. $k^2 \approx 1 - r^2/(4R_oR) \to 1$, the following result is obtained[1]

$$A_\phi \approx \frac{\mu_o I}{2\pi}[\ln(\frac{4}{(1-k^2)^{1/2}}) - 2] \approx \frac{\mu_o I}{2\pi}[\ln(\frac{8(R_oR)^{1/2}}{r}) - 2] \tag{13.16}$$

$$B_R = -\frac{\partial A_\phi}{\partial Z} \approx \frac{\mu_o I}{2\pi r}\frac{Z - Z_o}{r} \tag{13.17}$$

$$B_Z = \frac{1}{R}\frac{\partial(RA_\phi)}{\partial R} \approx -\frac{\mu_o I}{2\pi r}\frac{R - R_o}{r} \tag{13.18}$$

This expression is identical to the field of a straight wire. Specifically, close to the coil the magnetic field is independent of the coil radius R_o. In other words, close to the coil the effect of the curvature of the coil is negligible.

In the limit $(R^2 + Z^2)^{1/2} \to \infty (k \to 0)$, the following result is obtained

$$A_\phi \approx \frac{\mu_o M}{4\pi R}\frac{R^2}{(R^2 + Z^2)^{3/2}} \tag{13.19}$$

$$B_R = -\frac{\partial A_\phi}{\partial Z} \approx \frac{3\mu_o M}{4\pi R}\frac{ZR^2}{(R^2 + Z^2)^{5/2}} \tag{13.20}$$

$$B_Z = \frac{1}{R}\frac{\partial(RA_\phi)}{\partial R} \approx \frac{\mu_o M}{4\pi}\frac{2Z^2 - R^2}{(R^2 + Z^2)^{5/2}} \tag{13.21}$$

with $M = I\pi R_o^2$ being the magnetic dipole moment of the coil.

13.2 Rotational Transform and Large Aspect Ratio Tokamak Equilibrium

We have mentioned in Chap. 11 that plasma confinement along magnetic field lines can be realised by bending the magnetic field lines in such a way that a particle following the field lines never touches the wall of the reaction chamber. Bending the field can be made either around the large circumference of radius R_o *(major radius)* or around the small circumference of radius a *(minor radius)*. The first solution leads to the *toroidal field* whereas the second to the *poloidal field*. As it will be shown in Chap. 17 a toroidal equilibrium in general requires both components. We define the *aspect ratio* A as $A \equiv R_o/a$.

[1] Note that Eq. (13.16) includes the first order correction in r/R_o associated with toroidicity that will be used in Chap. 17.

A magnetic field line produced by the combination of toroidal and poloidal magnetic field wraps around a *magnetic surface*. Except for a numerable set of *rational surfaces* (the origin of the name will be explained below), the line has an ergodic coverage of the surface, i.e. a point following the magnetic field line comes arbitrarily close to any point on the surface. In the case of a rational surface a field line comes back to the point from which it has been originated after a m turns in the poloidal and n turns in the toroidal directions, with m and n integer numbers.

It is natural to ask if the existence of magnetic surfaces is always ensured. The simple answer is that configurations with at least one ignorable coordinate have always magnetic surfaces. This is the case e.g. of the tokamak that will be mostly considered in this paper. Tokamaks are axisymmetric configurations in which the toroidal angle ϕ is an ignorable coordinate: the equilibrium quantities do not depend on ϕ. Configurations such as the stellarator that do not have an ignorable coordinate display a more complex situations with regions where *good* magnetic surfaces exist interspersed with regions of stochastic magnetic field. For the time being we will assume that magnetic surfaces exist.

In a confined system the magnetic surfaces form a set of simply nested surfaces with the innermost surface that degenerates in a line called *magnetic axis*.

The way in which the field line wraps around the magnetic surface is described by the *rotational transform*. This can be defined by looking at the Poincaré plot of the field line crossing a plane at a fixed toroidal angle (Fig. 13.6). After the first turn in the toroidal direction the intersection point between the line and the plane will correspond to a rotation around the magnetic axis by $2\pi\iota_1$, by $2\pi\iota_2$ after the second turn and so on. The limit

$$\iota \equiv \frac{1}{N} lim_{N\to\infty} \sum \iota_k \tag{13.22}$$

defines the rotational transform ι. The value of the rotational transform depends on the specific magnetic surface. Rational surfaces are characterised by rational values of the rotational transform (i.e. $\iota = n/m$ in the example given above).

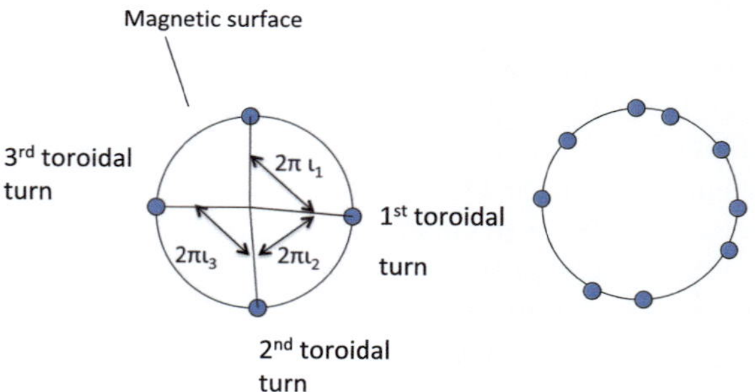

Fig. 13.6 Rotational transform. The Poincare plot of the magnetic field line shows the intersection of the line with a section at fixed toroidal angle

The inverse of the rotational transform $q \equiv 1/\iota$ is the *safety factor*. The safety factor and the rotational transform are equivalent concepts. In the tokamak community it is more common to use the safety factor, in the stellarator community the rotational transform.

The analysis presented above is valid for all the toroidal systems. It can be applied to stellarators, reversed field pinches, spherical tori as well as to tokamaks that will be the configuration mostly considered in this book. In the following we restrict the analysis to *tokamak* equilibria and introduce the large aspect ratio limit. A detailed analysis of the tokamak equilibrium will be given in Chap. 17.

For the purpose of describing the energetics of fusion reactors, it is convenient to introduce a simplified tokamak equilibrium that will be used in the next chapters for illustration. The simplified equilibrium is based on the large aspect ratio limit $A \gg 1$. In most of fusion experiments the aspect ratio evaluated at the plasma edge is typically $A \approx 3$ and the approximation is only marginally satisfied, with the approximation being more accurate closer to the magnetic axis. Nevertheless the large aspect ratio limit is a good approximation to order the various physical effects. Note however that noticeable exceptions exist such as spherical tori ($A \approx 1$ at the plasma edge). Furthermore, although the cross section of fusion plasmas is characterised by an elliptical shape with a finite amount of triangularity, in order to simplify the analysis a simple circular cross section will be used at first.

The large aspect ratio limit is equivalent to replacing the torus with a circular cylinder with periodicity $2\pi R_o$ as shown in Fig. 13.7.

In this limit:

- The variation of the toroidal field across the plasma is small

$$\frac{B_\phi(R_{in}) - B_\phi(R_{out})}{B_\phi(R_{in}) + B_\phi(R_{out})} = O\left(\frac{1}{A}\right) \tag{13.23}$$

- The poloidal field can be evaluated using the cylindrical approximation Eq. (13.6).

The poloidal field is a function of the coordinate r corresponding to the distance r from the cylinder axis that in this case coincide with the magnetic axis. The magnetic field lines are described by the equation

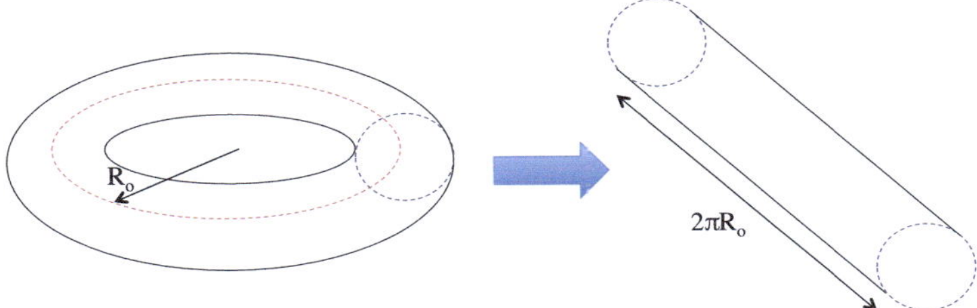

Fig. 13.7 The large aspect ratio limit corresponds to replacing the torus with a periodic cylinder with periodicity $2\pi R_o$

$$\frac{rd\theta}{B_\theta} = \frac{dz}{B_z} \tag{13.24}$$

upon integrating from $z = 0$ and $z = 2\pi R_o$ it is possible to find the poloidal angle made by a magnetic field line during a toroidal turn and therefore the value of the rotational transform

$$\iota = \frac{R_o B_\theta(r)}{r B_z} \tag{13.25}$$

Taking into account Eq. (13.6), it is possible to express the safety factor of the outermost plasma surface ($r = a$) as

$$q_a = \frac{2\pi a^2 B_z}{\mu_o R_o I_p} = \frac{5a(m)^2 B_z(T)}{R_o(m) I_p(MA)} \tag{13.26}$$

13.3 Plasma Shaping

As it will be shown in the following chapters tokamak magnetic surfaces have a shape characterised by a combination of vertical elongation and triangularity rather than by a circular cross section as illustrated above. The shaping is obtained through the superposition of the magnetic fields generated by the various coils and the plasma. To illustrate how shaping can be obtained, we consider, consistently with the large aspect ratio assumption, a pair of straight wires. Two possibilities exist: either the currents are parallel or anti parallel.

In the case of parallel currents the situation is that depicted in Fig. 13.8a with the two wires located at $\pm\mathbf{r}_j$. The two currents point towards the reader. They generate two magnetic fields with each magnetic field increasing as $1/|\mathbf{r} - \mathbf{r}_j|$ as the distance from the source decreases. Close to each wire the field of the other wire is small. Thus the total field is approximately given by the field of the closest wire. Moving along a field line in this region we encircle only one wire. If we move to a distance larger than the separation distance between the two wires, the observed field is, as a first approximation, equivalent to the field of a single wire located between the two wires and with a current given by the sum of the two individual currents. In this region, moving along a field line we encircle *both* wires. Thus, the magnetic field topology close to each wire is different from the topology at a distance much larger than the wire separation. The transition in topology takes place at a characteristic line called the *magnetic separatrix*, with a characteristic eight-shape. The point where the magnetic field lines on the separatrix cross is called the *X-point*. At the $X - point$ the sum of the magnetic fields of the two wires is zero. This can be easily understood for two wires with equal current. In the middle of the segment joining the two currents the azimuthal component of the field generated by each wire is equal in magnitude and opposite to that generated by the other, whereas the radial component for both is zero. As we move away from one wire the field lines are progressively "pulled" by the other wire, signalling an attractive force between the two wires.

The shape of the separatrix can be easily found in terms of the vector potential. In the case of a set of parallel wires along z considered here (the argument can however be generalised to more complex geometries) the vector potential can be written as $\mathbf{A} = A_z \mathbf{z}$ with

$$A_z = \sum_j \frac{\mu_o I_j}{2\pi} \ln \frac{a}{|\mathbf{r} - \mathbf{r}_j|} \tag{13.27}$$

The field line equation can be written as

$$0 = \mathbf{dl} \times \mathbf{B} = \mathbf{dl} \times (\nabla A_z \times \mathbf{z}) = -\mathbf{z}\mathbf{dl} \cdot \nabla A_z \tag{13.28}$$

where we have used the identity $\mathbf{dl} \cdot \mathbf{z} = 0$. From Eq. (13.28) it follows that the magnetic field lines correspond to the curve $A_z(x, y) = cost$. It can be shown that in the case of two equal currents at a distance $2r_1$ the field lines correspond to the curve

$$\frac{r^2}{r_1^2} = -1 + 2\sin^2\theta \pm (1 + 4\sin^4\theta - 4\sin^2\theta + c)^{1/2} \tag{13.29}$$

with the origin of the system taken at the mid point between the wires both placed on the vertical axis at $y = \pm r_1$.

In the case of anti-parallel currents (Fig. 13.8b) the field produced by each wire has the same sign in the region between the wires whereas, far from the wires, the two contributions tends to cancel each other. Two wires with antiparallel currents tend to repel each other. If the wires have different currents an $X - point$ is formed on the other side of the wire with the lower current. The location of the $X - point$ can be found by imposing that the component of the magnetic field normal to the line joining the two wires vanishes.

To elongate the plasma a pair of coils is needed above and below the plasma with a current with the same sign of the plasma current. The combination of the plasma current and of the elongation coils produces a pair of magnetic separatrices and two $X - points$. If there is a perfect up-down symmetry the two separatrices overlap. However, typically the plasma equilibrium is not up-down symmetric. The separatrix closer to the plasma defines the plasma boundary.

The presence of a separatrix is particularly important for the plasma exhaust. Particles crossing the separatrix and moving along field lines deposit their energy on the divertor (see Chap. 18).

As it will be shown later, it is convenient to produce magnetic configuration in which magnetic surfaces have a finite *elongation* κ and *triangularity* δ. The geometric definition of elongation and triangularity is shown in Fig. 13.9. The values of the upper and lower triangularity are usually different but in the rest of this book we will consider only the average triangularity. Elongation and triangularity are in general

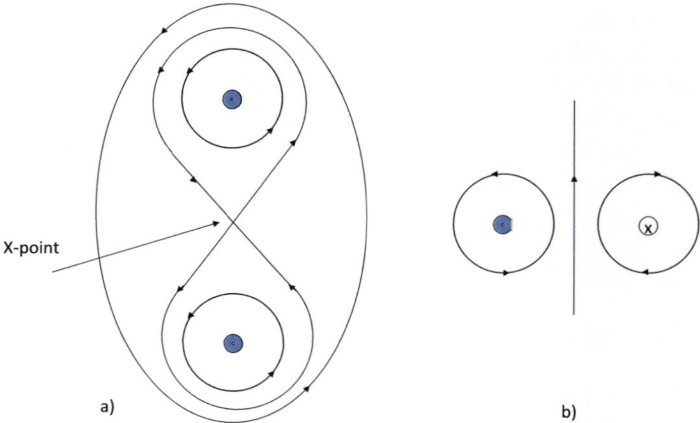

Fig. 13.8 Magnetic field generated by two wires carrying equal currents but with different directions: **a** parallel currents; **b** antiparallel currents

Fig. 13.9 Magnetic surfaces are typically characterised by a finite elongation κ in the vertical direction and a finite triangularity δ. The definition of these quantities is shown

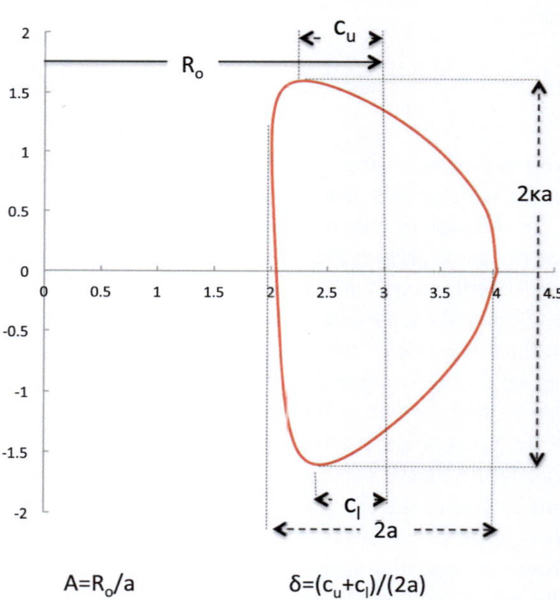

different on different magnetic surface. A special importance is associated to the value of these quantities at the magnetic surface that corresponds to the plasma boundary. Unless otherwise specified, the values of these quantities will correspond to the edge values.

The major radius $R_o = (R_{out} + R_{in})/2$ is defined as the average of the outermost R_{out} and innermost R_{in} radial coordinate of the magnetic surface corresponding to the plasma boundary and the minor radius as $a = (R_{out} - R_{in})/2$. The ratio $A \equiv R_o/a$ is the *aspect ratio* of the magnetic surface. In the following use will be made of the inverse aspect ratio parameter $\epsilon = a/R_o$. The aspect ratio measures how much a magnetic surface differs with respect to a cylindrical surface with the same cross section.

The shape of magnetic surfaces makes the poloidal angle no longer an ignorable coordinate. In particular the poloidal magnetic field becomes a function of both r and θ. This affects the expression of the safety factor /rotational transform that cannot be simply expressed as in Eq. (13.25). It can be shown (see Chap. 17) that the expression of the safety factor can be generalised by replacing the quantities evaluated for a given r with an average over the magnetic surface.

$$q(\rho) = \oint \frac{dl}{2\pi} \frac{B_z(r,\theta)}{RB_\theta(r,\theta)} \tag{13.30}$$

with the average performed on a given magnetic surface with label ρ. For finite aspect ratio equilibria with shaped plasma surfaces this expression of the safety factor must be generally evaluated by numerical methods. Note that as the poloidal magnetic field vanishes at the $X-point$, the safety factor has a singularity on the magnetic surface corresponding to the separatrix. For this reason in plasmas with X-point equilibria it is often defined an empirical value for the edge safety factor called q_{95} that corresponds to the value of q on a magnetic surface placed a slightly more internal location (about 95% of the minor radius—see Chap. 17).

A simple fit of the numerical results is given below

$$\begin{aligned} q_{95} &= \frac{5a(m)^2 B_z(T)}{R_o(m)I_p(MA)} S(\kappa, \delta, \epsilon) = \\ &= \frac{5a(m)^2 B_z(T)}{R_o(m)I_p(MA)} \frac{1+\kappa^2(1+2\delta^2-1.2\delta^3)}{2} \frac{1.17-0.65\epsilon}{(1-\epsilon^2)^2} \end{aligned} \tag{13.31}$$

Although this expression is an empirical fit to numerically generated equilibria some of its features are easily understandable. For example, the increase of q_{95} with elongation κ at fixed plasma current and minor radius reflects the fact that the average poloidal field is smaller on an elliptical surface of semi-axes a and κa as can be seen by applying Stokes theorem

$$4aE(\sqrt(1-\kappa^2)) < B_\theta >= \oint dl B_\theta(r,\theta) = \mu_o I_p \tag{13.32}$$

with E a complete elliptic integral. Taking into account that for $\kappa \approx 1$ $4aE(\sqrt(1-\kappa^2)) \approx 2\pi a(1+(\kappa^2-1)/4) > 2\pi a$ it follows that the average poloidal field is lower than the value obtained for circular magnetic surfaces and therefore the safety factor is correspondingly higher.

For tokamak equilibria the rotational transform takes values of order unity. This implies that the poloidal field is smaller than the toroidal field by a factor ϵ.

$$B_\theta \approx \epsilon B_z \tag{13.33}$$

Finally, as an example of magnetic fields in the poloidal plane that does not produce a rotational transform Fig. 13.10 shows an exapolar field generated by three pairs of straight wires.

13.4 Operational Limits and Safety Factor

The origin of the name *safety factor* comes from the fact that safe tokamak operation (i.e. operation with a low risk of onset of macroscopic instabilities leading to a loss of the equilibrium configuration—a disruption) require

$$q_{95} \geq 3 \tag{13.34}$$

Operation with $2 < q_{95} < 3$ are possible but they have a high occurrence of disruptions. The existence of a lower limit on the edge safety factor implies an upper limit in the plasma current that can be produced in a tokamak. As we will see in the next chapters, tokamak performance increase with plasma current therefore if high plasma currents must be achieved from Eq. (13.31) it follows that at fixed aspect ratio the value Ba must be increased which calls either for high magnetic fields/compact devices or low magnetic field large devices. As we will see in Chap. 16, the magnetic field intensity is limited by the superconducting material and by the electromagnetic stresses in the coils. These considerations have led to the ITER parameters. On the other hand the superconducting technology is evolving and high temperature superconductors operated at $4K$ have pushed the limit to much higher fields.

For a given Ba the maximum plasma current can be optimised by an appropriate choice of the shaping. In particular, elongations in the range $\kappa \approx 1.7$ allow roughly

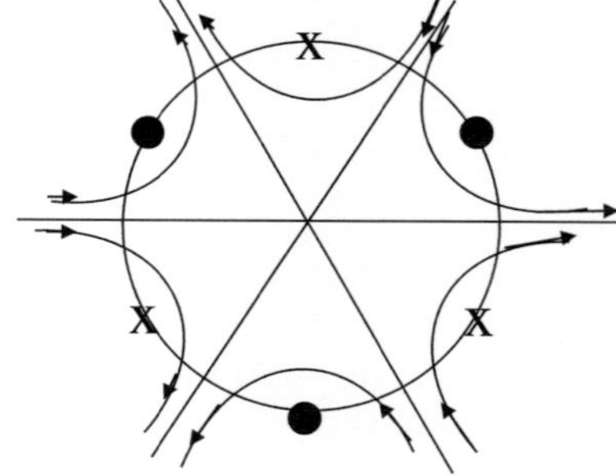

Fig. 13.10 Not all the magnetic fields in the poloidal plane produce a rotational transform. An example is the exapolar field generated by three pairs of straight wires shown here. Moving along a magnetic field line does not allow encircling the center. The field line remains always in the same sector of the plane

to double the maximum current. However, it is not possible to push much further the elongation as the plasma becomes unstable for perturbations corresponding to a vertical displacement.

13.5 Suggestions for Further Readings

The calculation of magnetic fields in complex geometry has been extensively discussed in Refs. [1, 2].

13.6 Exercises

Problem 13.1 Determine the explicit expression of the magnetic field for the current distribution shown in Fig. 13.1

Problem 13.2 Determine the explicit expression of the magnetic field for a finite cylindrical solenoid of length L.

Problem 13.3 Assuming that the poloidal field is generated by a current density along z $j_z = j_o$ for $r \leq a$ and $j_z = 0$ for $r \geq a$, evaluate the radial profile of the rotational transform and of the safety factor.

References

1. H.E. Knoepfel, *Magnetic Fields* (Wiley, 2000)
2. R.J. Thome, J.M. Tarrh, *MHD and Fusion Magnets. Field and Force Design Concepts* (Wiley, New York, 1982)

Chapter 14
Power Balance of Fusion Power Plants

Abstract *In this chapter a simple model for the fusion power plant is presented. The power balance equation is solved to determine the working parameters of a fusion power plant.*

14.1 Introduction

In this chapter we will determine the working conditions for the core plasma in order to ensure a net electricity production of a fusion power plant. To this aim, we first consider the overall balance of a fusion power plant as illustrated in the figure below for a DT reactor.

The power plant will be divided into the following domains:

- The plasma (in red). This is the region enclosed by the first wall and the divertor where fusion reactions occur. Charged particles (the plasma particles and the alpha particles produced in the fusion reactions) are confined whereas neutral particles (the 14.1 MeV neutrons produced by the fusion reaction) escape.
- The blanket (in yellow). This is the region where 14.1 MeV neutrons are absorbed, release their energy and breed tritium. The energy arriving on the first wall as radiation is assumed to contribute to the power balance of the blanket.
- The power-exhaust system (in purple)—to exhaust the power deposited in the divertor. The exhaust power is the net heating power i.e. the heating power reduced by the power radiated in the core.
- The auxiliary systems (in green). These are the systems that convert the thermal energy into electricity part of which is re-circulated in the plant.
- The fuelling system (in blue). This is the system to control the amount of deuterium and tritium in the plasma, extract tritium from the blanket and extract helium from both the plasma and the blanket.

Heating of the plasma region is provided both by the auxiliary power (P_{aux}) and by the alpha particles produced in the fusion reaction. The fusion power P_{fus} produced by the plasma divided by the auxiliary power defines the fusion gain Q

F. Romanelli, *Physics of Nuclear Energy*, Springer Series in Plasma Science and Technology, https://doi.org/10.1007/978-981-97-9609-0_14

$$Q = \frac{P_{fus}}{P_{aux}} \tag{14.1}$$

The 14.1 MeV neutrons are absorbed by the blanket where the reaction between the neutron and lithium, in addition to breeding tritium, produces additional energy (in the case of a pure ^6Li blanket 4.8 MeV for each neutron arriving corresponding to an amplification factor $f_{blanket} = 1 + 4.8/14.1 = 1.34$) and helium. Helium is removed from both the reaction chamber and the blanket. The thermal power P_{thb} and P_{thd} generated respectively in the blanket and the divertor is converted (with thermal conversion efficiencies η_{thb} and η_{thd}, respectively) into electric power P_e, of which a fraction $f_{rec,a}$ is re-circulated in the plant to operate the auxiliary power generator[1]. Assuming a conversion efficiency η_{aux}, the power balance illustrated above can be written as follows (f_{charge} is the fraction of the fusion power in charged products—20% for the DT case and f_R the fraction of heating power radiated to the first wall)

$$\begin{aligned}
P_{fus} &= Q P_{aux} = Q\eta_{aux} f_{rec,a} P_e = Q\eta_{aux} f_{rec,a}(\eta_{thb} P_{thb} + \eta_{thd} P_{thd}) = \\
&= Q\eta_{aux} f_{rec,a} \times \Big\{ \eta_{thb}[(P_{fus} - P_{charge}) f_{blanket} + f_R(P_{charge} + P_{aux})] + \\
&\qquad + \eta_{thd}(1 - f_R)(P_{charge} + P_{aux}) \Big\} = \\
&= Q\eta_{aux} f_{rec,a} \left\{ \eta_{thb} \left[(1 - f_{charge}) f_{blanket} + f_R \left(f_{charge} + \frac{1}{Q} \right) \right] + \right. \\
&\qquad \left. + \eta_{thd}(1 - f_R) \left(f_{charge} + \frac{1}{Q} \right) \right\} P_{fus}
\end{aligned} \tag{14.2}$$

which implies

$$\begin{aligned}
f_{rec,a} = \Bigg(Q\eta_{aux} \Big\{ \eta_{thb} \Big[(1 - f_{charge}) f_{blanket} + f_R \Big(f_{charge} + \frac{1}{Q} \Big) \Big] + \\
+ \eta_{thd}(1 - f_R) \Big(f_{charge} + \frac{1}{Q} \Big) \Big\} \Bigg)^{-1}
\end{aligned} \tag{14.3}$$

The global re-circulating power is the sum of the power needed to operate the auxiliary power generator, the control system and the pumping system. Since the re-circulating power fraction cannot exceed, say, 20%, and assuming, for the sake of illustration, that the dominant contribution is that associated with the auxiliary heating systems, the equation above sets a lower limit to the fusion gain Q. For $f_{blanket} = 1.3$, $f_{charge} = 0.2$, $\eta_{thb} = \eta_{thd} = 40\%$ and $\eta_{aux} = 50\%$ the fusion gain must be at least 20.

---

[1] We assume here that the thermal power of both blanket and divertor are used to produce electricity. In practical situations the temperature of the divertor coolant may be too low and the divertor thermal power can be used in a different way.

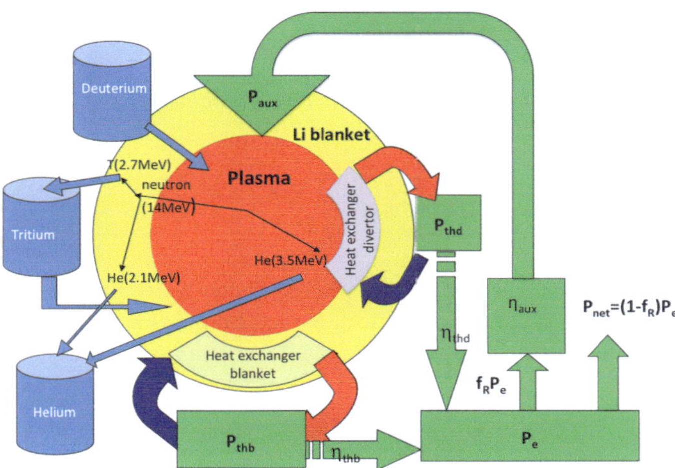

Fig. 14.1 Reactor geometry. Fusion reactions take place in the red core (plasma). Escaping neutrons release their energy in the yellow region (blanket) where tritium is also bred. Additional energy is released by the reactions between neutrons and ^6Li. A heat exchanger extract the heat from the blanket and converts it in electric power. A fraction f of the electric power is used for the auxiliary systems and recirculated in the plasma. He ashes are removed from the plasma and replaced with fresh DT fuel

In the rest of the chapter we will focus on the power balance equation in the reaction chamber (the red part of Fig. 14.1).

14.2 Equilibrium Geometry for the Power Balance Analysis

For the sake of simplicity the complicated equilibrium geometry of typical fusion plasmas will be replaced here with a idealised cylindrical configuration of length $2\pi R_o$ and minor radius a_p as explained in the previous chapter. The radius a_p is chosen in such a way that the total plasma volume is the same

$$V = 2\pi R_o \pi a_p^2 = 2\pi R_o \pi \kappa a^2 \tag{14.4}$$

The power balance equation involves a number of volume integrals. To make the calculation easy all the plasma profiles will be assumed to be given by generalized parabolas

$$n(r) = n_o\Big(1 - \frac{r^2}{a_p^2}\Big)^{\alpha_n} \qquad T(r) = T_o\Big(1 - \frac{r^2}{a_p^2}\Big)^{\alpha_T} \tag{14.5}$$

with the exponents α_n and α_T determined to fit a certain shape.

14.3 Power Balance Condition

The power balance condition is determined from the equation for the time evolution of the plasma internal energy W defined by

$$W = \frac{3}{2} \int dV \sum_i n_i T_i \tag{14.6}$$

with the summation extended to all the species (electrons and ions) and the integral extending over the plasma volume. With the assumptions on the plasma profiles made in the previous section the plasma energy can be written as

$$W = \frac{3}{2} V \frac{(n_{eo} + \sum_i' n_{io}) T_o}{1 + \alpha_n + \alpha_T} \tag{14.7}$$

with the summation $\sum'$ extended only to the ion species. We will assume that all the plasma species have the same temperature T. This is a good approximation for a burning plasma in which the equipartition time (the time required to bring all the species at the same temperature) is short compared with the characteristic time for the heat transport.

Charge neutrality must hold locally. Thus, the values of density at each point satisfy the following condition

$$n_e = \sum_i {}' Z_i n_i \tag{14.8}$$

The internal energy W evolves according to the following equation

$$\frac{dW}{dt} = P_{in} - P_{out} \tag{14.9}$$

with P_{in} the heating power and P_{out} the power lost through radiation, convection and conduction. The explicit form of P_{in} is

$$P_{in} = P_{charge} + P_{aux} + P_{ohm} \tag{14.10}$$

where

- P_{charge} is the fusion power released in the form of charged particles that are confined by the magnetic field and release their energy to the plasma by Coulomb collisions;
- P_{aux} is the auxiliary power;
- P_{ohm} is the power generated by the Joule effect in the plasma.

The explicit form of P_{out} is

$$P_{out} = P_{con} + P_{Brem} + P_{sync} + P_{line} \tag{14.11}$$

where

- P_{con} is the power lost through conduction and convection due to the effect of plasma turbulence and Coulomb collisions;
- P_{Brem} is the power lost via Bremsstrahlung;
- P_{sync} is the power lost by cyclotron and synchrotron radiation;
- P_{line} is the power lost by line radiation due to the transitions between different energy levels of the electrons belonging to the partially ionised species.

14.4 Fusion Power

The fusion power can be expressed in terms of the reactivity introduced in Chap. 3 and discussed for various fusion reactions in Chap. 8. For a generic fusion reaction between species i and j the expression for the fusion power is given by

$$P_{fus} = \int dV \frac{n_i n_j}{1+\delta_{ij}} E_{fus} < \sigma v >_{ij} \tag{14.12}$$

with the integral extending over the plasma volume. The symbol δ_{ij} is the Kronecker delta

$$\delta_{ij} = 1 \quad if\ i = j \qquad \delta_{ij} = 0 \quad if\ i \neq j \tag{14.13}$$

The factor $(1+\delta_{ij})^{-1}$ avoids counting twice the reaction between particles a and b and that between b and a. The fusion power entering in the energy balance equation Eq. (14.9) is only that associated with charged particles

$$P_{charge} = \int dV \frac{n_i n_j}{1+\delta_{ij}} E_{charge} < \sigma v >_{ij} \tag{14.14}$$

For example, for the DT reaction $E_{fus} = 3.5\,\mathrm{MeV} + 14.1\,\mathrm{MeV} = 17.6$ MeV whereas $E_{charge} = 3.5$ MeV. As shown in Chap. 10, the energy of charged particles is mostly absorbed by the electrons.

The calculation of the fusion power requires the explicit evaluation of the integral in Eq. (14.14). In general, this has to be done numerically. In the case in which the main contribution arises from regions with temperature in the range $10keV$ to $20keV$ the Maxwellian reactivity can be replaced with the simple expression $< \sigma v > \approx cT^2$ and the integral can be performed analytically yielding

$$P_{charge} = \frac{n_{io}n_{jo}}{1+\delta_{ij}} E_{charge} \frac{cT_o^2}{1+2\alpha_n+2\alpha_T} V \quad (14.15)$$

This is typically a good approximation for the DT reaction. In general we will approximate the fusion power as follows

$$P_{charge} = \frac{n_{io}n_{jo}}{1+\delta_{ij}} E_{charge} \frac{<\sigma v>_o}{1+2\alpha_n+2\alpha_T} V \quad (14.16)$$

with $<\sigma v>_o$ the reactivity evaluated with the temperature on axis, keeping in mind the limit of validity of the numerical factor associated to the profile shape.

14.5 Auxiliary Power

There are two main methods to heat the plasma from outside:

- the use of ion beams of high energy; and
- the use of electromagnetic waves. Three frequency ranges can be identified

  Ion cyclotron waves (ICW);
  Lower hybrid waves (LHW);
  Electron cyclotron waves (ECW).

For typical tokamak parameters ICWs fall in the radio frequency range (10–100 MHz), LHWs in the GHz range and ECW in the 100–200 GHz range. In order to evaluate the auxiliary heating power sophisticated codes must be used that determine the absorbed power on the basis of the plasma conditions. For the purpose of solving the energy balance equation P_{aux} will be considered a given number.

14.6 Ohmic Power

Fusion plasmas have a low but finite electrical resistivity. The Joule dissipation leads to a net input power to the electrons.

The ohmic power can be calculated from the following expression

$$P_{ohm} = \int dV \eta j^2 \quad (14.17)$$

with η being the plasma resistivity and j the toroidal plasma current driven by the toroidal electric field induced by the transformer.

The volume integral needs to be computed taking into account the exact expression of the plasma resistivity that has to include the effect of trapped particles as it will be explained in the next chapter. A simple approximation to the exact volume integral can be found e.g. in Ref. [1]

$$P_{ohm}(MW) = 0.15 I_p(MA)^2(1 - 0.14\,A)\frac{Z_{eff}}{< T(keV) >^{1.5}}\frac{2\pi R(m)}{\pi\kappa a^2} \tag{14.18}$$

14.7 Conduction and Convection

The heat transport in magnetic confinement devices is mainly associated with the effect of turbulence generated by the free-energy sources always present in a confined system. Significant progress has been made on the experimental side to reduce the heat and particle losses by accessing regimes with improved thermal confinement. The basic tokamak regime is the so called $L - mode$ (for low confinement mode). Its interest has been limited since it has been soon realised that a tokamak power plant based on such a regime of operation would be very large and expensive. In the $'80s$ it was discovered a high-confinement regime ($H - mode$) [2] corresponding to a local reduction of the heat flux at the plasma separatrix. This has been the first example of a number of self-organisation phenomena later observed in tokamaks leading to the formation of local transport barriers. The formation of a transport barrier leads to an increase of energy and particles in the region inside the barrier. In particular the reproduction of the H-mode regime has become so systematic that this regime is now considered the reference for ITER operation (see [3] for a review). In order to access H-mode conditions the power flowing via conduction and convection through the separatrix must be above a threshold value [4]

$$P_{L-H} = 2.15 \times e^{\pm 0.107} n_e(10^{20}m^{-3})^{0.782\pm 0.037} B_\phi(T)^{0.772\pm 0.031} \times \\ \times a(m)^{0.975\pm 0.08} R_o(m)^{0.999\pm 0.101} \tag{14.19}$$

The origin of this threshold has been the subject of several theoretical models. Although there is today a general consensus on the link to between the access to H-mode conditions and the onset of self-organised sheared flows at the plasma periphery thanks to the detailed measurements that have been made in the past twenty years, there is not yet evidence of this mechanism via first-principle numerical simulations.

In parallel, the theoretical understanding has substantially progressed. Today several codes have been developed to evaluate under different assumptions the turbulent heat flux (see e.g. Ref. [5–8]).

Although the understanding of plasma turbulence has made significant advances over the last few decades, for the purpose of our analysis, the effect of plasma turbulence on energy conduction and convection is captured through the so called *energy confinement time* τ_E.

The physical meaning of the energy confinement time can be understood by the following analogy. Let us consider a house that during winter is maintained at a certain temperature by its heating system and assume to stop heating and measure the decay in time of the internal temperature. The characteristic time of the decay is the energy confinement time of the house. It is important to stress that the energy confinement time has nothing to do with the age of the house but simply with the level of thermal insulation. The energy confinement time of a plasma can be defined in a similar way and, as in the example, it has nothing to do with the time a plasma is maintained. Plasmas lasting up to few hours have been already produced and controlled [9] but the energy confinement time was in the range of ms. The energy confinement time simply measures the degree of thermal insulation of the plasma. In terms of τ_E the energy lost through conduction and convection P_{cond} can be expressed as

$$P_{cond} = \frac{W}{\tau_E} \tag{14.20}$$

The energy confinement time can be experimentally determined as a function of the plasma parameters from the power balance equation and the measurement of the plasma energy. Assuming a non-nuclear plasma (no fusion reactions) and neglecting for the sake of simplicity the power lost through radiation and the ohmic power, the power balance condition yields

$$P_{aux} = P_{cond} = \frac{W}{\tau_E} \tag{14.21}$$

Thus, by knowing P_{aux} and measuring W, τ_E can be determined for different values of the plasma parameters. In this way it is possible to fit the data with a scaling law for τ_E i.e. an empirical expression of the form e.g.

$$\tau_E = C P_{aux}^{\alpha} I_p^{\beta} n^{\gamma} R^{\delta} \tag{14.22}$$

with C a constant. As the plasma dimensions cannot be easily changed on a single device, the dependence on the major radius R will require the comparison of machines of different size.

The exponents of this scaling law can be further constrained in order to satisfy the condition that the expression above obeys the invariance properties associated with the equations of plasma physics (or a more restricted model) [10, 11]. The idea is to exploit the invariance of the equations under scale transformations that can be illustrated using as an example the collision less Boltzmann equation

$$\frac{\partial f_i}{\partial t} + \mathbf{v} \cdot \nabla f_i + \frac{q}{m}(\mathbf{E} + \mathbf{v} \times \mathbf{B}) \cdot \nabla_v f_i = 0 \tag{14.23}$$

with the electric field determined by the charge neutrality condition

$$\sum_i q_i \int d^3 v f_i = 0 \tag{14.24}$$

A scaling transformation of the form

$$f \to \alpha_1 f \quad v \to \alpha_2 v \quad x \to \alpha_3 x \quad B \to \alpha_4 B \quad t \to \alpha_5 t \quad E \to \alpha_6 E \tag{14.25}$$

leaves invariant the model equations Eqs. (14.23) and (14.24) only in three cases

$$f \to \alpha_1 f$$
$$\begin{aligned} v \to \alpha_2 v \quad B \to \alpha_2 B \quad t \to \alpha_2^{-1} t \quad E \to \alpha_2^2 E \\ x \to \alpha_3 x \quad B \to \alpha_3^{-1} B \quad t \to \alpha_3 t \quad E \to \alpha_3^{-1} E \end{aligned} \tag{14.26}$$

The heat flux, defined as

$$\mathbf{Q} = \sum_i \int d^3 v f_i v^2 \mathbf{v} \tag{14.27}$$

the density n and the temperature T, scale as

$$Q \to \alpha_1 \alpha_2^6 Q \quad n \to \alpha_1 \alpha_2^3 n \quad T \to \alpha_2^2 T \tag{14.28}$$

If we look for a scaling law for the heat flux of the form $Q \propto n^p T^q B^r a^s$ the invariance requirements restricts the exponent to the following combination

$$p = 1 \quad 30 + 2q + r = 6 \quad r - s = 0 \tag{14.29}$$

or $Q \propto na^3 B^3 [T/(aB)^2]^q$. Thus, the energy confinement time $\tau_E \propto naT/Q$, with a the minor radius, scales as

$$B\tau_E \propto \left(\frac{T}{a^2 B^2}\right)^q \tag{14.30}$$

and, taking into account the power balance equation

$$\frac{nTa^3}{\tau_E} \propto P_{aux} \tag{14.31}$$

we can obtain a simple scaling for the energy confinement time

$$\tau_E \propto \left(\frac{n}{P_{aux}}\right)^{\frac{-q}{1-q}} a^{-\frac{-5q}{1-q}} B^{\frac{1+2q}{1-q}} \tag{14.32}$$

Equation (14.32) defines a relation between of the various exponents of the scaling law in Eq. (14.22) due to the constraints arising from the underlying model. For the case illustrated here the constraints are strong and only a single parameter (the exponent q) is free. For more general plasma models the constraints are less severe. For example, for the plasma model described by the Boltzmann equation and the complete set of Maxwell equation there is only one transformation that leaves the equations invariant. As a consequence, the energy confinement time has three free exponents

$$B\tau_E \propto (na^2)^{q_1}(Ta^{1/2})^{q_2}(Ba^{5/4})^{q_3} \tag{14.33}$$

Although the use of the energy confinement time to characterise the energy transport may sound simple, this approach requires:

- measurements of τ_E in the plasma regimes characterised by the same underlying physics;
- the assumption that when extrapolated to power plant conditions the underlying physics will remain the same.

The effort to determine a scaling expression for the energy confinement time has been one of the main successes of the ITER R&D programme over the 1990′s. It led to the formulation of the reference scaling law for the ITER design (the so-called $ITER98(y,2)$ scaling) given below [12] and compared in Fig. 14.2with the results of the tokamaks at that time

$$\tau_{ITER98(y,2)}(s) = 0.0562 I_p(MA)^{0.93} B_\phi(T)^{0.15} P_{heat}(MW)^{-0.69} \times n(10^{19}m^{-3})^{0.41} M^{0.19} R_o(m)^{1.97} \epsilon^{0.58} \kappa^{0.7} \tag{14.34}$$

with P_{heat} the total heating power (i.e. the sum of the auxiliary power and the fusion power deducted the power lost for radiation) and M the average mass number. The agreement between the experimental results and the scaling law is remarkable (Fig. 14.2). It covers three orders of magnitude in experimental data with ITER being

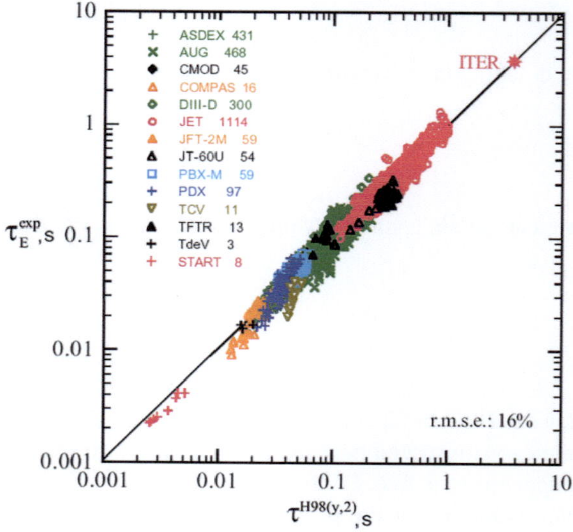

Fig. 14.2 ITER98(y,2) scaling [12]. The experimental value of the energy confinement time is shown on the vertical axis whereas the value coming from the scaling law Eq. (14.34) is shown on the horizontal axis. The ITER value foreseen using the scaling law is also shown. Reprinted from Ref. [12] with permission from IAEA

only a factor ≈ 3 above the JET points (corresponding to the largest energy confinement times achieved so far). In terms of dimensionless parameters the scaling law can be expressed as

$$\tau_{ITER98(y,2)}(s) \propto \tau_B \rho_*^{-0.70} \beta^{-0.90} \nu_*^{-0.01} \tag{14.35}$$

with $\tau_B \equiv a^2/D_B$ and $D_B \equiv T/(eB)$ the Bohm diffusion coefficient. In Eq. 14.35 $\rho_* = \rho_{Larmor}/a$, $\beta = 2\mu_o nT/B^2 and \nu_* = R\nu_{ee}/v_{the}$. However, it should be noted that the strong decrease of τ_E with β is not supported by the experimental findings of single device studies carried out e.g. on JET and DIIID [13]. This may not have a significant impact on the ITER predictions provided the range of beta is similar (actually the outcome of the JET/DIIID study would lead to better performance in ITER) however it tells that scaling laws should be used with a grain of salt!

The origin of the ITER98(y,2) scaling can be traced back to the characteristics of the small-scale turbulence responsible for the energy transport in fusion plasmas. Turbulence is characterized by a radial correlation length of the order of the ion Larmor radius and by a autocorrelation time of the order of the ion transit frequency. Therefore the thermal diffusivity χ is expected to scale as [14]

$$\chi \propto \rho_i^2 \frac{v_{ti}}{a} \tag{14.36}$$

In general, the thermal diffusivity will be also a function of a number of geometrical parameters such as the inverse aspect ratio ϵ and the elongation κ. Other dimensionless parameters can enter as the plasma collisionality and the plasma beta. For the sake of simplicity we will neglect such a dependence and replace the plasma dimension with a everywhere. Upon substituting in the energy balance equation, a simple relation for the plasma temperature is obtained

$$\frac{\chi n T}{a^2} \propto \rho_i^2 \frac{v_{ti}}{a} \frac{nT}{a^2} \propto \frac{P_{heat}}{a^3} \tag{14.37}$$

Thus, a scaling for plasma temperature in terms of the other parameters can be determined, which, in turn, yields a scaling for the energy confinement time

$$\tau_E \propto \frac{nTa^3}{P_{heat}} \propto B^{0.8} P_{heat}^{-0.6} n^{0.6} a^3 \tag{14.38}$$

Equation(14.38) displays several similarities with the ITER98(y,2) scaling. To see this, we note that the ITER98(y,2) scaling can be written in a slightly different form by taking the safety factor $q \propto Ba^2/I_p R$ as constant. Upon replacing I_p with Ba, Eq. (14.34) yields

$$\tau_{ITER98(y,2)} \propto B^{1.08} P_{heat}^{-0.69} n^{0.41} a^{2.9} \tag{14.39}$$

Therefore the core properties of the ITER98(y,2) scaling, namely the degradation of confinement with power, the linear dependence on the toroidal field (at fixed q) and the proportionality to the plasma volume can indeed be understood on the basis of the fundamental characteristics of plasma turbulence. This has motivated a large effort in the development of first principle turbulence codes.

14.8 Bremsstrahlung

The calculation of the stopping power due to Bremsstrahlung has been made in Chap. 10.

In order to have the radiation losses in a plasma it is convenient to start from Eq. (10.27) and sum over all the electron and ion distribution

$$P_{brem} = \int dV 2\pi \int dv_e dv_i f(v_e) f(v_i) E_{rad} v b db \tag{14.40}$$

Due to the different velocity of electron an ions the relative velocity v can be approximated with the electron velocity v_e. The integral over b can be made as shown in Chap. 10 and the expression for the power density becomes

$$\begin{aligned} P_{brem} &= \int dV \frac{\pi}{2} \frac{Z^2 e^6}{96\pi^3 \epsilon_o^3 c^3 m_e^2} 2\pi \int dv_e dv_i f(v_e) f(v_i) \frac{2\pi m_e v_e}{h} = \\ &= \int dV \frac{\pi}{2} \frac{e^6}{96\pi^3 \epsilon_o^3 c^3 m_e} \frac{4\pi^2}{h} Z^2 n_i \int dv_e f(v_e) v_e = \\ &= \int dV \frac{2^{1/2}}{24\pi^{1/2}} \frac{e^6}{\epsilon_o^3 c^3 h m_e^{3/2}} Z^2 n_i n_e T_e^{1/2} \end{aligned} \tag{14.41}$$

If multiple ion species are present in the plasma, the summation over the ion species yields the Z_{eff} value. Thus the final expression for the Bremsstrahlung

$$P_{brem} = \int dV \frac{2^{1/2}}{24\pi^{1/2}} \frac{e^6}{\epsilon_o^3 c^3 h m_e^{3/2}} Z_{eff} n_e^2 T_e^{1/2} \tag{14.42}$$

Upon using the expressions for the plasma profiles, the power lost may be expressed in practical units as follows

$$P_{brem}(MW) = k_{Br} Z_{eff} \frac{n_{eo}(10^{20} m^{-3})^2 T_{eo}(eV)^{1/2}}{1 + 2\alpha_n + 0.5\alpha_T} V \tag{14.43}$$

with $k_{Br} = 1.69 \times 10^{-4}$.

14.9 Cyclotron and Synchrotron Radiation

Even in the absence of collisions charged particles are accelerated due to the presence of the magnetic field. One would be tempted to evaluate the radiated power by replacing in Eq. (4.57) the acceleration with $a \approx v_{the}\Omega_e$. This would lead to a very large value that would make cyclotron losses already large in present laboratory plasmas. The correct evaluation [15, 16] is not reported here and takes into account the re-absorption of the emitted cyclotron power by the plasma that makes this contribution almost negligible in the power balance equation (but very useful for diagnostic reasons!).

This loss channel becomes significant only at very high temperature and must be included when dealing with advanced fusion reactions.

14.10 Ignition and Ideal Ignition Temperature

Upon solving the energy balance equation it is possible to determine the fusion power as a function of the auxiliary power. Their ratio is the fusion gain Q. In specific cases the energy balance equation admits stationary solutions ($dW/dt = 0$) with finite fusion power and no auxiliary power. This regime corresponds to the case in which the alpha particle heating is sufficient to heat the plasma and is called ignition. It should be noted that the term ignition is used in the literature with different meanings especially when dealing with transient conditions. We will restrict the use of ignition only to stationary conditions.

The most favourable situation to achieve ignition occurs when the conduction and convection losses are negligible (i.e. $\tau_E = \infty$). Upon neglecting the ohmic power and the cyclotron/synchrotron losses, the ideal ignition condition is obtained from the balance between the fusion power in the form of charged particles and Bremsstrahlung

$$1.6 \times 10^{15} \frac{n_i(10^{20}m^{-3})n_j(10^{20}m^{-3})}{1+\delta_{ij}} E_{charge}(MeV) < \sigma v >_{ij} (m^3/s) =$$

$$= k_{Br} Z_{eff} n_e (10^{20}m^{-3})^2 T_e(eV)^{1/2} \tag{14.44}$$

It should be noted that both members are quadratic in the plasma density. As a result, the ideal ignition temperature is independent of the density. The ideal ignition temperature T_{ideal} can be determined easily through graphic methods as shown in Fig. 14.3.

For the DT reaction $T_{ideal} = 4.3$ keV whereas for the D^3He reaction $T_{ideal} = 30$ keV. Below these values Bremsstrahlung losses are always larger than the fusion power.

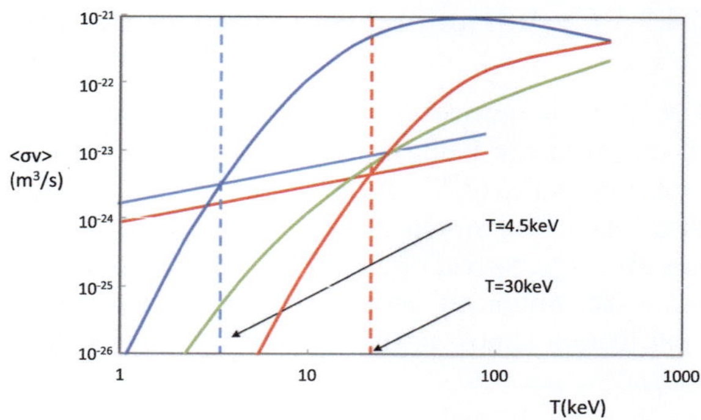

Fig. 14.3 Ideal ignition temperature. The ideal ignition temperature is the point at which the curve of the fusion reactivity crosses the line corresponding to Bremsstrahlung losses. The blue line corresponds to the DT reaction, the red line to the D^3He reaction

14.11 The Lawson Criterion

The power balance equation can be cast in a simple form involving only two quantities: the product $n\tau_E$ and the plasma temperature T [17]. Taking for the sake of simplicity the case $\alpha_n = \alpha_T = 0$ the energy balance equation Eq. (14.9) can be rearranged in the following form (see Fig. 14.4)

$$n_e\tau_E = \frac{3T}{2} \frac{1 + \sum_j c_j}{c_1 c_2 E_{charge} < \sigma v > -k_{Br} Z_{eff} T^{1/2}} = f(T) \tag{14.45}$$

the function $f(T)$ has a vertical asymptote for $T = T_{ideal}$ and it increases monotonically for $T \to \infty$. Therefore a minimum value exists corresponding to the minimum requirement for the product $n\tau_E$ in order to achieve ignition. The temperature at the minimum is the optimal temperature. As T approaches T_{ideal} the requirement on the product $n\tau_E$ becomes very stringent as ignition can be achieved only for negligible conductive losses. For temperatures larger than the optimal value, Bremsstrahlung

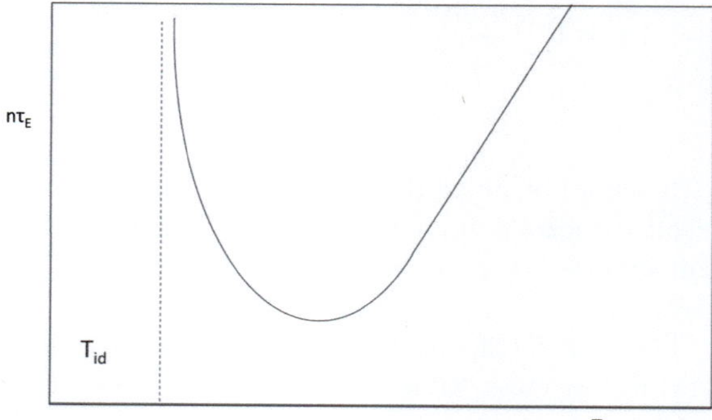

Fig. 14.4 Lawson criterion. The value of the product $n\tau_E$ is shown versus temperature. The optimal operation conditions correspond to the minimum of the curve

can be neglected in the denominator of Eq. (14.45) but the energy content of the plasma increases more rapidly than the fusion reactivity and again the required $n\tau_E$ value increases.

The ignition requirement is sometimes expressed in the form of the triple product $n\tau_E T$. Indeed, near the optimal temperature the fusion reactivity is well approximated by $< \sigma v > \approx < \sigma v > (T_{opt})(T/T_{opt})^2$. Upon neglecting Bremsstrahlung we obtain

$$n_e \tau_E T = \frac{3T_{opt}}{2} \frac{1 + \sum_j c_j}{c_1 c_2 < \sigma v > (T_{opt}) \frac{E_{charge}}{T_{opt}}} \tag{14.46}$$

The ignition condition expressed in the form of the triple product is meaningful only in the range of temperature in which the fusion reactivity has a quadratic dependence on temperature and Bremsstrahlung can be neglected ($T_{ideal} \to 0$).

14.12 Solution of the Power Balance Equation

In order to solve the power balance equation we need to determine the energy confinement time consistently with the amount of the plasma heating power. A simple way is to proceed as follows (see Exercises)

- fix a value for the central temperature;
- determine the corresponding value of the fusion, ohmic and radiated power;
- determine the net heating power;
- determine the value of τ_E and of the conduction/convection losses W/τ_E;
- determine the algebraic sum of the various power terms;
- repeat for a different temperature value until the algebraic sum is zero (power balance condition).

This procedure can be easily implemented with an Excel file.

The Excel file can then be used to analyze the behaviour of the solution in the (n, T) domain. The curves corresponding to the thermal equilibrium are shown for different P_{aux} values.

The r.h.s. of Eq. (14.9) as a function of temperature diverges as $T^{-3/2}$ for $T \to 0$ due to the ohmic heating and is always negative for sufficiently large temperatures. Therefore at least an equilibrium solution exists. More than one solution can exist if the r.h.s. is a non monotonic function of T as e.g. shown in Fig. 14.5. If three equilibrium solutions are found, two corresponds to stable equilibria whereas the third is an unstable solution. If the plasma is perturbed around such a point, it ends up in one of the two stable equilibria.

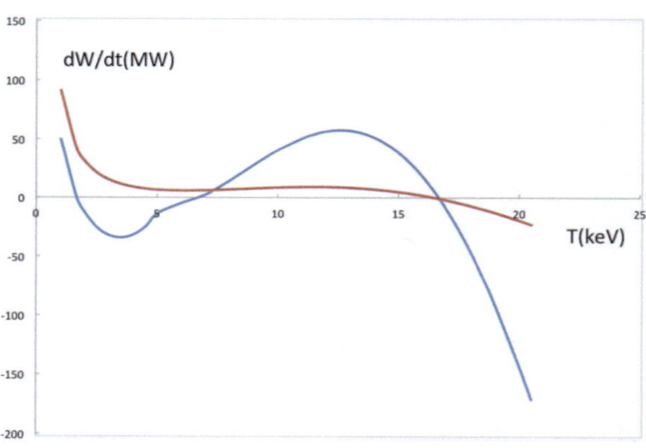

Fig. 14.5 Power balance. The quantity dW/dt is plotted versus the central temperature T. Two cases are shown with one (red curve) and three (blue curve) equilibrium points respectively. In the latter case the two extreme points are stable whereas the intermediate one is unstable

14.13 Operational Limits

The analysis presented in this chapter allows to determine the output of a fusion power plant given a set of input parameters. However, the outcome of the analysis must be consistent with a number of operational constraints as shown below.

- *Safety factor limit*—In Chap. 13 we have already seen that safe tokamak operation correspond to $q_{95} > 3$ (see Eq. (13.34)).
- *Vertical stability*—As mentioned in Chap. 13 it is not possible to exploit the beneficial effect of plasma elongation to arbitrary values of κ. If the plasma moves towards one of the coils that produce the elongation the attractive force increases (and the attractive force of the other coil decreases) leading to an instability. For finite aspect ratio plasmas a natural elongation exists below which the plasma is stable. Such a natural elongation increases as the aspect ratio decreases.
  Passive conducting structures around the plasma slow down the movement and allow the external coils to bring back the plasma in the correct position via active control algorithms provided the onset of the instability is detected at a sufficiently early stage.
  Therefore, the parameters of a tokamak fusion power plant must be carefully chosen without pushing simultaneously high values of elongation and aspect ratio.
- *Density limit*—Tokamaks exhibit a limit in the maximum line averaged density $\bar{n}_e$. This limit, known as *Greenwald limit* [18] is expressed as

$$\bar{n}_e = \frac{I_p(MA)}{\pi a(m)^2} \times 10^{20} m^{-3} \tag{14.47}$$

  The origin of the Greenwald limit is identified with the edge dynamics of the plasma. Interestingly, stellarators seem not to exhibit such a limit[2].

[2] In the case of stellarator (that have no net current in the plasma) the current must be replaced by an expression involving the poloidal field.

- *H-mode power threshold*—In order to access H-mode conditions the power flowing via conduction and convection through the separatrix must be above a threshold value given by Eq. (14.19).
- *Beta limit*—For sufficiently high value of the plasma pressure new classes of instabilities become likely. These instabilities strongly depend on the details of the safety factor profile and must be carefully investigated using sophisticated stability codes. For the purpose of the 0-D analysis described above it is convenient to formulate the limit as follows (see e.g. [19])

$$\frac{2\mu_o < p >}{B^2} \leq 0.01\beta_N \frac{I_p(MA)}{a(m)B(T)} \tag{14.48}$$

with β_N a number that depends on the safety factor profile and the aspect ratio. Values of $\beta_N = 2$ are considered safe. However, regimes with a substantial contribution of self-generated current require larger values.

To conclude this section, we underline the need for a sound design not to push all the parameters to their limits and, above all, to check the consistency of the assumptions with the appropriate numerical tools. To design a compact power plant by pushing all the parameters is easy but it has very little link with the everyday experience of tokamak operation!

14.14 Exercises

Problem 14.1 Prepare an excel file to solve the energy balance equation.

Problem 14.2 Using the excel file discuss the dependence of the equilibrium temperature on the auxiliary heating power for ITER parameters.

Problem 14.3 Using the excel file discuss the dependence of the volume averaged equilibrium temperature on the volume averaged density for ITER parameters.

Problem 14.4 Determine the ideal ignition temperature of a plasma made mostly of deuterium with a concentration $c_T = 10^{-2}$ of tritium. Quantify the relative importance of DD and DT reactions.

Problem 14.5 Determine the triple product of a DT plasma and compare with the value expected in ITER.

References

1. N. Uckan, ITER physics design manual 1989 ITER documentation series n.10, IAEA Vienna (1990)

2. F. Wagner et al., Regime of improved confinement and high beta in neutral-beam-heated divertor discharges of the ASDEX tokamak. Phys. Rev. Lett. **49**, 1408 (1982)
3. E. Doyle, et al., Chapter 2: plasma confinement and transport. Nucl. Fusion **47**, S18–S127 (2007)
4. Y.R. Martin, et al., Power requirement for accessing the H-mode in ITER. J. Phys.: Conf. Ser. **123**, 012033
5. C. Bourdelle et al., A new gyrokinetic quasilinear transport model applied to particle transport in tokamak plasmas. Phys. Plasmas **14**, 112501 (2007)
6. F. Jenko et al., Electron temperature gradient driven turbulence. Phys. Plasmas **7**, 1904 (2000)
7. M. Kotschenreuther, G. Rewoldt, W.M. Tang, Comput. Phys. Commun. **88**, 128 (1995)
8. A.G. Peeters et al., The nonlinear gyro-kinetic flux tube code GKW. Comput. Phys. Commun. **180**, 2650–2672 (2009)
9. H. Zushi et al., Nucl. Fusion **43**, 1600–1609 (2003)
10. B. Kadomtsev, So. Phys. J. - Plasma Phys. **1**, 295 (1975)
11. J.W. Connor, J.B. Taylor, Nucl. Fusion **17**, 1047 (1977)
12. ITER physics expert groups on confinement and transport and confinement modelling and database, ITER physics basis editors and ITER EDA. Nucl. Fusion **39**, 2175 (1999)
13. E.J. Doyle et al., Nucl. Fusion **47**, S18 (2007)
14. F. Romanelli, W.M. Tang, R.B. White, Anomalous thermal confinement in ohmically heated tokamaks. Nucl. Fusion **26**, 1515 (1986)
15. N.A. Trubnikov, in *Reviews of Plasma Physics*, vol. 7, ed. by A.M. Leontovich (Consultants Bureau, New York, 1979), p. 345
16. I. Fidone, R.L. Meyer, G. Giruzzi, G. Granata, Phys. Fluids B **4**, 4051 (1992)
17. J.D. Lawson, Proc. Phys. Soc. B **70**, 6 (1957)
18. M. Greenwald, Density limits in toroidal plasmas. Plasma Phys. Control Fusion **44**, R27–R53 (2002)
19. T.C. Hender, et al., Chapter 3: MHD stability, operational limits and disruptions. Nucl. Fusion **47** S128–S202 (2007)

Chapter 15
Particle Motion in Electric and Magnetic Fields

Abstract *In this chapter we investigate the orbits of particles in assigned electric and magnetic fields. The secular drifts associated with nonhomogeneous magnetic fields are classified. The particle orbits for a tokamak configuration are analyzed. The implication on particle confinement is discussed.*

A charged particle moves under the influence of the Lorentz force. In the non-relativistic limit the equation of motion is given by

$$m\frac{d\mathbf{v}}{dt} = m\frac{d^2\mathbf{x}}{dt^2} = e\mathbf{E}(t, \mathbf{x}) + e\mathbf{v} \times \mathbf{B}(t, \mathbf{x}) \tag{15.1}$$

This is a nonlinear differential equation and the solution is in general very complicated. However some general consequence can be easily deduced. If both sides of the equation are multiplied by $\mathbf{v} = d\mathbf{x}/dt$

$$m\mathbf{v} \cdot \frac{d\mathbf{v}}{dt} = \frac{d}{dt}\frac{mv^2}{2} = e\mathbf{v} \cdot \mathbf{E}(t, \mathbf{x}) \tag{15.2}$$

which expresses the fact that in the absence of electric fields the particle kinetic energy is conserved.

15.1 Particle Orbits in Static and Homogeneous Electric and Magnetic Fields

Let us consider first the case $\mathbf{E} = 0$ and $\mathbf{B}$ constant in time and homogeneous in space. The solution of Eq. (15.1) can be easily obtained by separating the velocity into its components along $\mathbf{B}$ ($v_{||}$) and perpendicular to $\mathbf{B}$ ($v_\perp$)yielding

$$\mathbf{v}(t) = v_{||}\frac{\mathbf{B}}{B} + v_\perp[\cos(\Omega t + \phi)\hat{\mathbf{x}} + \sin(\Omega t + \phi)\hat{\mathbf{y}}] \tag{15.3}$$

F. Romanelli, *Physics of Nuclear Energy*, Springer Series in Plasma Science and Technology, https://doi.org/10.1007/978-981-97-9609-0_15

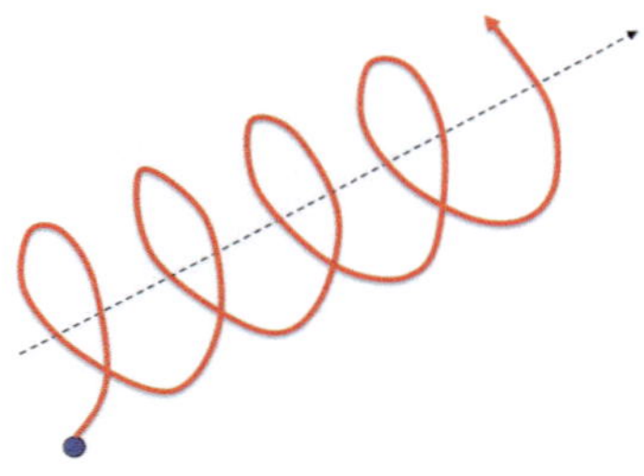

Fig. 15.1 Larmor orbit. In a homogenous magnetic field a charged particle orbit is an helix corresponding to the superposition of a uniform circular motion in the plane orthogonal to the magnetic field and a constant velocity liner motion along the magnetic field

with $\hat{\mathbf{x}}$ and $\hat{\mathbf{y}}$ being two orthogonal unit vectors, both orthogonal to $\mathbf{B}$, and ϕ is the initial constant. Note that both $v_{||}$ and $v_\perp$ are constants of motion in this case. The frequency Ω is the cyclotron frequency defined as

$$\Omega = \frac{eB}{m} \tag{15.4}$$

Note that the electron cyclotron frequency is larger than the proton cyclotron frequency by a factor m_p/m_e.

Equation (15.3) can be integrated in time to give the particle trajectory

$$\mathbf{x}(t) = \mathbf{const.} + v_{||}t\frac{\mathbf{B}}{B} + \frac{v_\perp}{\Omega}[\sin(\Omega t + \phi)\hat{\mathbf{x}} - \cos(\Omega t + \phi)\hat{\mathbf{y}}] \tag{15.5}$$

The quantity $\rho_L \equiv (v_\perp/\Omega)$ is the Larmor radius. A Larmor orbit corresponds to a helix (Fig. 15.1) around a magnetic field line.

If a static homogeneous electric field is added, the equation of motion (15.1) can be easily integrated with the following substitution of variable

$$\mathbf{v} = \mathbf{v}' + \frac{\mathbf{E} \times \mathbf{B}}{B^2} \tag{15.6}$$

with $\mathbf{v}'$ being given by Eq. (15.3) and the second term representing a constant (in space and time) drift independent of the charge and the mass of the particle.

The physical origin of the $\mathbf{E} \times \mathbf{B}$ drift can be easily understood by looking at Fig. 15.2.

During the Larmor motion a particle is accelerated (decelerated) by the electrostatic force $e\mathbf{E}$ and increases its perpendicular velocity $v_\perp$. This, in turn, increases (decreases) the radius of curvature of its orbit (that would be constant and equal to the Larmor radius for $E = 0$). The net effect of this change of $v_\perp$ is a net drift of the particle. Note that the drift is independent of charge because ions and electrons are accelerated in opposite directions but also the direction in which they follow a Larmor orbit is opposite.

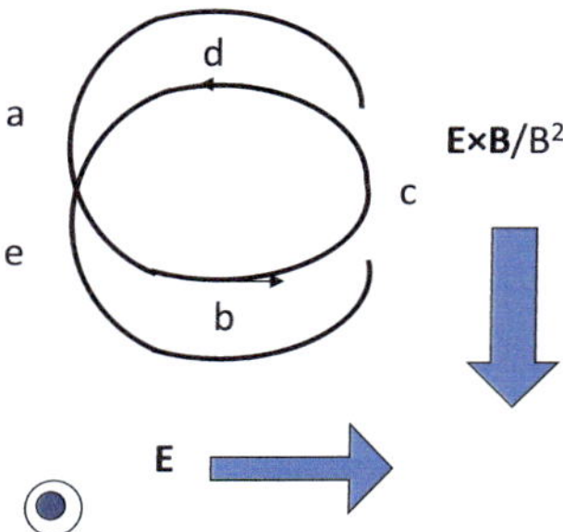

Fig. 15.2 The ExB drift illustrated in the case of an electron. The electron is decelerated during the semi orbit *abc* and accelerated during the semiorbit *cde*. The change in energy produces a change in the Larmor radius and as a consequence a uniform drift in the vertical direction. In the case of an ion the effect of the electric field is reversed but also the direction of the Larmor orbit is reversed, resulting in the same drift as for the electron

If the particle is subject to a constant force **F** perpendicular to **B** it will drift in the direction orthogonal to both **F** and **B**

$$\mathbf{v}_d = \frac{\mathbf{F} \times \mathbf{B}}{eB^2} \tag{15.7}$$

The origin of this drift is similar to the $E \times B$ drift but there is an important difference. In the case of Eq. (15.7) the drift depends on the charge and therefore is opposite for electrons and ions.

The center of the orbit at each time is the *guiding center*. The motion of the particle can be described as a slow drift of the guiding center with a superimposed Larmor motion.

15.2 Particle Orbits in Weakly Inhomogeneous Magnetic Fields

If the magnetic field is not homogeneous Eq. (15.1) cannot be solved exactly. It is however possible to obtain approximate solutions if the magnetic field is weakly inhomogeneous, i.e. if its characteristic length of variation $L_B \equiv (|\nabla B|/|B|)^{-1}$ is much larger than the Larmor radius. If this is the case, the equations can be solved by expanding the magnetic field around the initial position using as expansion parameter the ratio ρ_L/L_B. Three different cases can be considered:

- **B** varies in modulus but not in direction (∇B *drift*);
- **B** changes in direction (*curvature drift*);
- the magnetic field intensity B changes along **B**.

15.2.1 ∇B Drift

Making reference to Fig. 15.3, let us make a Taylor expansion of the modulus of the magnetic field around the initial position of the particle.

Without loss of generality we can define the coordinate system in such a way that the magnetic field is along the z axis and the magnetic field intensity depends only on the coordinate y: $B = B(y)\mathbf{z}$ and approximate $B(y)$ as

$$B(y) \approx B(y_0) + B'(y_0)(y - y_0) \tag{15.8}$$

The equations of motion become

$$m\frac{d^2x}{dt^2} = eB(y)\frac{dy}{dt} \tag{15.9}$$

$$m\frac{d^2y}{dt^2} = -eB(y)\frac{dx}{dt} \tag{15.10}$$

It is possible to look for solution of the above equations in series of the parameter $\delta \equiv \rho_L/L_B$.

$$x(t) = x^{(0)}(t) + \delta x^{(1)}(t) + O(\delta^2); \qquad y(t) = y^{(0)}(t) + \delta y^{(1)}(t) + O(\delta^2) \tag{15.11}$$

If the magnetic field inhomogeneity is weak (i.e. if $\delta \ll 1$), the second term in the expansion of $B(y)$ in Eq. (15.8) is $O(\delta)$. Upon substituting Eq. (15.11) in the equations of motion and equating terms of the same order in δ, the following system of equations is obtained

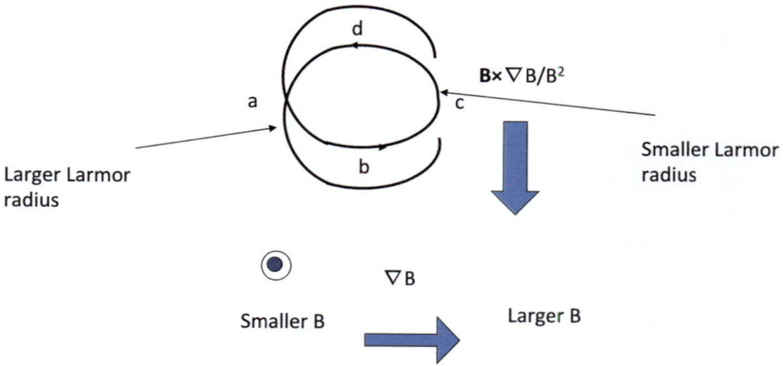

Fig. 15.3 Grad B drift. The instantaneous radius of curvature of the Larmor orbit is larger in a than in c producing a net vertical drift that is opposite for electrons and ions

$$m\frac{d^2x^{(0)}}{dt^2} = eB(y_0)\frac{dy^{(0)}}{dt} \tag{15.12}$$

$$m\frac{d^2y^{(0)}}{dt^2} = -eB(y_0)\frac{dx^{(0)}}{dt} \tag{15.13}$$

and

$$m\frac{d^2x^{(1)}}{dt^2} = eB(y_0)\frac{dy^{(1)}}{dt} + eB'(y_0)(y^{(0)} - y_0)\frac{dy^{(0)}}{dt} \tag{15.14}$$

$$m\frac{d^2y^{(1)}}{dt^2} = -eB(y_0)\frac{dx^{(1)}}{dt} + eB'(y_0)(y^{(0)} - y_0)\frac{dx^{(0)}}{dt} \tag{15.15}$$

At the lowest order in δ, the particle orbit $x^{(0)}(t)$ corresponds to a Larmor motion with frequency $\Omega_0 \equiv eB(y_0)/m$.

$$x^{(0)}(t) = x_0 + \frac{v_\perp}{\Omega_0}\sin(\Omega_0 t) \qquad y^{(0)}(t) = y_C + \frac{v_\perp}{\Omega_0}[1 - \cos(\Omega_0 t)] \tag{15.16}$$

Upon substituting this solution in Eqs. (15.14) and (15.15) we obtain

$$\begin{aligned} m\frac{d^2x^{(1)}}{dt^2} - eB(y_0)\frac{dy^{(1)}}{dt} &= eB'(y_0)\frac{v_\perp}{\Omega_0}[1 - \cos(\Omega_0 t)]v_\perp \sin(\Omega_0 t) = \\ &= eB'(y_0)\frac{v_\perp^2}{\Omega_0}\Big[\sin(\Omega_0 t) - \frac{1}{2}\sin(2\Omega_0 t)\Big] \end{aligned} \tag{15.17}$$

$$\begin{aligned} m\frac{d^2y^{(1)}}{dt^2} + eB(y_0)\frac{dx^{(1)}}{dt} &= -eB'(y_0)(\frac{v_\perp}{\Omega_0}[1 - \cos(\Omega_0 t)]v_\perp \cos(\Omega_0 t) = \\ &= -eB'(y_0)\frac{v_\perp^2}{\Omega_0}\Big[\cos(\Omega_0 t) - \frac{1}{2} - \frac{1}{2}\cos(2\Omega_0 t)\Big] \end{aligned} \tag{15.18}$$

Equations 15.17 and 15.18 show that the first order motion of the particle is the result of the Lorentz force and of an external forcing that has both periodic components and a constant component along y. The interesting contribution is the latter because it gives rise to a *secular* motion characterised by a constant drift along x

$$\mathbf{v}_d = \frac{v_\perp^2}{2\Omega_0}\frac{\mathbf{B} \times \nabla B}{B^2} \tag{15.19}$$

This drift is called the ∇B *drift* and it is opposite for electrons and ions.

15.2.2 Curvature Drift

The curvature of a magnetic field line is described by the curvature vector

$$\boldsymbol{\kappa} \equiv \mathbf{b} \cdot \nabla \mathbf{b} \tag{15.20}$$

with $\mathbf{b} \equiv \mathbf{B}/B$ the unit vector in the direction of the magnetic field. The curvature produces a centrifugal *acceleration* of a point moving along the magnetic field with constant velocity. Note that $\boldsymbol{\kappa}$ is always orthogonal to $\mathbf{b}$ as it can be seen also using the vector identity

$$\mathbf{b} \cdot \nabla \mathbf{b} = -\mathbf{b} \times \nabla \times \mathbf{b} \tag{15.21}$$

Thus in the presence of a curvature a particle moving with velocity $v_{||}$ along the magnetic field is subject to a centrifugal force given by

$$\mathbf{F}_c = -m v_{||}^2 \boldsymbol{\kappa} \tag{15.22}$$

The centrifugal force produces a drift in the direction normal to both the curvature and the magnetic field given by

$$\mathbf{v}_d = \frac{v_{||}^2}{\Omega_0} \mathbf{b} \times \boldsymbol{\kappa} \tag{15.23}$$

Again, this *curvature drift* is opposite for electron and ions.

15.2.3 Magnetic Field with Longitudinal Gradient. Magnetic Moment

The third case of magnetic field inhomogeneity corresponds to a field with a longitudinal gradient. In order to determine the effect of a longitudinal gradient on the particle orbit we need an expression for $\mathbf{B}$ valid in the region explored by the particle motion, region that is typically much smaller than the scale of variation of the field. It is convenient to separate the magnetic field in a dominant component along the direction of variation (that we will identify with z) and a small correction that must be introduced in order to preserve the solenoidal nature of B.

Taking a small circle of radius r around the z axis (see Fig. 15.4) and applying Gauss theorem to a cylinder of height dz, the condition $\nabla \cdot \mathbf{B} = 0$ yields

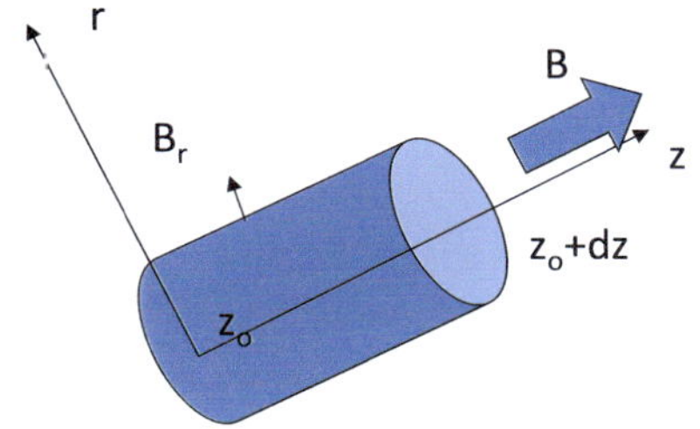

Fig. 15.4 Field with longitudinal gradient. To preserve the solenoidal nature of the magnetic field a small radial component must be added

$$2\pi r dz B_r + \pi r^2 B_z(z + dz) - \pi r^2 B_z(z) = 0 \tag{15.24}$$

that provides a relation between B_r and B_z

$$B_r = -\frac{r}{2}\frac{\partial B_z}{\partial z} \tag{15.25}$$

The radial component B_r vanishes if the magnetic field is homogeneous. However, in the presence of a longitudinal gradient $\partial B_z/\partial z \neq 0$ its presence is required to satisfy Maxwell's equations.

The effect of the radial component is to provide a net Lorentz force in the z direction. This can be seen by identifying the radius r with the Larmor radius of a particle with a gyro center placed in $r = 0$ and $z = 0$. The particle will feel a constant Lorentz force given by

$$e(\mathbf{v} \times \mathbf{B})_r = ev_\perp B_r = -\frac{mv_\perp^2}{2B_z}\frac{\partial B_z}{\partial z} = -\mu\frac{\partial B_z}{\partial z} \tag{15.26}$$

The quantity $\mu \equiv mv_\perp^2/(2B)$ is the *magnetic moment* and it is an approximate constant of motion. It can be easily seen that this definition of the magnetic moment is consistent with what has been given in Chap. 4 since $\mu = e(\Omega/2\pi)\pi\rho_L^2$. The magnetic moment describes the elementary magnetisation produced by the plasma particles in response to the presence of the magnetic field. The current associated with the Larmor motion produces an elementary field that far from the particles is a dipole field with a magnetic dipole moment given by μ. Such a magnetic field is directed in the direction opposite to the imposed magnetic field. Thus, the plasma is intrinsically a *diamagnetic* material. Note that the magnetic moment is independent of charge. Thus, as the average kinetic energy of electrons and ions are similar, the contribution to the magnetisation of electrons and ions is also of the same order.

The particle motion along the magnetic field can therefore be described by the following equation

$$m\frac{dv_{||}}{dt} = -\mu\frac{\partial B}{\partial s} \tag{15.27}$$

with $s \equiv \mathbf{b} \cdot \mathbf{r}$ the coordinate along B and $v_{||} = ds/dt$.

The invariance of the magnetic moment can be proved by taking the average over a Larmor period of the change in energy

$$\delta(\frac{mv^2}{2}) = \frac{e\|\Omega\|}{2\pi} \int_0^{2\pi/\|\Omega\|} dt \mathbf{E} \cdot \mathbf{v} \tag{15.28}$$

The line integral of **E** around the Larmor orbit that appears in the r.h.s. can be rearranged using Stokes theorem and Faraday's law as follows

$$\delta(\frac{m(v_\perp^2 + v_{||}^2)}{2}) = -\frac{e\|\Omega\|}{2\pi} \int \frac{\partial \mathbf{B}}{\partial t} \cdot \mathbf{n} dS \tag{15.29}$$

where the normal **n** has to be taken as shown in Fig. 15.5

Equation (15.29) can be rearranged by bringing the term related to the parallel energy to the r.h.s. and using Eq. (15.27)

$$\delta(\frac{mv_\perp^2}{2}) = \frac{e\Omega}{2\pi}(\pi\rho_L^2)\frac{\partial B}{\partial t} + \mu v_{||}\frac{\partial B}{\partial s} \tag{15.30}$$

The r.h.s. is proportional to the change of B along the particle trajectory. Upon substituting the definition of the magnetic moment we finally obtain

$$\delta(\mu B) = \mu \delta B \quad \rightarrow \quad \delta\mu = 0 \tag{15.31}$$

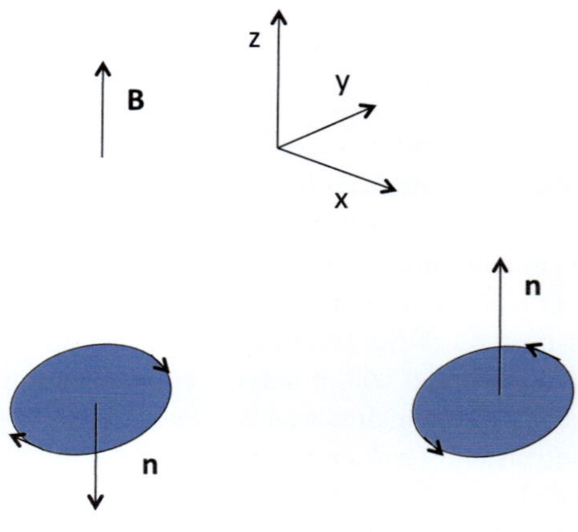

Fig. 15.5 The choice of the sign of the normal to the surface enclosed by the line integral in Stokes theorem depends on the direction of integration that follows the Larmor trajectory of a particle and is illustrated for electrons and ions

The magnetic moment is an example of *adiabatic invariant*, i.e. a quantity that is not an exact invariant (like the energy) but it remains invariant under slow change in the system (in this case exemplified by the weak inhomogeneity assumption).

15.3 Particle Confinement in Toroidal Magnetic Configurations

The drifts in the particle orbits determine the condition for particle confinement in a magnetic field. It is instructive to consider first the case of a purely toroidal magnetic field

$$B = B_\phi \hat{\phi} \tag{15.32}$$

With $\hat{\phi}$ the unit vector in the toroidal direction. Here a set of standard cylindrical coordinates (R, ϕ, Z) is used as shown in Fig. 15.6.

As shown in Sect. 13.1, the magnetic field is a solution of the Maxwell equations in vacuum: $B_\phi(R) = B_\phi(R_o)R_o/R$ with R_o being a generic radius (see Fig. 15.6). At the lowest order in the parameter δ the particle motion is a Larmor orbit around the toroidal magnetic field. However, the presence of drifts affects the particle confinement. Indeed, the combined effect of ∇B and curvature produce a drift given by Eqs. (15.19) and (15.23), i.e.

$$\mathbf{v}_D = (\frac{v_\perp^2}{2R\Omega} + \frac{v_\parallel^2}{R\Omega})\hat{\mathbf{Z}} \tag{15.33}$$

The drift is along the Z direction and is opposite for electrons and ions. Thus, a charge separation is produced that in turn produce an electric field E. We do not need to evaluate it. The sign of $\mathbf{E}$ can be inferred from the direction of the electron and ion drifts (see Fig. 15.7) and, using Eq. (15.6) it is possible to conclude that the

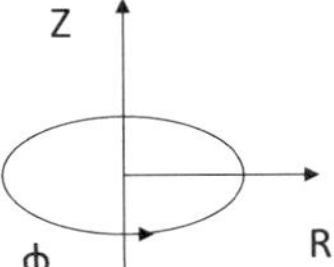

Fig. 15.6 Toroidal geometry. In this case the magnetic field is purely toroidal i.e. in the direction along ϕ

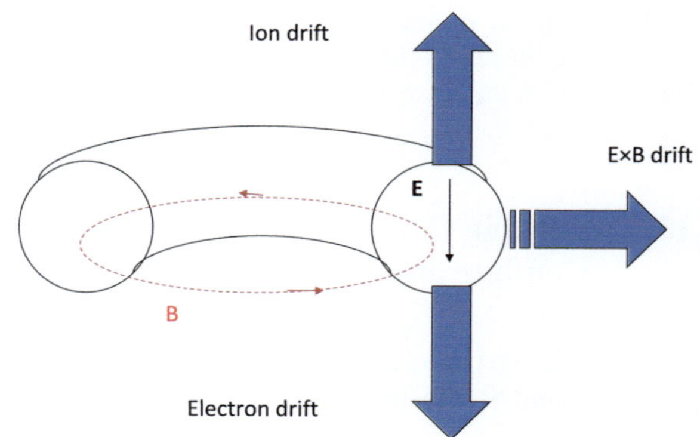

Fig. 15.7 A purely toroidal magnetic field does not confine a plasma. As shown, the curvature and ∇B drifts produce a charge separation and, in turn, an electric field that pushes the plasma out of the confinement region

drift is outwards in the radial direction. We therefore conclude that *a purely toroidal magnetic field does not confine a plasma*.

In order to confine a plasma the drift that produces the charge separation must be cancelled. This is possible by superimposing a poloidal component B_θ. In this way as soon as the charge separation is produced, the motion of the particles along the magnetic field line short circuit it, re-establishing charge neutrality and nullifying the $E \times B$ drift.

In order to demonstrate explicitly that particles are now confined, it is convenient to consider the large aspect ratio limit of the magnetic configuration in which we retain as a perturbation the effect of the drifts. In tokamaks the poloidal magnetic field is smaller than the toroidal field and can be described by a function $B_\theta = B_\theta(r)$ with $r = [(R - R_o)^2 + Z^2]^{1/2}$. The total magnetic field is now the sum of the poloidal and toroidal fields and a magnetic field line follows a trajectory on a surface $r = const.$. Thus, as the particle moves along B the magnitude of B_θ remains constant. In this limit the drift velocities can be evaluated using the inhomogeneity associated with the toroidal field only. Note that, although B_θ is constant along the field line, B_ϕ is not since it increases towards the torus axis. As a result, particles are subject to the longitudinal gradient force. Nevertheless, for the moment the effect of the longitudinal gradient is neglected and it is assumed that the particle moves at constant velocity $v_{||}$. The coordinate system r, θ, ϕ is a *quasi cylindrical* coordinate system that will be used also later in this book (see Fig. 15.8).

With these approximations the equation for the guiding center position becomes

$$\frac{d\mathbf{x}_{gc}}{dt} = v_D \hat{\mathbf{Z}} + v_{||}\mathbf{b} \tag{15.34}$$

with $v_D = v_\perp^2/(2R\Omega) + v_{||}^2/(R\Omega)$. The guiding center orbit can be determined by taking the components of Eq. (15.34) along R and Z. The magnetic field components are given by (see Fig. 15.8)

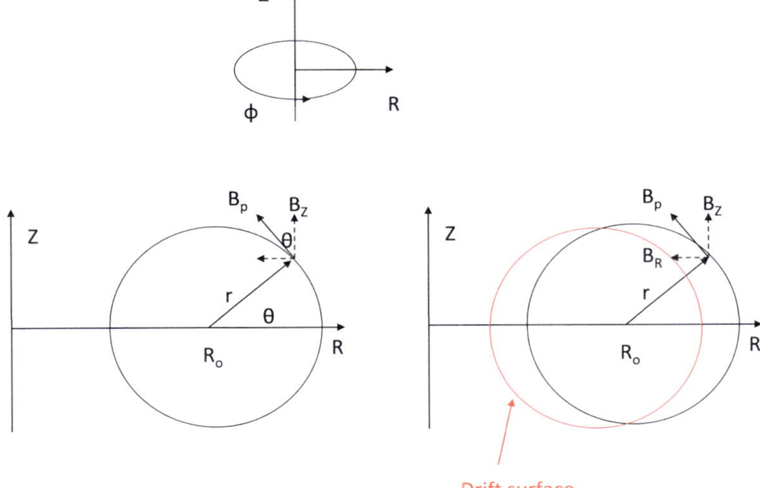

Fig. 15.8 The quasi cylindrical coordinate system r, θ, ϕ provides a convenient system to illustrate the effect of the drifts when a poloidal magnetic field is added to the toroidal magnetic field. A magnetic surface in this system corresponds to $r = const.$ and the poloidal field is assumed at the lowest order constant on the magnetic surface. The drift surface, i.e. the surface described by the guiding center motion, is shifted in the horizontal direction by an amount of the order of the Larmor radius

$$B_R = -B_p \sin\theta = -B_p \frac{Z}{r} \qquad B_Z = B_p \cos\theta = B_p \frac{R - R_o}{r} \tag{15.35}$$

The equation for the guiding center position then becomes

$$\frac{dR_{gc}}{dt} = \frac{B_R}{B} v_{||} = -\frac{B_p}{B} v_{||} \frac{Z_{gc}}{r} \tag{15.36}$$

$$\frac{dZ_{gc}}{dt} = \frac{B_Z}{B} v_{||} + v_D = \frac{B_p}{B} v_{||} \frac{R_{gc} - R_o}{r} + v_D \tag{15.37}$$

Assuming $v_{||} = const$ and assuming that the trajectory remains close to a surface $r = const$ the equations can be easily integrated to yield

$$(R_{gc} - R_o + \frac{v_D}{\omega_t})^2 + Z_{gc}^2 = const. \tag{15.38}$$

that corresponds to a circle shifted by an amount v_D/ω_t from the plasma center. The quantity $\omega_t \equiv (B_p/B)(v_{||}/r)$ is the *transit frequency*. Note that in Eqs. (15.36) and (15.37) we have also neglected the variation of B along the trajectory. Thus, the magnetic field inhomogeneity is accounted at the lowest order only by the drift velocity v_D.

It is possible to show that the shift of the magnetic surface with respect to the plasma center also holds for a generic magnetic field shape. The shift is inversely proportional to $v_{||}$ therefore it depends on its magnitude. We may wonder what happens for $v_{||} \to 0$ since this would give an infinite displacement. However, as shown below, the analysis above only holds for particles with non-vanishing parallel velocity. The amount of the shift can be estimated for thermal particles to be

$$\frac{v_D}{\omega_t} \approx \frac{v_t}{\Omega} \frac{rB}{RB_p} = q\rho_L \tag{15.39}$$

It is important to note that Eq. (15.39) is independent of the total magnetic field. The shift of the drift surface depends only on the poloidal field which is the real confining parameter in toroidal systems.

Finally, as the shift of the drift surface Eq. (15.38) with respect to the magnetic surface $r = const.$ is of order of the Larmor radius, the assumption that has been made in solving Eqs. (15.36)–(15.37) turns out to be *a posteriori* justified.

15.3.1 Trapped Particles

We now go back to the assumption $v_{||} = const.$ to understand under which condition it is not satisfied. Following a magnetic field line on a surface $r = const$ the magnetic field will vary as

$$B \approx B_\phi = B_o \frac{R_o}{R} \approx B_o(1 - \frac{r}{R_o} \cos\theta) \tag{15.40}$$

therefore the particle will be subject to a periodic force along the magnetic field. From the conservation of energy it is possible to obtain an equation for $v_{||}$.

$$E = \frac{m(v_\perp^2 + v_{||}^2)}{2} = \frac{mv_{||}^2}{2} + \mu B \tag{15.41}$$

or $v_{||} = \pm(2(E - \mu B)/m)^{1/2}$. The motion along the magnetic field is equivalent to the motion of a particle in a potential $V = \mu B$ as shown in Fig. 15.9.

If $E > \mu B_{max}$ the parallel velocity never vanishes and the particle moves along the field line always in the same direction. The velocity is modulated by the magnetic field variation but if the energy is sufficiently larger than μB_{max} such a modulation is small and can be neglected, as we have assumed in deriving Eq. (15.38). Particles for which the velocity never vanishes are called *circulating particles*.

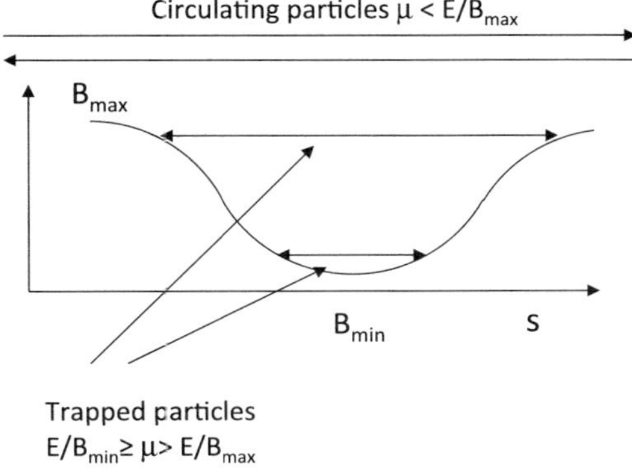

Fig. 15.9 Trapped particles

If $E < \mu B_{max}$ the parallel velocity vanishes at some point. The particle is then trapped between the points where $E = \mu B$. Note that by definition $E \geq \mu B_{min}$, therefore the trapped particle domain corresponds to

$$\frac{E}{B_{min}} \geq \mu \geq \frac{E}{B_{max}} \tag{15.42}$$

Trapped particles are not a negligible part of the plasma particles in a typical tokamak. The fraction of trapped particles can be determined by counting them at the point $B = B_{min}$[1] where $\mu = mv_\perp^2/(2B_{min})$ and Eq.15.42 yields

$$\lambda \equiv (\frac{B_{max}}{B_{min}} - 1)^{1/2} \geq \frac{|v_{||}|}{v_\perp} \geq 0 \tag{15.43}$$

Equation (15.43) defines the *trapping cone* and allows a simple evaluation of the trapped particle fraction f_{trap}

$$\begin{aligned} f_{trap} = & \int_{trapped} d^3u \frac{e^{-u^2}}{\pi^{3/2}} = \\ & = \frac{2\pi}{\pi^{3/2}} \int_0^\infty u_\perp du_\perp e^{-u_\perp^2} \int_{-\lambda u_\perp}^{\lambda u_\perp} du_{||} e^{-u_{||}^2} = \\ & = \frac{2}{\pi^{1/2}} \int_0^\infty dw e^{-w} \int_0^{\lambda w^{1/2}} du_{||} e^{-u_{||}^2} = \\ & = \frac{\lambda}{\pi^{1/2}} \int_0^\infty \frac{dw}{w^{1/2}} e^{-w(1+\lambda^2)} = \\ & = \frac{\lambda}{(1+\lambda^2)^{1/2}} = (1 - \frac{B_{min}}{B_{max}})^{1/2} \end{aligned} \tag{15.44}$$

[1] If we take any other point we miss the particles trapped at the bottom of the potential well.

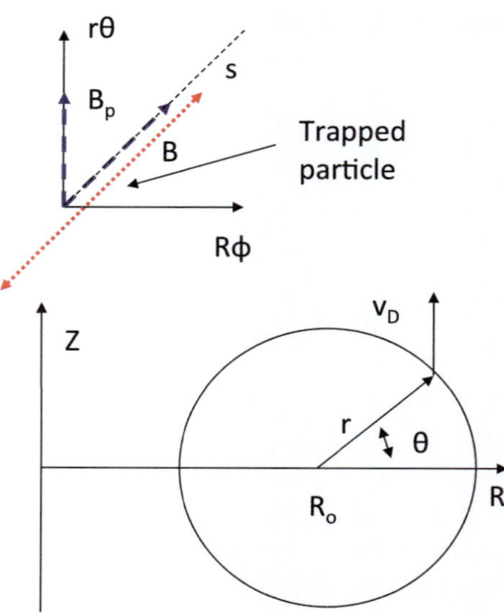

Fig. 15.10 Trapped particle geometry

For the tokamak equilibrium we have considered above $B_{max} = B_o/(1-\epsilon)$ and $B_{min} = B_o/(1+\epsilon)$ with $\epsilon = r/R$. Therefore, for a typical aspect ratio $A = 3$ the fraction of trapped particles at half radius ($r = a/2$) is $(2\epsilon/(1+\epsilon))^{1/2} \approx 50\%$. Note that at the plasma center the fraction of trapped particles vanishes.

The trapped particle orbit can be found again by integrating the equation of motion for the guiding center. In this case, rather than using the (R, Z) coordinates it is convenient to use the (r, θ) variables. Figure 15.10 shows the relation between the two set of coordinates

As before we cut the torus at a certain toroidal angle and straighten it to a cylinder. Next we open the cylinder on a line at $\theta = 0$ and flatten the surface into a plane with the horizontal axis corresponding to the toroidal position $R\phi$ and the vertical axis to the poloidal position $r\theta$. A magnetic field line is a straight line in the plane. A trapped particle moving along B will explore the portion of space between the two turning points.

The radial guiding center velocity is obtained from the component along r of the drift velocity v_D. The equation for the parallel velocity is simply Eq. (15.27) applied to the simplified tokamak equilibrium.

$$\frac{dr}{dt} = v_D \sin\theta \tag{15.45}$$

$$\frac{dv_{||}}{dt} = -\frac{B_p}{B}\frac{v_\perp^2}{2R_o}\sin\theta \tag{15.46}$$

which can be easily solved to give the following particle orbit

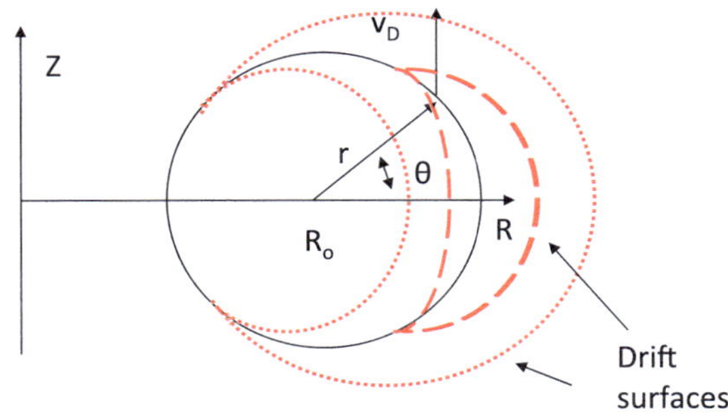

Fig. 15.11 Trapped particle orbits. The orbit has a characteristic banana shape

$$r = r_o - \frac{m}{eB_p} v_{||} \tag{15.47}$$

Note that in deriving Eq. (15.47) we have neglected the curvature drift with respect to the ∇B drift since, from Eq. (15.43) for trapped particles the parallel velocity is a factor $\epsilon^{1/2}$ smaller than the perpendicular velocity. As the particle oscillates between the two turning points it simultaneously drift outwards or inwards depending on the sign of the parallel velocity. The trajectory is shown in Fig. 15.11 and for obvious reason is called *banana* orbit.

Once again the orbit width is independent of the toroidal field and depends only on the poloidal field. By comparing Eq. (15.47) with Eq. (15.39) it is apparent that the former is a factor $\epsilon^{-1/2}$ larger

$$(r - r_o)_{max} \approx \frac{\epsilon^{1/2} v_\perp}{\Omega} \frac{rB}{RB_p} \frac{1}{\epsilon} \approx \frac{q}{\epsilon^{1/2}} \rho_L \tag{15.48}$$

15.4 Alpha Particle Confinement

In order to be confined, the $3.5 MeV$ alpha particles must have a displacement from the magnetic surface that is significantly smaller than the minor radius. Since trapped particles have the largest excursion it is reasonable to take as a condition $(r - r_o)_{max} < 0.3a$. From Eq. (15.48) the following condition is obtained

$$B_p(T)a(m) \geq \epsilon^{-1/2} \tag{15.49}$$

Taking $\epsilon = 1/3$ and recalling that $2\pi B_p a = \mu_o I_p$ Eq. (15.49) yields

$$I_p \geq 3MA \tag{15.50}$$

15.5 Effect of Trapped Particles on Spitzer Resistivity

A consequence of the existence of trapped particles is related to the electrical resistivity in tokamaks. Trapped particles are not on average accelerated by the electric field that is induced to generate and sustain the plasma current. This is equivalent to a local increase of the electrical resistivity over the Spitzer estimate given in Eq. (12.23) by a factor

$$\frac{\eta}{\eta_{Spitzer}} = \left(1 - \left(\frac{2\epsilon}{1+\epsilon}\right)^{1/2}\right)^{-1} \tag{15.51}$$

The average increase of resistivity is taken into account through the aspect ratio dependent factor in Eq. (14.18).

15.6 Exercises

Problem 15.1 Estimate the electric field generated by the charge separation in a purely toroidal magnetic field and the time to push the plasma out of the confinement region (hint: describe the charge separation as the charging of a capacitor with planar plates at the top and the bottom of the plasma)

Problem 15.2 Determine the shift of the drift orbits of an electron and an ion with energy equal to T and a ratio $v_{||}/v_{\perp}$ equal to 0.5 and 2 times the value corresponding to the boundary of the trapping cone for ITER parameters $T = 10keV$, $B = 5T$ and $R = 6.3m$.

Chapter 16
Forces Between Currents and Flux Balance

Abstract *In this chapter the interaction of a system of coils is studied. A lumped constant description is introduced first. The expression for the mutual and internal inductance is then used to determine the forces acting on a coil. Finally a model of the tokamak is introduced to determine the radial build of a device.*

We have shown in Chap. 13 how to evaluate the magnetic fields from various sources. In this chapter the concept of magnetic energy is introduced and it is shown how to evaluate the forces acting on the sources. Finally, using Faraday's law, we evaluate the e.m.f. in each circuit due to the variation of the magnetic flux. The last part of this chapter discusses a simplified model for the critical components of the tokamak magnetic system, namely the toroidal field (TF) magnet and the central solenoid (CS).

16.1 Magnetic Energy

We will now derive an expression for the magnetic energy of a set of interacting currents. The starting point is the expression of the magnetic energy given in Chap. 4

$$U = \int d^3r\prime \frac{B^2(\mathbf{r}\prime)}{2\mu_o} \tag{16.1}$$

The magnetic field is given by the linear superposition of the magnetic field generated by each element

$$\mathbf{B} = \sum_k \mathbf{B}_k \tag{16.2}$$

with $\nabla \times \mathbf{B}_k = \mu_o \mathbf{j}_k$. Upon substituting into Eq. (16.1) we obtain

$$U = \sum_k \int d^3r\prime \frac{B_k^2(\mathbf{r}\prime)}{2\mu_o} + \sum_{i>k} \int d^3r\prime \frac{\mathbf{B}_i(r') \cdot \mathbf{B}_k(r')}{\mu_o} \tag{16.3}$$

F. Romanelli, *Physics of Nuclear Energy*, Springer Series in Plasma Science and Technology, https://doi.org/10.1007/978-981-97-9609-0_16

We anticipate that the first term will be associated to the self-inductance of each element whereas the second term to the mutual inductance of each pair of elements.

Due to the symmetry of our problem, all the coils generating the poloidal field can be assumed to be circular coils. We might be also tempted to evaluate the integrals in the limit of arbitrarily thin coil. However it can be immediately seen that the self inductance contribution displays a divergence as the field increases as $1/r$ (with r the distance from the centre of the conductor) leading to a logarithmic singularity. Thus, the contribution in the first integral can be separated into two terms: the integral outside the conductor (vacuum region) and the integral inside the conductor.

The integral outside the conductor can be simplified using the vector identity

$$\mathbf{B} \cdot \nabla \times \mathbf{A} = \mathbf{A} \cdot \nabla \times \mathbf{B} + \nabla \cdot (\mathbf{A} \times \mathbf{B}) \tag{16.4}$$

This identity is valid for arbitrary vector fields $\mathbf{A}$ and $\mathbf{B}$. Applying the identity to the vector potential $\mathbf{A}_k$ and magnetic field $\mathbf{B}_k$ and observing that $\nabla \times \mathbf{B}_k = 0$ outside the conductor the following expression is obtained for the external contribution

$$\begin{aligned} \int_{ext} d^3r\prime \frac{B_k^2(\mathbf{r}\prime)}{2\mu_o} &= \int_{ext} d^3r\prime \frac{\mathbf{B}_k \cdot \nabla \times \mathbf{A}_k}{2\mu_o} = \\ &= \int_{ext} d^3r\prime \frac{\nabla \cdot (\mathbf{A}_k \times \mathbf{B}_k)}{2\mu_o} = \int_{S_{ext}} dS \frac{\mathbf{n} \cdot (\mathbf{A}_k \times \mathbf{B}_k)}{2\mu_o} \end{aligned} \tag{16.5}$$

where use has been made of Gauss theorem and $\mathbf{n}$ is the outgoing normal to the volume that corresponds to the vacuum region. This region can be considered bound by two surfaces: a sphere of radius R_s (with $R_s \to \infty$) and the surface of the conductor. Far from the coil it is possible to use the asymptotic expression corresponding to the dipole field (Eqs. (13.19)–(13.21))

$$A_k \approx R_s^{-2} \qquad B_k \approx R_s^{-3} \tag{16.6}$$

to show that the contribution on the sphere vanishes in the limit $R_s \to \infty$. Thus, the contribution outside the conductor reduces to an integral on the conductor surface. Note that the unit vector $\mathbf{n}$ is the *outgoing* normal to the external volume surface, i.e. it is the *ingoing* normal to the conductor surface.

To evaluate the surface integral Eq. (16.5) the exact expression of $\mathbf{A}_k$ on the conductor surface is needed that can be obtained from Eq. (13.2). For a circular coil of average major radius R_k made by a conductor with circular cross section of radius a_k the explicit expression can be obtained from the limiting form Eqs. (13.16)–(13.18)

$$\mathbf{A}_k \approx \frac{\mu_o I_k}{2\pi} \Big[\ln \Big(\frac{8R_k}{r} \Big) - 2 \Big] \hat{\phi} \tag{16.7}$$

with $r = [(R - R_k)^2 + Z^2]^{1/2}$. The magnetic field can be obtained from this limiting form taking into account that the leading contribution is associated with the derivative of the logarithmic term

$$\mathbf{B}_k = \nabla \times \mathbf{A}_k \approx -\frac{\mu_o I_k}{2\pi r} \nabla r \times \hat{\boldsymbol{\phi}} \tag{16.8}$$

Note that for $r \to 0$ the magnetic field corresponds to the field of a wire carrying a current I_k. The magnetic field lines are circles and the magnetic field must be evaluated at the conductor boundary $r = a_k$. Taking into account that for a circular conductor section $\mathbf{n} = -\nabla r$, the external contribution reduces to

$$\begin{aligned}\int_{ext} d^3r' \frac{B_k^2}{2\mu_o} &= \int_{S_{ext}} dS \frac{\mathbf{n} \cdot (\mathbf{A}_k \times \mathbf{B}_k)}{2\mu_o} = \\ &\approx \left(\frac{\mu_o I_k}{2\pi}\right)^2 \frac{1}{a_k} \ln\left(\frac{8R_k}{a_k} - 2\right) \int R_k d\phi a_k d\theta \frac{\nabla r \cdot \hat{\boldsymbol{\phi}} \times (\nabla r \times \hat{\boldsymbol{\phi}})}{2\mu_o} = \\ &= \frac{1}{2} L_{k,ext} I_k^2 \end{aligned} \tag{16.9}$$

with $L_{k,ext} = \mu_o R_k [\ln(8R_k/a_k) - 2]$ the external inductance of the coil.

The contribution to the magnetic energy from the region inside the conductor can be evaluated directly from Eq. (16.1). It is useful to introduce the internal inductance parameter l_{ik} as follows

$$\int d^3r' \frac{B_k^2}{2\mu_o} = \frac{B_k(r = a_k)^2}{2\mu_o} 2\pi^2 R_k a_k^2 l_{ik} = \frac{1}{2} \mu_o R_k \frac{l_{ik}}{2} I_k^2 \tag{16.10}$$

Thus the magnetic energy produced by each coil can be written as

$$\int d^3r' \frac{B_k^2}{2\mu_o} = \frac{1}{2} L_k I_k^2 = \frac{1}{2} \mu_o R_k \left(\ln\left(\frac{8R_k}{a_k}\right) - 2 + \frac{l_{ik}}{2} \right) I_k^2 \tag{16.11}$$

Note that the quantity l_{ik} is a dimensionless number related only to the distribution of current inside the conductor. It can be interpreted as the ratio between the internal magnetic energy and the magnetic energy associated with a constant magnetic field (equal to the edge value) over the volume of the conductor.

The contribution to the magnetic energy arising from the interaction between coils can be evaluated directly using Eq. (16.3) and the vector identity Eq. (16.4)

$$\begin{aligned}\int d^3r' \frac{\mathbf{B}_i \cdot \mathbf{B}_k}{\mu_o} &= \int d^3r' \frac{\mathbf{B}_i \cdot \nabla \times \mathbf{A}_k}{\mu_o} = \\ &= \int d^3r' \frac{(\mathbf{B}_k \cdot \nabla \times \mathbf{A}_i + \nabla \cdot (\mathbf{A}_k \times \mathbf{B}_i))}{\mu_o} = \\ &= \frac{\mu_o}{4\pi} \int d^3r' \int d^3r'' \frac{\mathbf{j}_k(\mathbf{r}'') \cdot \mathbf{j}_i(\mathbf{r}')}{|\mathbf{r}' - \mathbf{r}''|} \end{aligned} \tag{16.12}$$

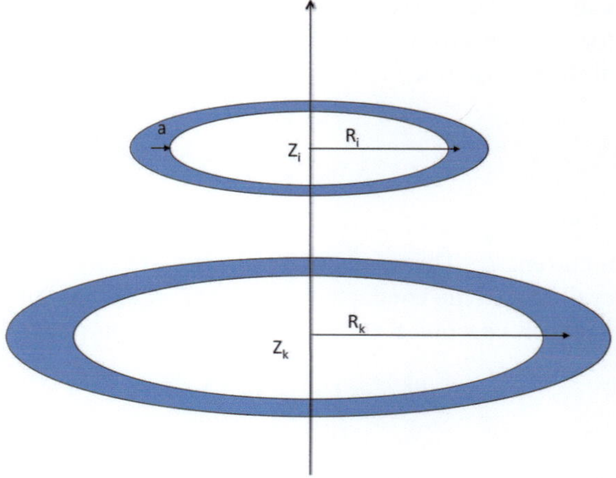

Fig. 16.1 Magnetic field coil geometry. The conductor systems can be approximated as a set of coaxial coils

where again the divergence term provides a vanishing contribution when integrated on a sphere with radius $R_s \to \infty$. The expression can be further simplified by assuming that the coils have a thickness negligible with respect to their radius and to the distance between them. This allows us to perform the integration over each conductor cross section separately and is equivalent to making the following replacement

$$\mathbf{j}_i(\mathbf{r}\prime)d^3r\prime \to I_i d\mathbf{l}_i \tag{16.13}$$

with $d\mathbf{l}_i$ being the line element along the $i-th$ coil, yielding

$$\int d^3r\prime \frac{\mathbf{B}_i \cdot \mathbf{B}_k}{\mu_o} = \frac{\mu_o}{4\pi} I_i I_k \int\int \frac{d\mathbf{l}_k(\mathbf{r}\prime\prime) \cdot d\mathbf{l}_i(\mathbf{r}\prime)}{|\mathbf{r}\prime - \mathbf{r}\prime\prime|} \equiv M_{ik} I_i I_k \tag{16.14}$$

A simple application of the expression above for the mutual inductance is the case of coaxial circular coils (Fig. 16.1)) of radius R_i and R_k located on the plane $Z = Z_i$ and $Z = Z_k$, respectively. The line element can be written as $d\mathbf{l}_i = (-R_i \sin\phi, R_i \cos\phi, 0)$, yielding

$$\begin{aligned} M_{ik} =& \frac{\mu_o}{4\pi} 2\pi R_i R_k \int_0^{2\pi} \cos\phi d\phi (R_i^2 + R_k^2 + (Z_i - Z_k)^2 - 2R_i R_k \cos\phi)^{-1/2} = \\ &= \mu_o[(R_i + R_k)^2 + (Z_i - Z_k)^2]^{1/2} \left[\left(1 - \frac{\kappa^2}{2}\right) K(\kappa^2) - E(\kappa^2)\right] \end{aligned} \tag{16.15}$$

with

$$\kappa^2 = \frac{4R_i R_k}{(R_i + R_k)^2 + (Z_i - Z_k)^2} \tag{16.16}$$

A second illustrative example is a coil (denoted by i) inside a cylindrical solenoid (denoted by k). The coil radius R_i is much smaller than the solenoid radius R_k. For the sake of simplicity we assume that the solenoid width Δ_s is small and use the relation $\mu_o j_{k\phi}\Delta_s = B_z$. In this case it is easier to use Eq. (16.4) to give

$$\int d^3r\prime \frac{\mathbf{B}_i \cdot \mathbf{B}_k}{\mu_o} = \int d^3r\prime \frac{\mathbf{A}_i \cdot \nabla \times \mathbf{B}_k}{\mu_o} = 2\pi R_k B_z \int_{-\infty}^{+\infty} dZ \frac{A_{i\phi}(R = R_k)}{\mu_o} \tag{16.17}$$

The integral can be evaluated using the asymptotic form for $A_{i\phi}$ given in Eq. (13.16) yielding

$$\int_{-\infty}^{+\infty} dZ \frac{A_{i\phi}(R = R_k)}{\mu_o} = \frac{I\pi R_o^2}{4\pi R_k} R_k^2 \int_{-\infty}^{+\infty} \frac{dZ}{(R_k^2 + Z^2)^{3/2}} = \frac{I\pi R_o^2}{2\pi R_k} \tag{16.18}$$

Thus

$$\int d^3r\prime \frac{\mathbf{B}_i \cdot \mathbf{B}_k}{\mu_o} = (I\pi R_o^2)B_z \tag{16.19}$$

Similarly, it is possible to evaluate the self-inductance of a cylindrical solenoid of radius R_{CS} and length L_{CS}. We assume that the magnetic field is non-zero only inside the solenoid volume. The magnetic field is given by Eq. (13.8) and is constant inside the volume. Thus

$$U = \frac{B^2}{2\mu_o}\pi R_{CS}^2 L_{CS} = \frac{1}{2}\frac{\mu_o \pi R_{CS}^2}{L_{CS}} I^2 \tag{16.20}$$

16.2 Relation Between Magnetic Energy and Magnetic Fluxes

In the previous section we have introduced the concept of self and mutual inductance starting from the definition of magnetic energy. It is possible to define the same quantities in terms of the magnetic fluxes.

The flux Φ_{ik} across the $i-th$ coil of the magnetic field generated by the $k-th$ coil is defined as

$$\Phi_{ik} = \int dS_i \mathbf{n} \cdot \mathbf{B}_k = \int dS_i \mathbf{n} \cdot \nabla \times \mathbf{A}_k = \oint d\mathbf{l}_i \cdot \mathbf{A}_k =$$
$$= \frac{\mu_o}{4\pi} I_k \int\int \frac{d\mathbf{l}_i(\mathbf{r}\prime\prime) \cdot d\mathbf{l}_k(\mathbf{r}\prime)}{|\mathbf{r}\prime - \mathbf{r}\prime\prime|} = M_{ik} I_k \tag{16.21}$$

The same definition can be applied to the case $i = k$ with the caveat about the logarithmic singularity of the integral for the self-inductance.

Using matrix notation we can write

$$\Phi_i = \sum_k \Phi_{ik} = \sum_k M_{ik} I_k \quad or \quad I_i = \sum_k M_{ik}^{-1} \Phi_k \tag{16.22}$$

The magnetic energy can be expressed either in terms of the currents or in terms of the fluxes

$$U = \frac{1}{2} \sum_{ik} M_{ik} I_i I_k = \frac{1}{2} \sum_i I_i \Phi_i \tag{16.23}$$

16.3 Force Between Coils

A conductor carrying a current density **j** immersed in a magnetic field **B** is subject to a force per unit volume $\mathbf{j} \times \mathbf{B}$ with **B** being the total magnetic field (i.e. the magnetic field generated by all the currents). The calculation of this force requires the self-consistent solution of Eq. (13.2) for **A**. A complete analysis can be found in Refs. [1, 2]. However there is a simpler method to evaluate the force on a coil by using the expression of the magnetic energy together with the principle of virtual works.

The change of the magnetic energy in time is given by two terms

$$\frac{dU}{dt} = \frac{1}{2} \sum_k \left[I_k \frac{d\Phi_k}{dt} + \frac{dI_k}{dt} \Phi_k \right] \tag{16.24}$$

using Eq. (16.22) the change in magnetic energy can also be expressed as follows

$$\begin{aligned}
\frac{dU}{dt} &= \frac{1}{2} \sum_k I_k \frac{d\Phi_k}{dt} + \frac{1}{2} \sum_{kj} \Phi_k (M^{-1})_{kj} \frac{d\Phi_j}{dt} + \frac{1}{2} \sum_{kj} \Phi_k \frac{d(M^{-1})_{kj}}{dt} \Phi_j = \\
&= \frac{1}{2} \sum_k I_k \frac{d\Phi_k}{dt} + \frac{1}{2} \sum_{kjlm} \Phi_k [-(M^{-1})_{kl} \frac{dM_{lm}}{dt} (M^{-1})_{mj}] \Phi_j = \\
&= \frac{1}{2} \sum_k I_k \frac{d\Phi_k}{dt} - \frac{1}{2} \sum_{lm} \frac{dM_{lm}}{dt} I_l I_m
\end{aligned} \tag{16.25}$$

the first term corresponds to the input electrical power whereas the second term to the output mechanical power, with the latter that can be identified with $-\mathbf{F} \cdot (d\mathbf{r}dt)$. Therefore we can write e.g. for the force in the direction x acting on the $j - th$ coil

$$F_{x_j} = \frac{1}{2} \sum_k \frac{\partial M_{jk}}{\partial x_j} I_j I_k \tag{16.26}$$

where the derivative is performed with respect to the space coordinates of the $j - th$ coil.

16.4 Equilibrium of a Current Carrying Loop

The problem of the equilibrium of a current carrying loop is a first approximation to the equilibrium of a toroidal plasma that will be considered in the next chapter.

The force can be determined from the expression of the magnetic energy of a coil. Making the derivative at constant flux as shown above we can determine the component F_R in the direction of the major radius R_o and the component F_a in the direction of the minor radius a.

$$F_R = \frac{I_{loop}^2}{2}\frac{\partial L}{\partial R_o} \approx \frac{I_{loop}^2}{2}\mu_o\left[\ln\left(\frac{8R_o}{a}\right) + \frac{l_i}{2} - 1\right] \tag{16.27}$$

$$F_a = \frac{I_{loop}^2}{2}\frac{\partial L}{\partial a} \approx -\frac{I_{loop}^2}{2}\mu_o\frac{R_o}{a} \tag{16.28}$$

Thus a current carrying loop is subject to an outward force that tends to increase its major radius and a compression force in the direction of the minor radius. As we will see in the next chapter the latter balances the expansion force associated with the plasma pressure and the poloidal current. For the moment we will indicate the quantities with the suffix *cl* (current loop) to indicate that these refer to the present simplified equilibrium.

To maintain the loop at a constant radius R_o we need to impose a vertical field. For a pure vertical field the equilibrium condition can be determined by including in the expression of the magnetic energy the interaction energy between the plasma (assumed to be a current carrying loop) and the vertical field. This term has been determined above in Eq. (16.19) and corresponds to the energy of a magnetic dipole $\mu = \pi R_o^2 I_{loop}$.

$$U_1 = \frac{L I_{loop}^2}{2} + \mu B_z^{cl} \tag{16.29}$$

Taking the derivative of U_1 with respect to R_o (at constant magnetic flux Φ) and setting the resulting force to zero we obtain

$$F_R = \frac{I_{loop}^2}{2}\frac{\partial L}{\partial R_o} + 2\pi R_o I_{loop} B_z^{cl} = 0 \tag{16.30}$$

or

$$B_z^{cl} = \mu_o\frac{I_{loop}}{4\pi R_o}\left[\ln\left(\frac{8R_o}{a}\right) + \frac{l_i}{2} - 1\right] \tag{16.31}$$

In the next chapter we will determine a more precise expression of the vertical field by solving the equilibrium equation.

In order to generate the vertical field with the minimal set of coils it is possible to use a pair of Helmoltz coils with the coil current flowing in the direction opposite to that of the plasma. The field generated in this way has a vertical component that maintains the horizontal position of the plasma.

16.5 Forces and Stresses on a Cylindrical Solenoid

As a first application, we study the case of a cylindrical solenoid. The force in the radial direction is given by

$$F_R = -\frac{\partial U}{\partial R_{CS}} = -\frac{\partial}{\partial R_{CS}} \frac{\Phi^2/2}{(\mu_o \pi R_{CS}^2/L_{CS})} = \\ = \frac{\Phi^2}{\mu_o \pi R_{CS}^3/L_{CS}} = \mu_o \pi \frac{R_{CS}}{L_{CS}} I^2 \tag{16.32}$$

where the magnetic energy of the solenoid has been expressed in term of the magnetic flux. The force tends to open the solenoid (hoop force) and produces a radial outward displacement of the solenoid that, in turn, produces a radial compression and a tension along ϕ.

A simple evaluation of the stresses can be made by taking a sector of angle $\Delta\phi$ of a solenoid of finite width $|R_{out} - R_{in}|$ and assuming that the dominant contribution comes from the hoop stress that acts on the two sections of area $L_{CS}(R_{out} - R_{in})$

$$2\sigma_\phi L_{CS}(R_{out} - R_{in})\frac{\Delta\phi}{2} \approx \frac{J_{CS}B_{CS}}{2} R_{CS}\Delta\phi L_{CS}(R_{out} - R_{in}) \tag{16.33}$$

or, taking $B_{CS} \approx B_{CS}(R_{in})/2, \sigma_\phi \approx (B_{CS}(R_{in})^2/2\mu_o)R_{CS}/(R_{out} - R_{in})$ (Fig. 16.2).

It is instructive to compare this simple estimate with a proper evaluation of the stresses that in this case can be made analytically. The stress-strain relation for an elastic isotropic material can be written as

$$\sigma_i = \frac{Y}{(1+\nu)}\left(\frac{\partial u_i}{\partial x_i} + \frac{\nu e}{1-2\nu}\right) \tag{16.34}$$

$$\tau_{ij} = \frac{Y}{2(1+\nu)}\left(\frac{\partial u_i}{\partial x_j} + \frac{\partial u_j}{\partial x_i}\right) \tag{16.35}$$

where σ_i are the elements of the principal diagonal of the stress tensor, τ_{ij} are the shear stresses, Y is the Young module, ν is the Poisson ratio and $e \equiv \sum_i \partial u_i/\partial x_i$ with $\mathbf{u}$ the displacement vector.

For a cylinder, the choice of standard cylindrical coordinates makes the shear stresses to vanish. The equilibrium condition

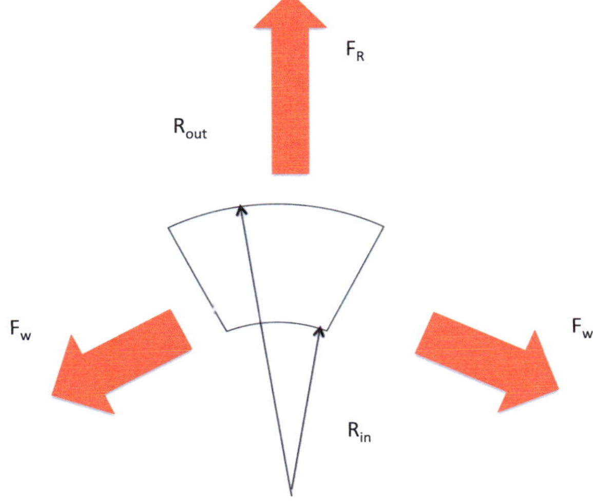

Fig. 16.2 Forces on a cylindrical solenoid. The radial expansion force (hoop force) is mostly balanced by the hoop stress

$$\nabla \cdot \sigma + \mathbf{j} \times \mathbf{B} = 0 \tag{16.36}$$

has only the radial component as non trivial condition

$$\frac{d\sigma_R}{dR} + \frac{\sigma_R - \sigma_\phi}{R} + (\mathbf{j} \times \mathbf{B})_R = 0 \tag{16.37}$$

Upon substituting the expression of σ_R and σ_ϕ in terms of $\mathbf{u}$, Eq. (16.37) becomes a differential equation for the radial displacement u_R

$$\frac{d}{dR}\frac{1}{R}\frac{d(Ru_R)}{dR} + \frac{1-\nu^2}{Y}(\mathbf{j} \times \mathbf{B})_R = 0 \tag{16.38}$$

and we have used the plane stress condition $\sigma_Z = 0$. Equation (16.38) can be analytically integrated to give the following expression for the hoop stress $\sigma_\phi$and the radial stress σ_R

$$\sigma_\phi = \frac{Y}{1-\nu^2}\left(\frac{u_R}{R} + \nu\frac{du_R}{dR}\right) \tag{16.39}$$

$$\sigma_R = \frac{Y}{1-\nu^2}\left(\nu\frac{u_R}{R} + \frac{du_R}{dR}\right) \tag{16.40}$$

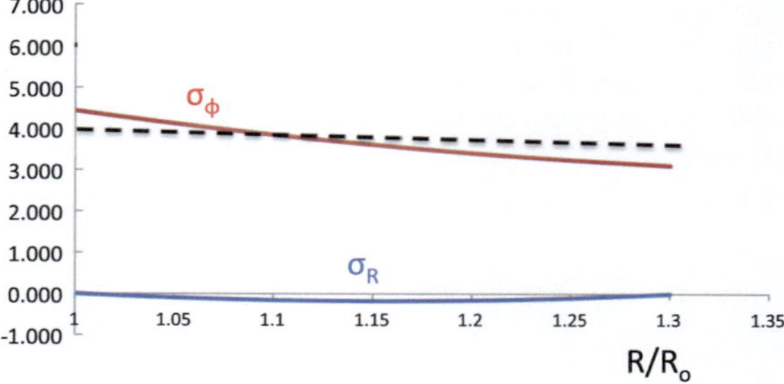

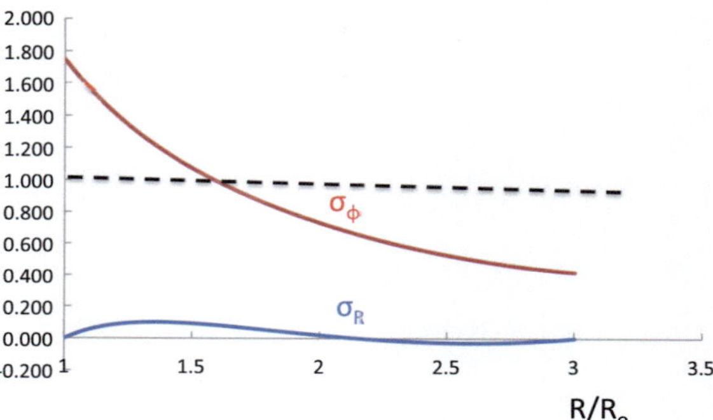

Fig. 16.3 Stresses in a cylindrical solenoid: σ_ϕ (red), σ_R (blu) and the approximation Eq. (16.33) as a function of R/R_o with R_o the solenoid inner radius. The stresses are normalised to $B^2/(4\mu_o)$. The upper figure corresponds to a thin solenoid whereas the lower figure to a thick solenoid

with the radial displacement u_R given by

$$Ru_R = -\frac{1-\nu^2}{Y}\int_{R_{in}}^{R} R\prime dR\prime \int_{R_{in}}^{R\prime} dR\prime\prime(\mathbf{j}\times\mathbf{B})_R + CR^2 + D \tag{16.41}$$

and the two constants C and D determined from the boundary conditions $\sigma_R(R_{in}) = \sigma_R(R_{out}) = 0$. The solution is plotted in Fig. 16.3 for a thin solenoid and a thick solenoid. It is clear from Fig. 16.3 that the hoop stress is the main component of the stress tensor. The radial $\mathbf{j} \times \mathbf{B}$ force is equilibrated by the tensile stress of the coil.

The stresses inside the solenoid attain the largest value at the inner radius and are proportional to $B^2/2\mu_o$. Thus the production of high magnetic field presents important challenges. The stresses from Eqs. (16.39)–(16.40) must be compared with the yield strength of the material. Typical values for copper alloys are $80MPa$ (note that $B^2/(4\mu_o) = 80MPa$ for $B = 20T$). To be noted that the main challenge typically arises from the shear stresses on the conductor insulator.

From the equations above it is possible to derive also the expressions for σ_R and σ_ϕ in the case of no volume force but with a pressure imposed at the two boundaries $\sigma_R(R_{in}) = -p_{in}$ and $\sigma_R(R_{out}) = -p_{out}$. In particular

$$\sigma_\phi = \frac{p_{in}R_{in}^2 - p_{out}R_{out}^2}{R_{out}^2 - R_{in}^2} + (p_{in} - p_{out})\frac{R_{in}^2 R_{out}^2}{R^2(R_{out}^2 - R_{in}^2)} \tag{16.42}$$

Use of this expression will be made later in this chapter when discussing the stresses on the central leg of the toroidal field magnet.

16.6 Bending-Free Toroidal Magnet

In this section we address the problem of determining the forces acting on a toroidal magnet. In general the problem has to be solved numerically but there is a simple limiting case that allows an analytic solution, namely the case of a magnet with a radial width small compared with its main dimension. Following the analysis of the cylindrical solenoid we may assume that the main contribution to the stress tensor is associated with the tensile force. We can model this case either with a continuous toroidal conductor with the current flowing in the poloidal direction or with a discrete set of coils. We will focus on the latter (more realistic) case and determine the optimal shape of the coils. There is a second contribution to the stresses that arise from the centering force on the coils. This force is assumed to be balanced by *wedging* the inner part of the coil to form a vault (Fig. 16.4). The alternative method *(bucking)* of supporting the centering force by the CS or by a bucking cylinder is also used.

Thus, the central leg of the toroidal field magnet can be modelled as a cylinder subject to a radial compression force on the outer radius and an axial tension.

The natural cross section of a cylindrical solenoid is a circle due to the symmetry of the problem. We may ask what happens if the cylindrical solenoid is bent into a

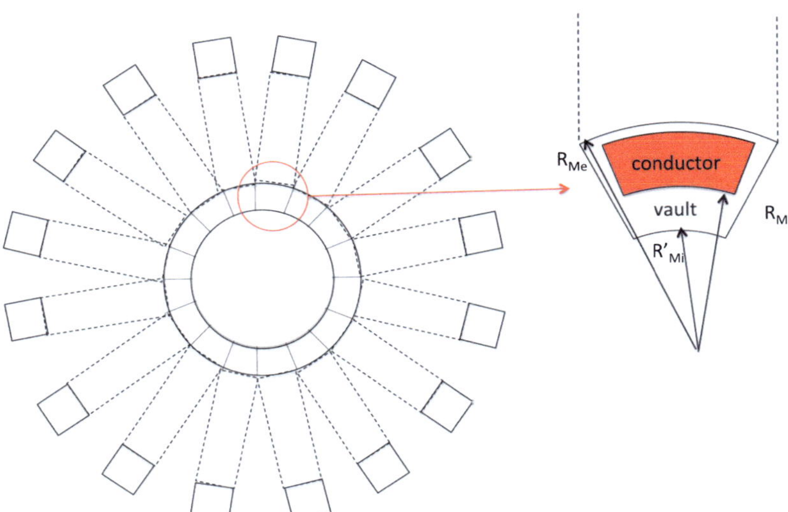

Fig. 16.4 The centering force can be reacted by wedging the inner part of the toroidal field coils to form a vault

torus. The natural shape for a toroidal magnet corresponds to the condition that the force due to the magnetic pressure is balanced by a constant tension. This problem is similar to that of finding the natural shape of a bubble in which the pressure difference is balanced by the surface tension. In the case of a soap bubble we know that the natural shape is a sphere. What is the solution for a toroidal magnet?

From symmetry considerations we know that the shape must be symmetric around the torus axis, i.e. must be independent of the toroidal angle ϕ. What we have to determine is the expression of the function $Z = Z_M(R)$ describing the shape on a $\phi = const.$ plane.

Let us consider a toroidal magnet made by a discrete set of N_{coil} filamentary coils. The current I_{coil} in each coil is $I_{coil} = I_M/N_{coil}$ with $\mu_o I_M = 2\pi B_o R_o$. Integrating the equilibrium equation Eq. (16.36) on a volume of length $\rho_c d\theta$ of the filamentary coil and using Gauss theorem we obtain

$$T d\theta = \frac{I_{coil} B(R) \rho_c}{2} d\theta \tag{16.43}$$

with the factor 2 inserted to take into account the average value of the magnetic field in the coil and the radius of curvature ρ_c of the curve $Z = Z_M(R)$ given by[1]

$$\rho_c = \frac{[1 + (\frac{dR}{dZ_M})^2]^{3/2}}{\frac{d^2R}{dZ_M^2}} = -\frac{(1+u^2)^{3/2}}{du/dR} \tag{16.44}$$

where $u = dZ_M/dR$ and in the stress tensor we have considered only the effect of the tension T along the filament. The above conditions can be cast in a simple equation for u by taking $B(R) = B_M R_M/R$

$$u = \frac{dZ_M}{dR} = \pm \frac{\ln(\frac{R}{R_M})}{[\kappa^2 - \ln^2(\frac{R}{R_M})]^{1/2}} \tag{16.45}$$

with $\kappa = 2T/(I_{coil} R_M B_M)$ and R_M the radius at which $u = 0$. Note that since the bending-free shape is a D-shape, R_M is different from the major radius of the plasma, i.e. the quantity R_o that we have used so far. It is convenient to express κ and R_M in terms of the two radii R_{out} and R_{in} at which $u = \infty$, yielding $\kappa = (1/2)\ln(R_{out}/R_{in})$ and $R_M = (R_{out} R_{in})^{1/2}$. We can identify R_o with the arithmetic average of R_{out} and R_{in}, yielding $R_o = R_M \cosh(\kappa)$ and the magnet minor radius a_M with $(R_{out} - R_{in})/2 = R_M \sinh(\kappa)$. The shape of the coil is obtained by integrating Eq. (16.45) from $R = R_{out}$ ($Z_M = 0$) and $R = R_{in}$. At $R = R_{in}$

[1] This expression can be easily obtained by Taylor expanding the curve around $R = R_o$ and $Z = Z_o = Z_M(R_o)$ and imposing that up to the second order in $(R - R_o)$ the curve is identical to that of a circle $(R - R_c)^2 + (Z - Z_c)^2 = \rho_c^2$. Imposing that the two expressions are identical the expressions for R_c, Z_c and ρ_c can be found.

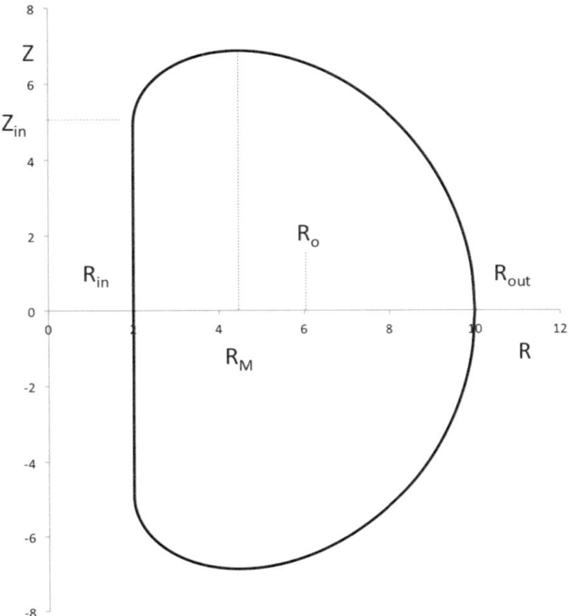

Fig. 16.5 Bending-free shape. The solution of Eq. (16.45) is shown up to $R = R_{in}$, $Z = \pm Z_{in}$. These two points are connected by a vertical line

we find $Z_M(R_{in}) = Z_{in} = \pi R_M \kappa I_1(\kappa)$, with $I_n(\kappa)$ being modified Bessel functions [3], and in order to close the coil we have to join the two symmetric points ($R = R_{in}$, $Z = \pm Z_{in}$) with a straight vertical line yielding the D shape shown in Fig. 16.5. The vertical line is the inner leg of the magnet. It is subject to a vertically directed tension T but it is not bending free. It is also subject to the centering force F_c (see below) that must be properly equilibrated.

The tension T can be written as

$$T = \frac{1}{4} I_{coil} R_M B_M \ln\left(\frac{R_{out}}{R_{in}}\right) = \frac{1}{4} I_{coil} B_M (R_o^2 - a_M^2)^{1/2} \ln\left(\frac{R_o + a_M}{R_o - a_M}\right) \quad (16.46)$$

Taking as an example the ITER magnet the resulting force is $120MN$.

Let us consider a toroidal solenoid with a shape given by the curve $Z = Z_M(R)$. The magnetic field decreases as R^{-1} (see Eq.13.11) thus the magnetic energy is given by [2]

$$U = \frac{1}{2}\frac{\mu_o}{\pi} \int Z_M(R) \frac{dR}{R} I_M^2 = \frac{1}{2}\left[\frac{\mu_o R_M \kappa^2}{2} (I_0(\kappa) + 2I_1(\kappa) + I_2(\kappa)) N_{coil}^2\right] I_{coil}^2 \quad (16.47)$$

with the integral extended between the R_{in} and R_{out} and I_M the magnet current. The expression in square bracket is the inductance of each coil.

For comparison we report also the inductance for a magnet with rectangular and elliptical shape. For a rectangular shape $Z_M(R) = b_M$ and we obtain

$$U = \frac{1}{2}\frac{\mu_o}{\pi} b_M \ln\left(\frac{R_o + a_M}{R_o - a_M}\right) I_M^2 \tag{16.48}$$

whereas for an elliptical shape $Z_M(R) = b_M[1 - (R - R_o)^2/a_M^2]^{1/2}$ we have

$$U = \frac{1}{2}\mu_o \frac{b_M}{a_M}(R_o - (R_o^2 - a_M^2)^{1/2}) I_M^2 \tag{16.49}$$

On each coil acts a centripetal force given by

$$F_c = -\frac{1}{N_{coil}}\left(\frac{\partial U}{\partial R_o}\right)_\Phi = \frac{I_M^2}{N_{coil}}\left(\frac{\partial L}{\partial R_o}\right)_{a_M} \tag{16.50}$$

In making the derivative with respect to R_o one must be careful to take constant a_M. This is easy for the rectangular and the elliptic shape since the inductance is written explicitly in terms of R_o and a_M whereas for the bending free shape it requires some care. In the latter case

$$\left(\frac{\partial L}{\partial R_o}\right)_{a_M} = \left(\frac{\partial L}{\partial R_M}\right)_\kappa \frac{\partial R_M}{\partial R_o} + \left(\frac{\partial L}{\partial \kappa}\right)_{R_M} \frac{\partial \kappa}{\partial R_o} =$$
$$= \frac{\mu_o \kappa^2}{2} \frac{I_0(\kappa) + 2I_1(\kappa) + I_2(\kappa)}{\cosh(\kappa)} + \frac{\mu_o}{2\sinh(\kappa)} \frac{d}{d\kappa}[\kappa^2(I_0(\kappa) + 2I_1(\kappa) + I_2(\kappa))] \tag{16.51}$$

Taking as an example the rectangular shape we obtain

$$F_c \approx \frac{\mu_o}{\pi} \frac{a_M b_M}{R_o^2 - a_M^2} N_{coil} I_{coil}^2 \tag{16.52}$$

Taking as a reference the ITER parameters $B_o = 5.3T$, $R_o = 6.2m$, $I_M = 157MA$, $N_{coil} = 18$, $I_{coil} = 8.7MA$, $a_M = 4m$, $b_M = 8m$ we obtain $F_c \approx 360MN$.

The force acting on the central leg alone (the expressions above correspond to the balance between the inward force on the inner leg and the outward force on the outer leg) can be evaluated directly from the Lorentz force assuming a constant current density $j(R) = j_M$ for $R_{Mi} \le R \le R_{Me}$ yielding (see Fig. 16.4)

$$F_c = \int \frac{j \times B dV}{N_{coil}} = \int \mu_o j_M^2 \frac{R^2 - R_{Mi}^2}{2R} 2\pi R dR \frac{L_{TF}}{N_{coil}} =$$
$$= \frac{4\pi}{3} L_{TF} \frac{B(R_{Me})^2}{\mu_o} \frac{R_{Me}^2(R_{Me} + 2R_{Mi})}{(R_{Me} + R_{Mi})^2 N_{coil}} \tag{16.53}$$

with L_{TF} the length of the central leg. The resulting stress can be evaluated by dividing the centering force on all the coils by the inner surface of the coil $2\pi R_{Mi} L_{TF}$, yielding

$$\sigma_{R,center} = \frac{4}{3} \frac{B(R_{Me})^2}{2\mu_o} \frac{R_{Me}^2(R_{Me} + 2R_{Mi})}{R_{Mi}(R_{Me} + R_{Mi})^2} \tag{16.54}$$

The centering force is equilibrated by the vault formed by the nose of the TF coils. The solution for the ϕ component of the stress is determined by considering a cylinder subject to an uniform external pressure due to the centering force and zero internal pressure. Using Eq. (16.42) the following expression for the maximum value of the hoop stress in the vault is obtained

$$\sigma_{\phi,vault} = 2\sigma_{R,center} \frac{R_{Mi}^2}{R_{Mi}^2 - R_{Mi}'^2} \tag{16.55}$$

with $R'_{Mi} < R_{Mi}$ the inner vault radius.

Finally, it must be noted that in addition to the centering and vertical forces, the toroidal field magnet is subject to *out of plane forces* arising from the combination of the magnet current and the poloidal field. These forces produce a tilting moment that must be equilibrated by adequate supporting structures joining adjacent coils.

16.7 Production of Magnetic Fields

The production of the external magnetic field can be made via copper or superconducting coils. Copper magnets have been in use for many decades. The problem with the use of copper is that the power dissipated in the coils may be very large. The dissipated power heats the coils that, in turn, increase their resistivity dissipating even more power. Therefore, for fusion energy application the preferred solution is the use of superconducting magnets. Copper coils are used only when the conditions do not allow the use of superconducting materials, such as for coils mounted inside the reaction chamber.

Superconductivity is the property, discovered in 1911 by Kamerlingh Onnes [4], of certain materials that exhibit zero electrical resistivity below a critical temperature. A related property is the *Meissner effect* [5]: a superconducting material immersed in a magnetic field produces a magnetisation equal and opposite to the applied magnetic field in such a way that the total magnetic field is zero inside the material. There is a critical value of the magnetic field above which the Meissner effect disappears. More precisely, it is possible to distinguish between two types of superconducting materials:

- Type I superconductors exhibit a sharp threshold in the Meissner effect that disappears completely above a critical value.

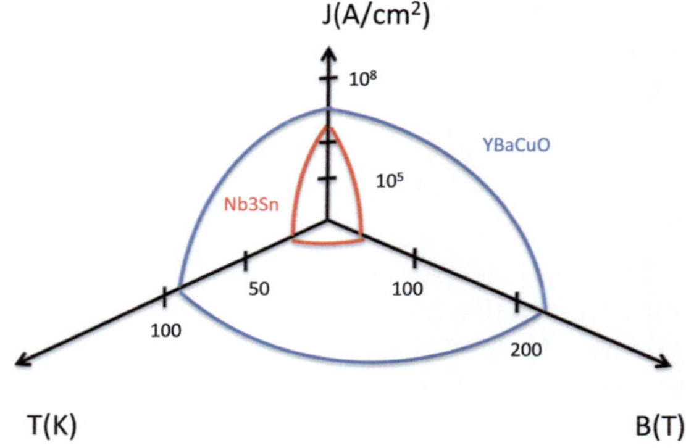

Fig. 16.6 Region of existence of superconducting materials for the case of YBaCuO and Nb3Sn. The operating point must be chosen on the basis of a trade off between the three quantities

- Type II superconductors exhibit two critical values of the magnetic field (H_{c1} and H_{c2}). In this case the Meissner effect disappears totally only above H_{c2} whereas between H_{c1} and H_{c2} the material is in the so called *vortex state*. In the vortex state the magnetic field does not penetrate inside the entire superconducting material but is instead concentrated in regions called *fluxons*. Inside these regions the material is no longer superconductive.

When current flows inside a Type II superconducting material, a Lorenz force is produced that tends to move the fluxons that are kept in place by the *pinning force*. Above a critical current density J_c the Lorentz force overcomes the pinning force and the fluxons start to move. This generate a variable magnetic flux (thus an electric field) that dissipates energy and destroys the superconducting state. The critical current density depends on the impurity content of the material.

Therefore, a superconducting material displays a region of existence in a 3D space temperature-magnetic field-current density. An example of this region for two superconducting materials is given in Fig. 16.6. The design of superconducting fusion magnets requires an optimal choice between these parameters since they have to operate at significant magnetic fields (especially on the inner leg of the toroidal field magnet) and at sufficiently high values of the current density to contain the magnet dimensions within reasonable limits.

For the use in magnetic fusion experiments the superconducting materials that are employed are mostly two: $NbTi$ and Nb_3Sn. More recently, there has been a growing interest on high temperature superconductors such as $YBCO$ (made by $Y - Ba - Cu - O$) and $BSCCO$ ($Bi - Sr - Ca - Cu - O$). The interest in high-T superconductors is mostly related with their capability of maintaining the superconducting state at much higher values of magnetic field and current density than $NbTi$ and Nb_3Sn, provided they are cooled down to liquid He temperature. This may lead to a reduction of the magnet dimensions at fixed plasma current and safety factor provided high yield steels are employed to contain magnetic stress [6].

As an example, the critical current density in the ITER strand can be modelled as follows [7]

$$J_{crit} = \frac{C}{B} s(\epsilon)[1 - (\frac{T}{T^*_{crit}(0,\epsilon)})^{1.52}][1 - (\frac{T}{T^*_{crit}(0,\epsilon)})^2](\frac{B}{B^*_{c2}(T,\epsilon)})^p(1 - (\frac{B}{B^*_{c2}(T,\epsilon)}))^q \tag{16.56}$$

where

$$T^*_{crit}(B,\epsilon) = T^*_{c0max}[s(\epsilon)]^{1/3}(1 - \frac{B}{B^*_{c2}(0,\epsilon)})^{1/1.52} \tag{16.57}$$

$$B^*_{c2}(T,\epsilon) = B^*_{c20max}[s(\epsilon)][1 - (\frac{T}{T^*_{crit}(0,\epsilon)})^{1.52}] \tag{16.58}$$

$$s(\epsilon) = 1 + \frac{1}{1 - C_{a1}\epsilon_{0,a}}[C_{a1}((\epsilon^2_{sh} + \epsilon^2_{0,a})^{1/2} - ((\epsilon - \epsilon_{sh})^2 + \epsilon^2_{0,a})^{1/2}) - C_{a2}\epsilon] \tag{16.59}$$

$$\epsilon_{sh} = \frac{C_{a2}\epsilon_{0,a}}{(C^2_{a1} - C^2_{a2})^{1/2}} \tag{16.60}$$

with ϵ being the strain on the cable. For the ITER TF Nb_3Sn conductor the fitting constants are $C = 16500$, $B^*_{c20max} = 32.97T$, $T^*_{c0max} = 16.06K$, $p = 0.63$, $q = 2.1$, $C_{a1} = 44$, $C_{a2} = 4$, $\epsilon_{0,a} = 0.00256$.

Superconducting cables can experience off normal events *(quench)* in which superconductivity is lost at some location, energy start to be dissipated and the entire magnet makes a transition to a normal conductor state. Since the energy stored in fusion magnets is very high, the rapid dissipation of this energy may damage the component. For this reason, superconducting cables are designed to dissipate the energy during a quench through copper filaments that are part of the cable. When a hot spot is detected, the power supply is disconnected and the current is short circuited on a resistor. The resistor is chosen in such a way that the characteristic decay time L/R is sufficiently long that the voltage in the coil remains below a certain value again to avoid possible breakdowns in the insulator.

The cable is enclosed in a steel jacket. For the sake of simplicity, in the following the conductor will be modelled as shown in Fig. 16.7 with a square steel casing. The conductor is immersed in a bath of liquid He that circulate to remove the heat generated e.g. by residual nuclear heating. To become a superconducting material the cable must be heat treated first at 680^oC. The Nb_3Sn becomes brittle after heat treatment and needs to be wound in its final shape before heat treatment. After heat treatment the conductor is insulated by wrapping around it a suitable material (e.g. kapton). The conductor is then impregnated with resin to form a winding pack that is inserted in a steel casing to contain the large forces acting on the coil.

In the design of a superconducting cable the non-copper section is determined by the required temperature margin ΔT_{margin} whereas the copper section is determined from the hot-spot temperature T_{max}, the maximum temperature that can be reached

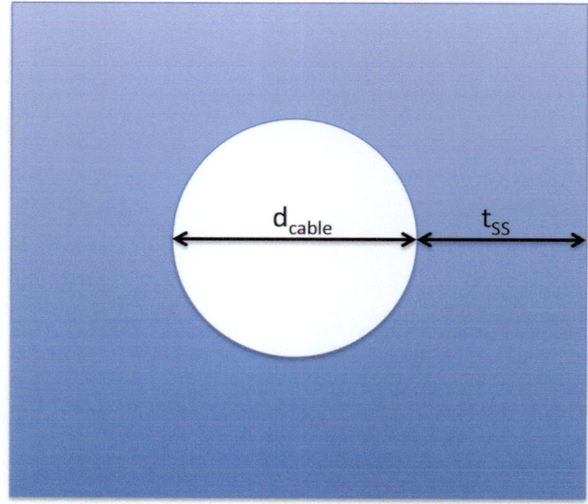

Fig. 16.7 Model for a square SC conductor. The steel jacket thickness is determined from the stresses in the magnet. The circular cable section is subdivided in a He section A_{He}, a non-copper section $A_{noncopper}$ and a copper section A_{Cu}

in the conductor during a fast safety discharge. Furthermore the critical current must be evaluated taking into account the strain in the magnet.

In the following discussion we will simply assume that the current density in the superconducting cable is fixed at the value of $J_{cable} = 50MA/m^2$ and determine the dimensions of the steel jacket and casing on the basis of the allowable stresses.

The average current in the conductor can be determined from geometric consideration as

$$J_{conductor} = J_{cable}\frac{\pi}{4(1+\frac{2t_{ss}}{d_{cable}})^2} \tag{16.61}$$

16.8 Flux Balance

The minimal set of coils needed to keep in equilibrium a tokamak plasma is made of:

- A pair of coils for the horizontal position control. These coils produce a vertical field and have a radius larger than the outer plasma radius. The current has the opposite sign of the plasma current.
- A pair of coils above and below the plasma to produce a finite elongation. Thus the current has the same sign of the plasma current. The combination of the plasma current and of the elongation coils produces a pair of magnetic separatrices and two $X-points$. If there is a perfect up-down symmetry the two separatrices overlap. However, typically the plasma equilibrium is not up-down symmetric. The separatrix closer to the plasma defines the plasma boundary.

- A vertical solenoid to produce a time varying magnetic flux and therefore an e.m.f. that drives the plasma current. After bringing the plasma current from zero to the desired value, an additional e.m.f. is necessary to maintain the current against the dissipation due to the finite plasma resistivity.

In this section we evaluate the requirements on the central solenoid to generate the plasma current and to maintain it for a certain amount of time.

During the plasma current ramp up the current in each coil changes in time in order to match the equilibrium conditions. This produces an e.m.f. that has in principle to be accounted for the plasma current evolution. For the sake of simplicity we will consider the contribution to the e.m.f. coming only from the central solenoid that we will assume to be infinite in the vertical direction. Therefore the flux balance will be determined on the basis of a single coil (the plasma) encircling a cylindrical solenoid.

Faraday's law, written in integral form, yields

$$\frac{d(LI_{plasma})}{dt} + R_{plasma} I_{plasma} = \frac{d\Phi_{CS}}{dt} \tag{16.62}$$

with R_{plasma} the plasma resistance and Φ_{CS} the flux generated by the central solenoid. The plasma resistance can be obtained from the resistivity η and the plasma geometry

$$R_{plasma} = \frac{2\pi R_o}{\pi \kappa a^2}\eta \tag{16.63}$$

whereas the central solenoid flux is obtained from Eq. (13.8) (see also Exercise 2 at the end of Chap. 13)

$$\Phi_{CS} = \mu_o \pi j_{CS} \frac{R_{CSe}^3 - R_{CSi}^3}{3} = \pi B_{CS}(R_{CSi}) \frac{R_{CSe}^2 + R_{CSe} R_{CSi} + R_{CSi}^2}{3} \tag{16.64}$$

with R_{CSe} and R_{CSi} the external and internal radius of the central solenoid as shown in Fig. 16.8.

The plasma current evolution is schematically shown in Fig. 16.9. It increases linearly during the ramp-up phase from $I = 0$ to the target value achieved at the start of the flat-top phase ($t = t_{SOF}$). The end of the current flat-top is at $t = t_{EOF}$. After the flat top the current is ramped down to zero in a controlled way.

We can integrate Eq. (16.62) from the start-up time ($t = 0$) to the end of the plasma current flat-top time t_{EOF}, yielding

$$\begin{aligned} &LI_{plasma}(t_{SOF}) + \int_0^{t_{SOF}} dt\, R_{plasma} I_{plasma} + \\ &+ R_{plasma} I_{plasma}(t_{SOF})(t_{EOF} - t_{SOF}) = \Phi_{CS}(t_{EOF}) - \Phi_{CS}(0) \end{aligned} \tag{16.65}$$

The first term is the so-called inductive flux, the second and the third terms are the resistive flux, associated with the plasma dissipation, during the ramp=up and flat top phase, respectively. For a machine like ITER $L \approx 9\mu H$, $I_{plasma} = 15MA$,

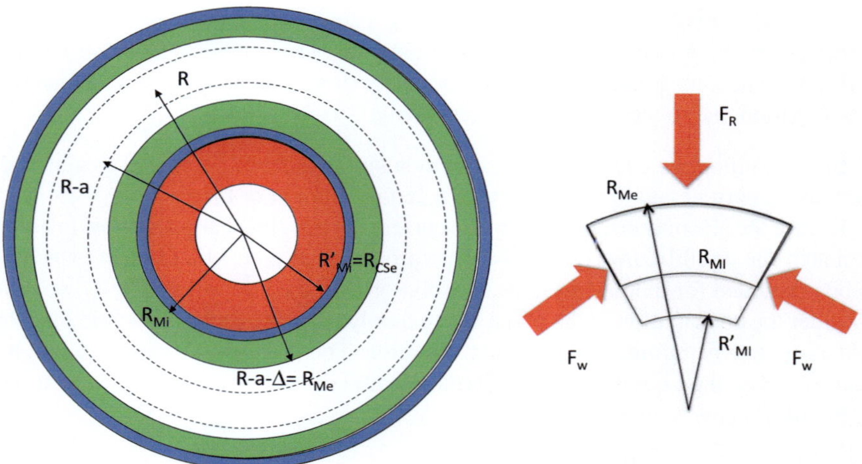

Fig. 16.8 Model of the tokamak system. The centering force acting on the inner leg of the toroidal field magnet (inner green annulus) is balanced by a vault forming the inner blue annulus. The central solenoid (red) is characterised by an outer and an inner radius determined on the basis of the flux requirements

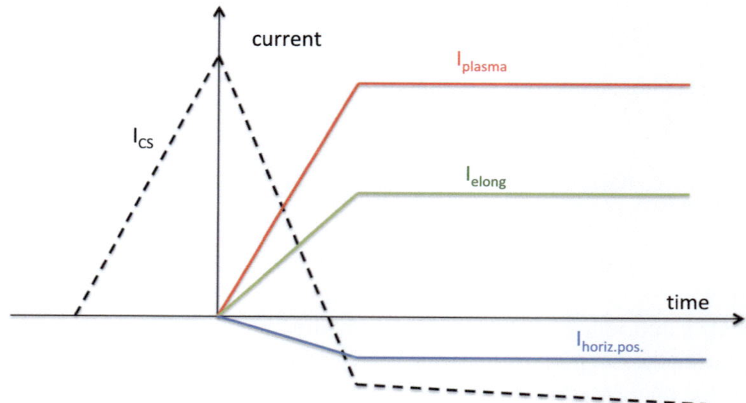

Fig. 16.9 Example of current evolution in the tokamak windings

and $V = R_{plasma} I_{plasma} \approx 0.08V$ for $< T >\approx 10keV$ (see exercise). Therefore the inductive flux is of the order $135Wb$ whereas the resistive flux for a flat-top duration of $300s$ is $24Wb$ (often the resistive flux is measured in the equivalent unit volt seconds). The resistive flux consumption during the ramp-up phase is estimated empirically as

$$\int_0^{t_{SOF}} dt R_{plasma} I_{plasma} = C_{Ejima} \mu_o R_o I_{plasma}(t_{SOF}) \tag{16.66}$$

with the Ejima coefficient estimated as $C_{Ejima} = 0.3$. The sum of the inductive and resistive flux must be equal to the flux variation of the central solenoid. For a given dimension, this is limited by the maximum current density and magnetic field of the central solenoid. The maximum variation is obtained by starting at $t = 0$ with the solenoid carrying the maximum possible current and ending with the maximum possible current in the opposite direction (double swing). The maximum magnetic field B_{CSmax} is set by the properties of the superconducting cable. Therefore the size of the central solenoid is determined by the flux balance condition Eq. (16.65).

The flux variation due to the central solenoid can be compared with that associated with the horizontal position coils. The latter can be estimated as the change Φ_V in the flux of the vertical magnetic field across the plasma from $t = 0$ ($I = 0$) and the end of the ramp-up ($I = I_{plasma}(t_{SOF})$). From Eq. (16.31) the following estimate is obtained

$$\Phi_V = \frac{\mu_o I_{plasma}(t_{SOF})}{4\pi R_o}\left[\ln\left(\frac{8R_o}{a}\right) + \frac{l_i}{2} - 1\right]\pi R_o^2 \tag{16.67}$$

whereas the inductive flux is given by

$$L I_{plasma}(t_{SOF}) \approx I_{plasma}(t_{SOF})\mu_o R_o\left[\ln\left(\frac{8R_o}{a}\right) + \frac{l_i}{2} - 1\right] \tag{16.68}$$

Therefore the vertical field flux variation is approximately 25% of the inductive flux variation and, as a first approximation, will be neglected

16.9 Model for the Tokamak Magnet System

We are now in the position to formulate a simple model for the superconducting toroidal field magnet and central solenoid (the two more critical components in a machine). This follows the analysis given in Refs. [8, 9].

The toroidal field coils are modelled as a bending-free shape plus a straight central leg. The critical part in the analysis is the central leg that we assume divided in two regions:

- the winding pack that extends between the two radii R_{Me} and R_{Mi} that are determined by the design value for the magnetic field in the centre and the current density of the magnet conductor;
- the vault that extends between the radii R_{Mi} and R'_{Mi}. It is assumed that the centering force is entirely supported by the vault.

The central solenoid extends between the radii $R_{CSe} = R'_{Mi}$ and R_{CSi}. The latter is determined by the maximum magnetic field and the current density in the central solenoid.

The design parameters are the plasma current I_{plasma} (the main control parameter of the physics performance as we have seen in Chap. 13) and plasma shape (aspect ratio, elongation and triangularity), the maximum current density and magnetic field in the cable and the mechanical parameters of steel. The plasma current is related to the magnetic field on the magnetic axis through Eq. (16.69)

$$q_{95} = \frac{2\pi B_\phi a^2}{R_o \mu_o I_{plasma}} S(\kappa, \delta, \epsilon) \tag{16.69}$$

with $q_{95} \geq 3$.

The design starts with a guess for the plasma major radius R_o. From the plasma aspect ratio the plasma minor radius a is determined. The external radius R_{Me} of the TF magnet internal leg is

$$R_{Me} = R_o - a - \Delta_b \tag{16.70}$$

with the width Δ_b accounting for the plasma-wall distance, the blanket width (this is the main contribution), the vacuum vessel width and the thermal shield width and is determined by other considerations. We will assume $\Delta_b = 1m$. At this point we have to check that the maximum toroidal field (achieved in $R = R_{Me}$) is lower than the maximum field allowed by the superconductor

$$B_{\phi max} = \frac{B_o R_o}{R_{Me}} < B^*_{c2} \tag{16.71}$$

if $B_{\phi max} > B^*_{c2}$, the radius R_o must be increased. The radius R_{Mi} is obtained from the condition

$$I_M = \pi J_{conductor}(R^2_{Me} - R^2_{Mi}) = \frac{2\pi B_o R_o}{\mu_o} \tag{16.72}$$

with the current density $J_{conductor}$ determined from Eq. (16.61). The quantity t_{ss}/d_{cable} is determined by imposing the Tresca condition. The cable of the toroidal magnet and of the central solenoid have the same geometry but in general with different values of t_{ss} and d_{cable}. The stress arises from the combined effect of the centering stress and the hoop stress on the coil.

The stresses are concentrated on the steel jacket with the radial stresses amplified by a factor $(2t^{TF}_{ss} + d^{TF}_{cable})/(2t^{TF}_{ss})$ over the average radial stresses and can be determined from the average value given in Eq. (16.54) by imposing that it is supported by the steel in the jacket

$$\sigma_{R,eff} = \frac{4}{3}\frac{R^2_{Me}(R_{Me} + 2R_{Mi})}{R_{Mi}(R_{Me} + R_{Mi})^2}\frac{B^2_{\phi max}}{2\mu_o}\frac{2t^{TF}_{ss} + d^{TF}_{cable}}{2t^{TF}_{ss}} \tag{16.73}$$

The hoop force on the magnet from Eq. (16.50) produces a tension directed along Z and given by

$$T = \frac{\pi}{\mu_o}(R_{Me}B_{\phi max})^2 \ln\left(\frac{R_2}{R_{Me}}\right) \tag{16.74}$$

To determine the tensional stress we take the average $(T/2)$ and apply it on the steel section S of the coil given by the sum of the section of the vault and the steel section of the conductor

$$S = \pi(R_{Mi}^2 - R_{Mi}'^2) + \pi(R_{Me}^2 - R_{Mi}^2)\Big[1 - \frac{\pi}{4}\Big(1 - \frac{2t_{ss}^{TF}}{2t_{ss}^{TF} + d_{cable}^{TF}}\Big)^2\Big] \tag{16.75}$$

yielding a peak tensional stress

$$\sigma_{Z,eff} = -\frac{B_{\phi max}^2}{2\mu_o}\frac{\pi R_{Me}^2 \ln(\frac{R_2}{R_{Me}})}{S} \tag{16.76}$$

Thus, the Tresca condition can be written as

$$\sigma_{R,eff} - \sigma_{Z,eff} \leq \sigma_o \tag{16.77}$$

with σ_o being the yield stress for steel.

The inner vault radius R'_{Mi} is determined by considering a cylinder subject to an uniform external pressure due to the centering force Eq. (16.55), yielding

$$\frac{4}{3}\frac{R_{Me}^2(R_{Me} + 2R_{Mi})}{R_{Mi}(R_{Me} + R_{Mi})^2}\frac{B_{\phi max}^2}{2\mu_o}\frac{2R_{Mi}^2}{R_{Mi}^2 - R_{Mi}'^2} \leq \sigma_o \tag{16.78}$$

Upon combining Eqs. (16.72)–(16.78) it is possible to determine R_{Mi}, R'_{Mi} and $t_{ss}^{TF}/d_{cable}^{TF}$ for a given R_{Me}. From these values and Eq. (16.61) it is possible to determine the conductor current density. If the magnet is made of N_{coil} coils each with N_{wire} wires it is possible to determine the (square) conductor area and therefore separately d_{cable}^{TF} and t_{ss}^{TF}.

After having determined R_{Mi} and R'_{Mi}, and so the external solenoid radius R_{CSe}, we can determine the internal solenoid radius R_{CSi} from the condition

$$B_{CSmax} = \mu_o j_{CS}(R_{CSe} - R_{CSi}) < B_{c2}^* \tag{16.79}$$

The central solenoid must again satisfy the condition that the maximum hoop stress is below σ_o. The hoop stress is obtained from Eq. (16.33) multiplying for a correction factor to account for the effective area of the steel (a better estimate of the effective stress can be found in Ref. [8]). The expression for the conductor current density is again obtained from Eq. (16.61). These two conditions can be explicitly solved to give

$$\frac{2t_{ss}^{CS}}{d_{cable}^{CS}} = \left(\frac{2\alpha + \pi\mu/4}{\mu(1+\alpha)}\right)^{1/2} - 1 \tag{16.80}$$

with $\alpha \equiv (B_{c2}^*)^2/(4\mu_o\sigma_o)$ and $\mu \equiv 4B_{c2}^*/(\pi\mu_o J_{cable} R_{CSe})$, and

$$R_{CSi} = R_{CSe}\left[1 - \mu\left(1 + \frac{2t_{ss}^{CS}}{d_{cable}^{CS}}\right)^2\right] \tag{16.81}$$

At this point we have all the elements to evaluate the central solenoid flux swing Φ_{CS} (Eq. (16.64)) and check if the flux balance equation is satisfied. If the central solenoid radius is too small the plasma radius R_o must be increased until the flux balance is satisfied.

It is important to stress that the design of the magnet system critically depends on the steel yield stress σ_o. Conventional austenitic steels such as AISI 316 LN have an yield strength at $4K$ that can be as high as $900MPa$. Taking some margin a value of $650MPa$ can be used in the expressions above. To increase further the magnetic field would not produce a reduction of the magnet dimension at fixed yield strength because a larger volume of steel would be required due to the larger magnetic forces [10]. To produce a significant reduction, higher yield strength materials must be considered as, e.g., Nitronic 50 [11] with yield strengths that exceed $1GPa$. It should be also noted that a further obstacle to the reduction of the toroidal field magnet dimensions comes from the finite extension of the blanket that is included in the quantity Δ_b in Eq. (16.70). Also in this case, materials with high shielding capability are required to reduce the radial build without compromising the shielding function of the blanket.

16.10 Suggestions for Further Readings

Many of the calculations presented in this chapter are extensively discussed in Refs. [1, 2]. The tokamak magnet system is described in more detail in [8, 9].

16.11 Exercises

Problem 16.1 Derive Eq. (16.47) using the properties of the modified Bessel functions of integer order.

Problem 16.2 Using Eq. 16.15 determine the forces between two coils with coil currents $1MA$ with radius $1m$ and an axial distance of $1m$.

Problem 16.3 Evaluate the inductance of a coil with dimensions equal to the ITER plasma and an internal inductance $l_i = 0.5$.

Problem 16.4 Derive Eq. (16.53)

Problem 16.5 Using an excel file to solve the equations for the model of the tokamak magnet system discuss the dependence of the machine radius on the parameter J_{cable}.

References

1. H.E. Knoepfel, *Magnetic Fields* (Wiley, 2000)
2. R.J. Thome, J.M. Tarrh, *MHD and Fusion Magnets. Field and Force Design Concepts* (Wiley, New York, 1982)
3. M. Abramowitz, I.A. Stegun, (eds.), Handbook of Mathematical Functions with Formulas, Graphs, and Mathematical Tables. Applied Mathematics Series, vol. 55 (Ninth reprint with additional corrections of tenth original printing with corrections (December 1972); first ed.). Washington D.C.; New York: United States Department of Commerce, National Bureau of Standards; Dover Publications (1983)
4. H. Kamerlingh Onnes, Further experiments with liquid helium. D. On the change of electric resistance of pure metals at very low temperatures, etc. V. The disappearance of the resistance of mercury, Communications from the Physical Laboratory of the University of Leiden; No. 122b, (1911)
5. W. Meissner, R. Ochsenfeld, Ein neuer Effekt bei Eintritt der Supraleitfähigkeit. Naturwissenschaften **21**(44), 787–788 (1933)
6. A.J. Creely et al., Overview of the SPARC tokamak. Publish. Online J. Plasma Phys. **86**, 865860502 (2020)
7. L. Bottura, B. Bordini, $J_C(B, T, \epsilon)$ parameterisation for the ITER Nb_3Sn production. IEEE Trans. Appl. Superconduct. **19**, 1521 (2009)
8. J.-L. Duchateau, Conceptual integrated approach for the magnet system of a tokamak reactor. Fusion Eng. Des. **89**, 2606–2620 (2014)
9. M. Kovari et al., "Process": a systems code for fusion power plants - part 2: engineering. Fusion Eng. Des. **104**, 9–20 (2016)
10. G. Federici et al., Relationship between magnetic field and tokamak size-a system engineering perspective and implications to fusion development. Nucl. Fusion **64**, 036025 (2024)
11. C.G. Fountzoulas, E.M. Klier, J.E. Catalano, D. Casem, L. Lamberson, J. Kimberley, (eds.), *Dynamic Behavior of Materials, Volume 1: Dynamic Characterization of Nitronic 30*, 40 and 50 Series Stainless Steels (Springer, 2016), p. 22. ISBN 978-3319411323

Chapter 17
Plasma Equilibrium

Abstract *In this chapter the analysis of axisymmetric equilibria is reviewed. The one dimensional equilibria are discussed first. The Grad-Shafranov equation is derived and simple solutions such as the Solovev equilibria are discussed.*

17.1 Introduction

The question we address in this chapter is the following: what is the distribution of plasma current in which all the forces acting on the plasma are balanced? The forces acting on plasmas at rest (i.e. without stationary flows) are associated with the plasma pressure p (that drives the plasma expansion) and the Lorenz force $\mathbf{j} \times \mathbf{B}$. Since the magnetic field is produced by the current density distribution, this is an example of a problem that must be solved self-consistently.

The force balance is expressed by the following equation

$$\nabla p = \mathbf{j} \times \mathbf{B} \tag{17.1}$$

together with Maxwell equations

$$\nabla \times \mathbf{B} = \mu_o \mathbf{j} \qquad \nabla \cdot \mathbf{B} = 0 \tag{17.2}$$

supplemented with appropriate boundary conditions that we will discuss in a moment. The output of the equilibrium calculation is the equilibrium magnetic field. But what is the input? We may assume that the pressure p is provided but this is generally not enough. Ampére's law provides an expression for the current density as a function of the magnetic field. The three components of the magnetic field, in turn, are constrained to satisfy $\nabla \cdot \mathbf{B} = 0$ which reduces the number of unknowns from three to two. The equilibrium equation provides three conditions but, in general, only one of them is non trivial. Thus, we need to provide a second function (e.g. one of the magnetic field component) in order to close the system.

F. Romanelli, *Physics of Nuclear Energy*, Springer Series in Plasma Science and Technology, https://doi.org/10.1007/978-981-97-9609-0_17

17.2 One Dimensional Configurations

As a first illustration of the equilibrium problem we consider a straight cylinder with circular cross section (Fig. 17.1). This problem is characterized by both azimuthal (i.e. along θ) and translational (i.e. along z) invariance. All the equilibrium quantities must be therefore functions only of the distance r from the cylinder axis. The condition $\nabla \cdot \mathbf{B} = 0$ together with $\partial/\partial\theta = 0$ and $\partial/\partial z = 0$ yields

$$\frac{1}{r}\frac{\partial(rB_r)}{\partial r} = 0 \tag{17.3}$$

that has the solution $B_r = A/r$. Since the field must be regular at $r = 0$ we need to set $B_r = 0$. Thus, the magnetic field has only two components B_θ and B_z. The current density, in turn, can be determined from Ampére's law

$$\mu_o j_\theta = -\frac{\partial B_z}{\partial r} \tag{17.4}$$

$$\mu_o j_z = \frac{1}{r}\frac{\partial(rB_\theta)}{\partial r} \tag{17.5}$$

$$\mu_o j_r = \frac{1}{r}\frac{\partial B_z}{\partial\theta} - \frac{\partial B_\theta}{\partial z} = 0 \tag{17.6}$$

with the condition $j_r = 0$ following from the symmetry of the configuration. Since both B_r and j_r are identically 0, the components of the equilibrium equation along θ and z are trivially satisfied. Only the component along r provides a non trivial condition that can be written as

$$\mu_o\frac{dp}{dr} = -B_z\frac{dB_z}{dr} - \frac{B_\theta}{r}\frac{d(rB_\theta)}{dr} \tag{17.7}$$

The term on the l.h.s. of this equation is the radial component of the pressure gradient, whereas on the r.h.s. the first term is the force associated with the axial field and the azimuthal current and the second term is the force associated with the azimuthal field and the axial current. Given e.g. the two functions $p(r)$ and $B_z(r)$ the poloidal magnetic field can be determined by multiplying by r^2 and integrating

$$\int_0^r r\prime^2 dr\prime \frac{d}{dr\prime}\left[p(r\prime) + \frac{B_z(r\prime)^2}{2\mu_o}\right] = -\frac{(rB_\theta)^2}{2\mu_o} \tag{17.8}$$

or, integrating by parts,

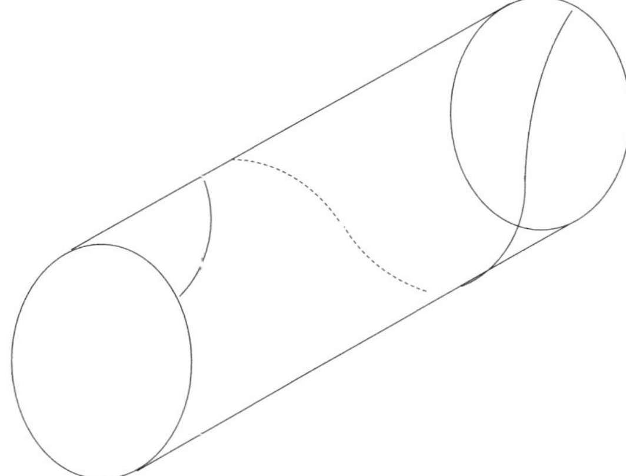

Fig. 17.1 A screw pinch is an example of one dimensional equilibrium. The magnetic field is the sum of a component along the cylinder axis and an azimuthal component

$$\frac{(rB_\theta)^2}{2\mu_o} = \int_0^r 2r\prime dr\prime \left[p(r\prime) + \frac{B_z(r\prime)^2}{2\mu_o}\right] - r^2\left[p(r) + \frac{B_z(r)^2}{2\mu_o}\right] \tag{17.9}$$

Equation 17.9 applied to $r = a$, the plasma radius at which the pressure is assumed to vanish and the axial magnetic field component matches its vacuum solution (i.e. the field produced by external coils), provides an interesting condition

$$\frac{(\mu_o I_p)^2}{8\pi^2} = \int_0^a 2rdr\left[\mu_o p(r) + \frac{B_z(r)^2}{2}\right] - a^2\frac{B_z(a)^2}{2} \tag{17.10}$$

with I_p being the total axial current and use has been made of the relation $B_\theta(a) = \mu_o I_p/(2\pi a)$. Upon introducing the poloidal beta defined as

$$\beta_p \equiv \frac{16\pi^2 \int_0^a rdrp(r)}{\mu_o I_p^2} \tag{17.11}$$

Equation (17.9) can be written as

$$< B_z^2 > -B_z(a)^2 = B_\theta(a)^2(1 - \beta_p) \tag{17.12}$$

with the brackets indicating the volume average of a quantity

$$< f >= \frac{\int_0^a 2rdrf(r)}{a^2} \tag{17.13}$$

Thus, if $\beta_p < 1$ the average axial field is larger than the vacuum field whereas if $\beta_p > 1$ the average axial field is smaller. We can say that *from the point of view of the axial magnetic field component* for $\beta_p < 1$ ($\beta_p > 1$) the equilibrium is paramagnetic (diamagnetic). However this should not be misleading. If we consider the effect on the total magnetic field, the plasma has always a diamagnetic behaviour.

17.3 Two Dimensional Configurations

The simple example of the straight cylinder shows that to solve the equilibrium equation two functions (e.g. $p(r)$ and $B_z(r)$) are needed, supplemented by the condition of regular behaviour of the solution near the axis (that has been used in solving Eq. (17.3) by setting $A = 0$) and a condition at the plasma boundary (that in this specific case could be chosen at $r = a$).

The above example can be easily generalised to toroidal geometry in the case of axisymmetry in which an ignorable coordinate exists ($\partial/\partial\phi = 0$) to derive the Grad-Shafranov equation [1, 2]. In this case, all the quantities are functions of the two variables R and Z only.

The equilibrium condition Eq. (17.1) has some general consequences. Taking the scalar product with $\mathbf{B}$, it follows that

$$\mathbf{B} \cdot \nabla p = 0 \tag{17.14}$$

This equation *per se* simply tells that the pressure is constant along the magnetic field line. However, if magnetic surfaces exist, following a field line that covers in an ergodic way a surface, this condition implies that the pressure is constant on the magnetic surface, i.e. ∇p is orthogonal to the magnetic surface.

On the other hand, taking the scalar product of Eq. (17.1) with $\mathbf{j}$, it follows that

$$\mathbf{j} \cdot \nabla p = 0 \tag{17.15}$$

Since ∇p is orthogonal to the magnetic surface, it follows that $\mathbf{j}$ must lie on the magnetic surface.

The existence of magnetic surfaces can be proved rigorously only for configurations with at least one degree of symmetry. In this case a smooth function f of the non-ignorable coordinates u_1, u_2 can be explicitly found such that the equation $f(u_1, u_2) = const.$ define the surface. Thus, for axisymmetric equilibria magnetic surfaces do exist. In this case the ignorable coordinate is the toroidal angle ϕ. An example of simply nested magnetic surfaces is given in Fig. 17.2.

To proceed further we need a suitable representation of the axisymmetric magnetic field. It can be shown that the more general form is

$$\mathbf{B} = F\nabla\phi + \nabla\psi \times \nabla\phi \tag{17.16}$$

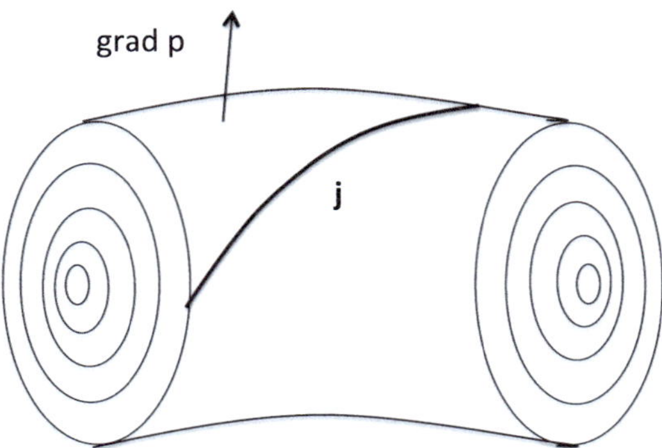

Fig. 17.2 Example of axisymmetric configuration with a system of simply nested magnetic surfaces. The degenerate surface is the magnetic axis. The pressure gradient is orthogonal to the magnetic surface and the current lies on the magnetic surface. Each surface corresponds to $\psi(R, Z) = const.$ with ψ the flux function

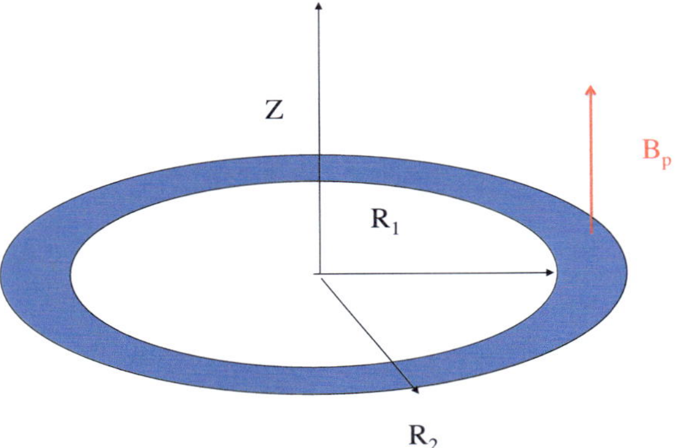

Fig. 17.3 The poloidal flux ψ is the flux of the poloidal magnetic field across a circular corona with two radii R_1 and R_2

with $\nabla\phi = \hat{\boldsymbol{\phi}}/R$ and $\hat{\boldsymbol{\phi}}$ being the unit vector along the ϕ coordinate. The functions F and ψ are unknown at the moment except that the axisymmetry requires $\partial F/\partial\phi = 0$ and $\partial\psi/\partial\phi = 0$. The two components of Eq. (17.16) can be obviously identified with the toroidal and the poloidal component, respectively. The function ψ takes the name of poloidal flux function as it can be shown to coincide with the flux of the poloidal field across a circular corona (Fig. 17.3).

We can now use the condition Eq. (17.14) together with $\partial p/\partial\phi = 0$ to obtain

$$\nabla\psi \times \nabla\phi \cdot \nabla p = 0 \tag{17.17}$$

This condition can be identically satisfied provided $p = p(\psi)$. The poloidal flux is therefore a label for the magnetic surface $\psi(R, Z) = const.$

From the general form of the magnetic field we can also determine an expression for the current density

$$\mu_o \mathbf{j} = \nabla \times \mathbf{B} = \nabla F \times \nabla\phi + \nabla \times (\nabla\psi \times \nabla\phi) = \\ = \nabla F \times \nabla\phi - \Delta\psi\nabla\phi - (\nabla\psi \cdot \nabla)\nabla\phi = \nabla F \times \nabla\phi - \Delta^*\psi\nabla\phi \quad (17.18)$$

where use has been made of the vector identities $\nabla \times \nabla f = 0$ and

$$\nabla \times (\mathbf{A} \times \mathbf{B}) = \mathbf{A}\nabla \cdot \mathbf{B} - \mathbf{B}\nabla \cdot \mathbf{A} + (\mathbf{B} \cdot \nabla)\mathbf{A} - (\mathbf{A} \cdot \nabla)\mathbf{B} \quad (17.19)$$

together, again, with the symmetry condition $(\nabla\phi \cdot \nabla) f = 0$. The second term in Eq. (17.18) can be also written in cylindrical coordinates as

$$\Delta\psi\nabla\phi + (\nabla\psi \cdot \nabla)\nabla\phi = [\frac{1}{R}\frac{\partial}{\partial R}(R\frac{\partial\psi}{\partial R}) + \frac{\partial^2\psi}{\partial Z^2} - \frac{1}{R}\frac{\partial\psi}{\partial R}]\nabla\phi = \\ = [R\frac{\partial}{\partial R}(\frac{1}{R}\frac{\partial\psi}{\partial R}) + \frac{\partial^2\psi}{\partial Z^2}]\nabla\phi \equiv \Delta^*\psi\nabla\phi \quad (17.20)$$

and the Grad-Shafranov operator Δ^* has been introduced. Again, the two terms in Eq. (17.18) can be easily identified as the components of the current density in the poloidal and toroidal directions, respectively. Using the expression for $\mathbf{j}$ Eq. (17.18), we can apply the condition Eq. (17.15) that, similarly to what has been shown above, yields

$$\nabla F \times \nabla\phi \cdot \nabla p = \nabla F \times \nabla\phi \cdot \nabla\psi\frac{dp}{d\psi} = 0 \quad (17.21)$$

or $F = F(\psi)$. Finally, the component of the equilibrium condition along the normal to the magnetic surface (i.e. along $\nabla\psi$) yields

$$\Delta^*\psi + \mu_o R^2\frac{dp}{d\psi} + F\frac{dF}{d\psi} = 0 \quad (17.22)$$

Equation (17.22) is the Grad-Shafranov equation. As in Eq. (17.7) the three terms represent the effect of the toroidal current and of the poloidal magnetic field, the effect of the pressure gradient and the effect of the poloidal current and of the toroidal field. It is an elliptic second-order partial differential equation. It can be solved for ψ once the two function $p(\psi)$ and $F(\psi)$ are assigned together with appropriate boundary conditions. In addition to the regularity of the function ψ at the origin, the second boundary condition depends on the problem. If the shape of the plasma boundary is assigned (fixed boundary problem) the value of ψ (or its normal derivative or a linear combination of the two) is assigned on that boundary. In practical situations, the form of the plasma boundary is unknown and has to be determined self consistently from the combination of the field produced by the plasma current and that produced by the external coils (free boundary problem). In the latter case the second boundary condition is the regularity of the solution for $r \to \infty$ and the matching of ψ at the plasma vacuum interface.

17.4 Analytic Solutions of the Grad-Shafranov Equation

The solution of the Grad-Shafranov equation generally requires a numerical integration. There are however simple choices for the function $p(\psi)$ and $F(\psi)$ that allow an analytical treatment.

The simplest choice is [3, 4]

$$p(\psi) = -\frac{A_{Sol}(\psi - \psi_b)}{\mu_o} \qquad F(\psi)^2 = F_o^2 - 2B_{Sol}(\psi - \psi_b) \tag{17.23}$$

that leads to the following linear, non-homogeneous equation

$$\Delta^* \psi = A_{Sol} R^2 + B_{Sol} \tag{17.24}$$

The general solution is the combination of a particular solution plus the solution of the homogeneous equation. The particular solution that we consider here is an up-down symmetric solution

$$\psi = f_o(R) + f_2(R) Z^2 \tag{17.25}$$

Upon inserting Eq. (17.25) into Eq. (17.24) the following system of equations is obtained for the unknown functions $f_n(R)$

$$R \frac{d}{dR} \frac{1}{R} \frac{df_2}{dR} = 0 \tag{17.26}$$

$$R \frac{d}{dR} \frac{1}{R} \frac{df_o}{dR} + 2f_2 = A_{Sol} R^2 + B_{Sol} \tag{17.27}$$

The equations can be straightforwardly solved, yielding

$$f_o = a_o + b_o \frac{R^2}{2} + (A_{Sol} - b_2) \frac{R^4}{8} + (B_{Sol} - 2a_2) \frac{R^2}{4} (\ln(R^2) - 1)$$
$$f_2 = a_2 + b_2 \frac{R^2}{2} \tag{17.28}$$

The constants a_o, b_o, a_2, b_2 can be determined by imposing some geometrical constraints. For example, it is possible to fix the $X - point$ position at (R_X, Z_X)

and determine two of the constants from the condition that the the poloidal field vanishes at the $X - point$, yielding

$$0 = \left(\frac{\partial \psi}{\partial Z}\right)|_{(R_X, Z_X)} = Z_X \left(a_2 + b_2 \frac{R_X^2}{2}\right) \tag{17.29}$$

$$0 = \left(\frac{\partial \psi}{\partial R}\right)|_{(R_X, Z_X)} = R_X [b_o + (A_{Sol} - b_2)\frac{R_X^2}{2} + (B_{Sol} - 2a_2)\ln(R_X) + b_2 Z_X^2] \tag{17.30}$$

Equations (17.29) and (17.30) can be used e.g. to determine b_2 and b_o in terms of the other constants and the $X - point$ coordinates. The constant a_o can be obtained by setting the boundary $\psi = \psi_b$ at the magnetic separatrix. Since the $X - point$ is part of the magnetic separatrix this condition yields the following expression for the particular solution

$$\psi = \psi_b + a_2\left(1 - \frac{R^2}{R_X^2}\right)(Z^2 - Z_X^2) + \frac{B_{Sol} - 2a_2}{4}\left[R^2 \ln\left(\frac{R^2}{R_X^2}\right) + R_X^2 - R^2\right] + \\ + \left(A_{Sol} + \frac{2a_2}{R_X^2}\right)\frac{(R^2 - R_X^2)^2}{8} \equiv \zeta_2(R)(Z^2 - Z_X^2) + \zeta_o(R) \tag{17.31}$$

Finally, the constant a_2 can be obtained by imposing the point at which the separatrix crosses the equatorial plane $Z = 0$. The inner point is $R = R_X$ whereas the outer point $R = R_S$ must satisfy Eq. (17.31) with $Z = 0$ yielding

$$a_2 = \frac{\frac{B_{Sol}}{4}[R_S^2 \ln(\frac{R_S^2}{R_X^2}) + R_X^2 - R_S^2] + A_{Sol}\frac{(R_S^2 - R_X^2)^2}{8}}{(1 - \frac{R_S^2}{R_X^2})Z_X^2 + \frac{1}{2}[R_S^2 \ln(\frac{R_S^2}{R_X^2}) + R_X^2 - R_S^2] - \frac{2}{R_X^2}\frac{(R_S^2 - R_X^2)^2}{8}} \tag{17.32}$$

To assign R_S is equivalent to assign the inverse aspect ratio of the configuration $\epsilon = (R_S - R_X)/(R_S + R_X)$. Thus, the four constants appearing into Eq. (17.28) have been determined in terms of the two equilibrium parameters A_{Sol} and B_{Sol}, the X-point position and the aspect ratio. The vertical $X - point$ position is related to the elongation by the relation

$$\kappa_X = \frac{2Z_X}{(R_S - R_X)} \tag{17.33}$$

The magnetic axis position $R = R_o$ can be determined by the condition of vanishing poloidal field for $Z = 0$

$$a_2 Z_X^2 / R_X^2 + \frac{B_{Sol} - 2a_2}{4}\ln\left(\frac{R_o^2}{R_X^2}\right) + \left(A_{Sol} + \frac{2a_2}{R_X^2}\right)\frac{R_o^2 - R_X^2}{4} = 0 \tag{17.34}$$

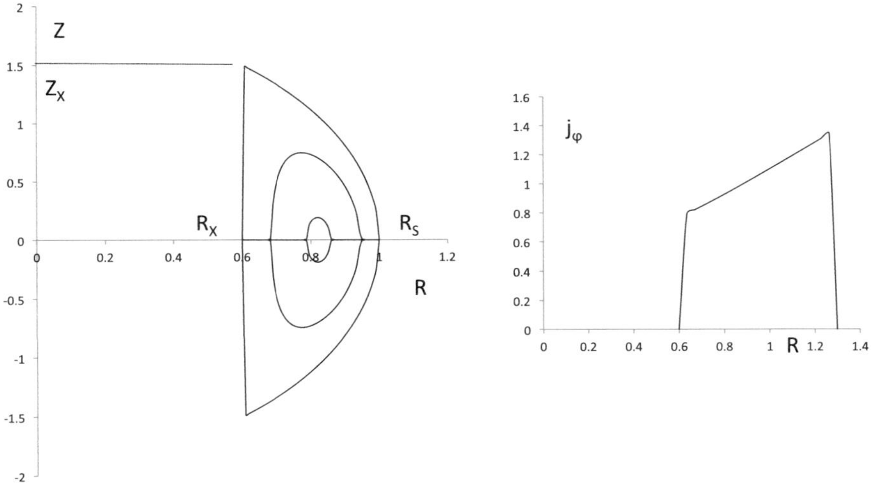

Fig. 17.4 Solovev equilibrium. Left: contour plot of the flux surfaces. Right: profile of the toroidal current along R. Note the discontinuity of the toroidal current at the separatrix location

The solution is plotted in Fig. 17.4 for a representative set of parameters. The magnetic axis is shifted with respect to the middle point between R_X and R_S by a quantity called *Shafranov shift*.

From the expression of ψ it is also possible to obtain the safety factor profile

$$q = \frac{F(\psi)}{2\pi} \oint \frac{dl}{R^2 B_p} \tag{17.35}$$

The line integral has to be performed over a contour $\psi = const$ that from Eq. (17.31) corresponds to the following curve

$$Z = \hat{Z} \equiv \pm \left(Z_X^2 + \frac{\psi - \zeta_o}{\zeta_2} \right)^{1/2} \tag{17.36}$$

Similarly, from the equilibrium solutions it is possible to evaluate the volume averaged density and the plasma current.The integrals have to be evaluated numerically.

The simple solution discussed so far has a number of limitations. The simple expression of the two equilibrium functions leads to a toroidal current with the profile shown in Fig. 17.4. The shape of the separatrix is also constrained by the limited numbers of parameters. While the former limitation is intrinsic to the approximations Eq. (17.23), the latter can be removed by including a suitable homogeneous solution [5]. The homogeneous solution can be found as a truncated series of the form

$$\psi_h = \sum_n h_n(R) Z^n \tag{17.37}$$

with the functions h_n satisfying the recurrence relation

$$R \frac{d}{dR} \frac{1}{R} \frac{dh_n}{dR} + (n+2)(n+1) h_{n+2} = 0 \tag{17.38}$$

that can be solved iteratively starting from the largest non vanishing function. Since the recurrence relation does not mix odd and even n-numbers, two chains of relations must be solved, one for n-odd and the other for n-even. In this way it is possible to describe also up-down asymmetric equilibria (as in ITER). We refer to Ref. [5] for a comprehensive analysis of this case.

17.5 The Low-Beta, Large Aspect Ratio Tokamak

As already noted the Grad-Shafranov equation requires a numerical integration. However, it is instructive to see how the effect of toroidicity modifies the screw-pinch equilibrium discussed at the beginning of this chapter. To this aim it is possible to make use of the large aspect ratio limit introduced in Chap. 13 and look for solutions of the Grad-Shafranov equation through an expansion in the inverse aspect ratio ϵ. To make the analysis as simple as possible we assume that the flux surface corresponding to the plasma boundary is circular and impose the fixed-boundary condition $\psi = \psi_a$ at $r = a$ where again the quasi cylindrical coordinate system $R = R_o + r\cos\theta$ and $Z = r\sin\theta$ is used.

Up to the first order in ϵ the Grad-Shafranov operator can be expressed as

$$\Delta^* \psi = \nabla^2 \psi - \frac{1}{R_o}\left(\cos\theta \frac{\partial \psi}{\partial r} + \frac{\sin\theta}{r}\frac{\partial \psi}{\partial \theta}\right) + O(\epsilon^2) \tag{17.39}$$

Seeking a solution of the form $\psi = \psi_o(r) + \epsilon\psi_1(r)\cos\theta$, replacing in the Grad-Shafranov equation and equating terms of the same order in ϵ the following system of equations is obtained

$$\nabla^2 \psi_o + \mu_o R_o^2 \left.\frac{dp}{d\psi}\right|_{\psi=\psi_o} + \frac{1}{2}\left.\frac{dF^2}{d\psi}\right|_{\psi=\psi_o} = 0 \tag{17.40}$$

which is exactly the screw pinch equilibrium equation Eq. (17.7) with the position $B_\theta = (1/R_o) d\psi_o/dr$. The expression of ψ_o can be obtained by integration yielding

$$\psi_o = \psi_a + B_o \ln\frac{r}{a} - \int_r^a dr' R_o B_\theta(r') \tag{17.41}$$

with $B_o = 0$ to ensure the regularity of the solution at the origin. Equation 17.41 takes also into account the boundary condition at the plasma boundary.

The correction due to toroidicity comes at the first order in the expansion of the Grad-Shafranov equation that yields

$$\nabla^2(\psi_1 \cos\theta) - \frac{1}{R_o}\cos\theta\frac{d\psi_o}{dr} + 2\mu_o R_o r\cos\theta\left.\frac{dp}{d\psi}\right|_{\psi=\psi_o} + \left[\mu_o R_o^2 \left.\frac{d^2p}{d\psi^2}\right|_{\psi=\psi_o} + \frac{1}{2}\left.\frac{d^2F^2}{d\psi^2}\right|_{\psi=\psi_o}\right]\psi_1\cos\theta = 0 \quad (17.42)$$

The last term of the l.h.s. of Eq. (17.42) can be obtained directly from Eq. (17.40)

$$\mu_o R_o^2 \left.\frac{d^2p}{d\psi^2}\right|_{\psi=\psi_o} + \frac{1}{2}\left.\frac{d^2F^2}{d\psi^2}\right|_{\psi=\psi_o} = -\frac{dr}{d\psi_o}\frac{d}{dr}\nabla^2\psi_o = -\frac{1}{B_\theta}\frac{d}{dr}\left(\frac{1}{r}\frac{d(rB_\theta)}{dr}\right) \quad (17.43)$$

Upon substituting Eq. (17.43) into Eq. (17.42) the equation for ψ_1 becomes

$$\frac{d}{dr}\left[(rB_\theta^2)\frac{d}{dr}\frac{\psi_1}{B_\theta}\right] = -2\mu_o r^2\frac{dp}{dr} + rB_\theta^2 \quad (17.44)$$

to be solved with the boundary condition $\psi_1(a) = 0$. The integration is straightforward yielding

$$\psi_1(r) = B_\theta \int_r^a dr\prime \frac{1}{r\prime B_\theta(r\prime)^2}\int_0^{r\prime} dr\prime\prime(2\mu_o r\prime\prime^2\frac{dp}{dr\prime\prime} - r\prime\prime B_\theta(r\prime\prime)^2) \quad (17.45)$$

Equations (17.41) and (17.45) are the solutions in the plasma and have to be matched to the solution in the vacuum region ψ^{vac}. The solution can be obtained from $\Delta^*\psi^{vac} = 0$ using the expansion in the inverse aspect ratio. Upon taking the asymptotic solution for the vector potential A_ϕ given in Eq. (13.16) (including the first order correction i ϵ) and noting that $\psi = RA_\phi$, we obtain

$$\psi_o^{vac} = \frac{\mu_o I_p R_o}{2\pi}(\ln\frac{8R_o}{r} - 2) \quad (17.46)$$

and

$$\psi_1^{vac} = \frac{\mu_o I_p}{4\pi}[Cr + \frac{D}{r} + r(\ln\frac{8R_o}{r} - 1)] \quad (17.47)$$

Matching the function ψ_1 with the solution Eq. (17.45) at the plasma boundary $r = a$ ($\psi_1(a) = 0$) yields

$$D = -Ca^2 - a^2\left(\ln\frac{8R_o}{a} - 1\right) \quad (17.48)$$

whereas, matching the derivative of ψ_1 in $r = a$ yields

$$C + \ln \frac{8R_o}{a} - \frac{3}{2} = -\left(\beta_p + \frac{l_i}{2}\right) \tag{17.49}$$

The vacuum solution corresponds to a vertical field given by

$$B_v = -\mu_o \frac{I_p}{4\pi R_o}\left[\ln\left(\frac{8R_o}{a}\right) + \frac{l_i - 3}{2} + \beta_p\right] \tag{17.50}$$

with β_p given in Eq. (17.11) and l_i in Eq. (16.10). The expression for B_v in Eq. (17.50) differs from the simplified expression B_v^{cl} in Eq. (16.31) due to the effect of the plasma pressure and of the poloidal current. A physical description of these effects can be found in Ref. [6].

The magnetic axis can be determined from the condition $\partial\psi/\partial R = 0$ for $Z = 0$. By construction the plasma boundary corresponds to $r = a$.

To conclude this section it is important to note the ordering of the various quantity that we have implicitly assumed. As already noted in Chap. 13 in tokamak plasmas the safety factor is of order unity. This implies $B_\theta = O(\epsilon)B_\phi$. Furthermore, if the plasma pressure enters the equilibrium equation only at the first order it means that $\beta_p = O(1)$ or that $\beta = O(\epsilon^2)$. Similarly, if the effect of the poloidal current competes at the same order with the other two effects it follows that $F = RB_\phi^{vacuum}(1 + O(\epsilon^2))$. This ordering is the so-called low-beta tokamak ordering.

17.6 The Virial Theorem

A general consequence of the equilibrium condition Eq. (17.1) is a form of the virial theorem that can be expressed as follows: it is not possible to confine a plasma with magnetic fields generated only by the plasma itself [2]. In other words, to confine a plasma we need to generate magnetic fields with an appropriate set of active (coils) or passive conductors.

The theorem can be demonstrated as follows. Equation (17.1) can be written as

$$\nabla \cdot T = 0 \tag{17.51}$$

with T the stress tensor given by

$$T \equiv (p + \frac{B^2}{2\mu_o})(I - \mathbf{bb}) + (p - \frac{B^2}{2\mu_o})\mathbf{bb} \tag{17.52}$$

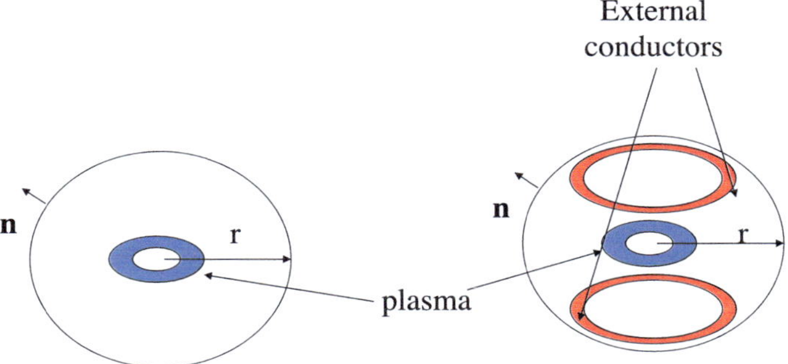

Fig. 17.5 The virial theorem states that a plasma cannot be confined by magnetic fields autogenerated only. In the absence of external conductor the equilibrium equation leads to an inconsistency that is removed by the presence of external conductor

Using the tensor identity

$$\nabla \cdot (\mathbf{r} \cdot T) = \mathbf{r} \cdot \nabla \cdot T + Tr(T) = 3p + \frac{B^2}{2\mu_o} \tag{17.53}$$

where use has been made of Eq. (17.51), and integrating Eq. (17.53) over a sphere of radius R_s enclosing the plasma, the following result is obtained

$$\int dV(3p + \frac{B^2}{2\mu_o}) = \int dV \nabla \cdot (\mathbf{r} \cdot T) = \int dS[(p + \frac{B^2}{2\mu_o})\mathbf{n} \cdot \mathbf{r} - \frac{B^2}{2\mu_o}\mathbf{n} \cdot \mathbf{b}\ \mathbf{r} \cdot \mathbf{b}] \tag{17.54}$$

Since the magnetic field far from the plasma is asymptotically equivalent to a dipole field and the plasma pressure outside the confinement region is 0, the r.h.s. of Eq. (17.54) vanishes as $R_s \to \infty$. On the other hand the l.h.s. is always a finite (positive) number, which gives an inconsistency. The only way to resolve the inconsistency is to consider external conductors as shown in Fig. 17.5.

In the presence of external conductor the r.h.s. of Eq. (17.54) does not vanish and its contribution can be evaluated (using Gauss theorem) from the contribution at the conductor surface (as done in Chap. 16 for the calculation of the inductance in the region outside the conductor).

17.7 Non-axisymmetric Equilibria

So far we have considered axisymmetric equilibria in which an ignorable coordinate (the toroidal angle) exists. The main limitation of these configuration is the fact that in order to create a rotational transform a net current must flow in the toroidal direction in the plasma. In tokamaks the configuration can be maintained only for a finite amount of time determined by the flux available in the transformer (see Chap. 16) unless the plasma current is driven via non-inductive methods. Furthermore, the plasma current provides a free-energy source that can drive instabilities with a rapid loss of the equilibrium configuration *(disruptions)*.

A solution to this problem is provided by the use of non-axisymmetric configurations. In this case the rotational transform is produced by a set of helical coils wound around the vacuum vessel. These may be supplemented by a set of coils to produce a toroidal field as in a tokamak to ensure maximum flexibility in the achievable configuration. The helical coils can be replaced with sets of non-planar modular coils that allow an easier assembly of the machine.

In the case of 3D configurations no net plasma current must be produced but a number of challenges must be faced as summarised below.

- In the case of fully 3D equilibria the existence of magnetic surfaces is not guaranteed and the equilibrium configuration will correspond to regions of good magnetic surfaces (and therefore good confinement) interspersed with regions of stochastic field. The volume of the stochastic regions can be minimised by an optimisation of the imposed field.
- The orbits of plasma particles are affected by further inhomogeneities of the equilibrium field that lead to very large transport at low collisionalities. This again need a proper control of the equilibrium configuration.
- The insertion of a divertor and of a blanket is more difficult and is still matter of investigation although significant progress in this respect have been made.

Stellarators were one of the first configuration proposed for plasma confinement, they were partially abandoned at the end of the '60 s when the tokamak configuration demonstrated superior confinement but, after a substantial improvement in the theoretical basis, they emerged again as the best long-term option for toroidal confinement.

Today there are two major experiments in operation: Wendelstein-7X [7] in Germany and the Large helical Device (LHD) [8] in Japan. A recent review of stellarator physics can be found in Ref. [9]

17.8 Suggestions for Further Readings

There are several books and journal papers on the subject of tokamak equilibrium. An intuitive introduction to tokamak equilibrium can be found in [6] and [10]. In addition we signal the original Shafranov paper Ref. [2].

17.9 Exercises

Problem 17.1 Using Eq. (17.35) determine the dependence of the safety factor at 95% of the separatrix flux on elongation and aspect ratio and compare with the fit shown in Eq. (13.31)

Problem 17.2 Using Eq. (17.34) determine the dependence of the Shafranov shift on the plasma beta.

References

1. H. Grad, H. Rubin, Hydromagnetic Equilibria and Force-Free Fields, in *Proceedings of the 2nd UN Conference on the Peaceful Uses of Atomic Energy*, vol. 31 (IAEA, Geneva, 1958), p. 190
2. V.D. Shafranov, Plasma Equilibrium in a Magnetic Field, Reviews of Plasma Physics, vol. 2 (Consultants Bureau, New York, 1966), p. 103
3. L.S. Solovev, Sov. Phys. JETP **26**, 400 (1968); N.M. Zueva, L.S. Solovev, Atomnaya Energia **24**, 453 (1968)
4. R.H. Weening, Phys. Plasmas **7**, 3654 (2000)
5. A.J. Cerfon, J.P. Freidberg, "One size fits all" analytic solutions to the Grad-Shafranov equation. Phys. Plasmas **17**, 032502 (2010)
6. J.P. Freidberg, *Plasma Physics and Fusion Energy* (Cambridge University Press, 2007)
7. T. Klinger et al., Overview of first Wendelstein 7-X high-performance operation. Nucl. Fusion **59**, 112004 (2019)
8. M. Fujiwara et al., Overview of LHD experiments. Nucl. Fusion **41**, 1355 (2001)
9. P. Helander et al., Stellarator and tokamak plasmas: a comparison. Plasma Phys. Control. Fusion **54**, 124009 (2012)
10. J. Wesson, *Tokamaks* (Clarendon press, Oxford, 2004)

Chapter 18
Scrape-Off Layer Dynamics

Abstract *This chapter presents an introduction to the heat and particle exhaust in tokamaks. The equation for the scrape-off layer dynamics are introduced and solved in some limiting cases. The regimes of operation of the divertor are classified. The engineering challenges of the divertor due to the high heat loads are discussed and possible solutions based on advanced divertor concepts or high core radiation are briefly reviewed.*

18.1 Introduction

In stationary conditions, the heat that is generated in the plasma core by fusion reactions and maintains the plasma at high temperature must be eventually exhausted somewhere. At the same time, the fusion reaction ashes (the charged fusion products that have transferred all their energy to the plasma) cross the magnetic separatrix due to convection and diffusion processes and must be efficiently removed to avoid diluting the plasma. All these processes involve the interaction of plasma particles with material objects and must take place with negligible extraction of the plasma facing material atoms to avoid impurity contamination of the plasma and fast erosion of the plasma facing components (PFC).

This is a non trivial task as the heat load on the PFC can achieve values of the order of the load on the Sun surface (60 MW/m^2).

If the heat generated inside the plasma were lost entirely via isotropic radiation the resulting load would be of the order of only a few hundreds of kW/m^2. The challenge of large heat loads arises because the heat that crosses the separatrix flows along the magnetic field down to the component on which it is exhausted. The radial width of this region is narrow (a few *mm*) and, as a result, the surface wetted by the plasma is small, leading to high localised heat loads. One may argue that by increasing the fraction of radiated power the problem can be reduced or even eliminated. However, there are two obstacles to overcome. First, the radiated power must be originated by the edge plasma and not by the plasma core to avoid cooling the region where fusion reactions take place. To produce a radiative layer that remains localised at the plasma edge is not easy. Second, we have seen that to access H-mode conditions the power

F. Romanelli, *Physics of Nuclear Energy*, Springer Series in Plasma Science and Technology, https://doi.org/10.1007/978-981-97-9609-0_18

crossing the separatrix via conduction and convection must exceed a certain value, setting a lower limit to the power that ultimately flows to the exhaust component.

The heat exhaust component is generally a niche in the reaction chamber (usually at the bottom) called *divertor* (see Fig. 18.1). A divertor configuration is naturally formed in the presence of a magnetic separatrix. The divertor has the advantage that the plasma-wall interaction takes place far from the main plasma. Furthermore, with appropriate design of the divertor geometry, ashes coming from the main plasma can be contained inside the divertor, allowing an efficient pumping. If a separatrix is not present (as e.g. in the phase of plasma formation) the heat can be exhausted on a *limiter*, i.e. a component protruding inside the reaction chamber. The limiter wetted surface must be large enough to reduce the loads to manageable levels at least for limited durations although the price to pay is that the interaction takes place close to the hot plasma and can lead to impurities contamination. Furthermore, due to its geometry, a limiter cannot ensure an adequate pumping.

In order to understand how the interaction between plasma and limiter/divertor takes place it is necessary to analyze first the plasma dynamics in the so called *scrape-off layer (SOL)*, the region immediately beyond the last closed magnetic surface. As a general reference for more detailed discussions we refer to the monograph of Stangeby [1] and to the review papers [2–4] that cover most of the material presented here.

18.2 SOL Dynamics

The SOL geometry is shown in Fig. 18.2. Since the SOL width is small, it is possible to describe the SOL dynamics in slab geometry. The SOL can be divided in a main SOL region, located between the magnetic separatrix and the wall, and the divertor regions between the divertor plates and the X-point region. Shown in Fig. 18.1 is also the *private region*. The main difference between the SOL and the region inside the separatrix is that the SOL is characterised by open field lines connecting the two divertor plates. Therefore, charged particles moving along B will eventually impinge

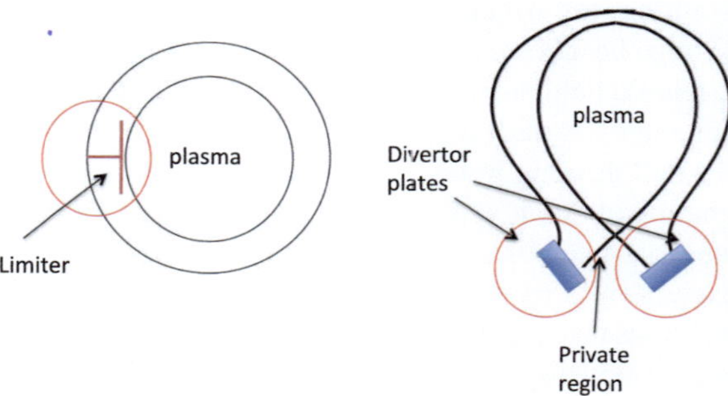

Fig. 18.1 Divertor and limiter geometry

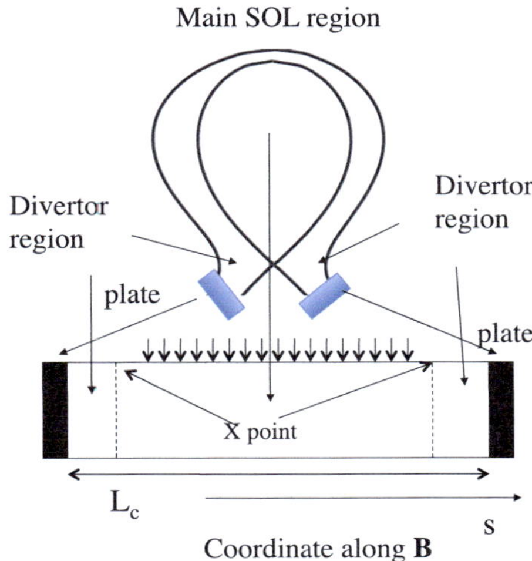

Fig. 18.2 Simplified SOL geometry. The SOL boundary is mapped into a box with the upper face corresponding to the magnetic separatrix, the lower face corresponding to the wall and the two lateral faces corresponding to the divertor plates. Power and particles are flowing through the upper face in the SOL

on the plate. This is exactly what we want in order to extract the heat but it opens the possibility to the extraction of the plate material, contamination of the main plasma and rapid erosion of the divertor plates. The aim of this chapter is to understand if it is possible to find plasma conditions that minimise the side effects while preserving the role of the divertor for heat extraction.

It is convenient to start with what happens close to the divertor plate. Charged particles impinge on the plate but, as the electrons have a much larger thermal velocity than ions, they will be initially preferentially lost in such a way that the plate becomes negatively charged. It will be shown below that the negative plate potential is $\Phi_p \approx -3T_{ep}/e$ with T_{ep} the electron temperature of the plasma in front of the plate. This negative potential has two effects: on one hand it decelerates the incoming electrons in such a way that only the electrons with kinetic energy sufficiently large to overcome the potential arrive on the plate; on the other hand it accelerates the ions. A stationary condition is rapidly established that allows to equalise the fluxes to the plate of negatively and positively charged particles maintaining charge neutrality in the SOL. As it will be shown below, the main plasma ions are accelerated by the electrostatic field up to the local value of the sound velocity $C_s \equiv [(T_e + T_i)/m_i]^{1/2}$ [5]. This flow of plasma ions on one hand contributes to plate erosion and impurity production, on the other hand produces a friction force on the impurity ions that tends to keep impurities emitted by the plate and the He ashes confined in the divertor region facilitating their extraction. Since the ion average velocity is directed in opposite directions at the two plates, a point exists along B at which $v_{\|i} = 0$ (the *stagnation point*).

Within a region of few Debye lengths close to the plate, the plasma does not satisfy the charge neutrality condition. This region (the *sheath*) is characterised by large electric field values along the magnetic field that substantially distort the distribution functions and determine the effective heat transmission coefficient across the sheath.

Since the flux of ions to the plate is given by $n_p C_{sp}$ (with the suffix p denoting the values at the plate) and the flux of electrons must be the same to maintain charge neutrality in the SOL, the parallel heat flux $q_{||}$ can be written as

$$q_{||p} = n_p C_{sp}(\gamma T_p + \epsilon_{pot}) \tag{18.1}$$

where electron and ion temperature have been assumed, for the sake of simplicity, equal and $\gamma \approx 7$ is the sheath transmission coefficient [4]. In Eq. (18.1) ϵ_{pot} is the energy released on the plate by the neutralisation of a hydrogen ion and the subsequent recombination in a molecule ($\epsilon_{pot} \approx 13.6\,\text{eV} + 2.2\,\text{eV}/2$).

In steady state conditions the heat that flows via convection and conduction from the main plasma to the divertor plates (minus the fraction radiated in the SOL) must be transmitted through the sheath via Eq. (18.1) that provides a constraint on the density and temperature *of the plasma in front of the plate*.[1]

The SOL length, measured along a magnetic field line between the two plates, is the *connection length* L_c. It is important to remember that the connection length is measured *along* the total magnetic field and therefore should not be confused with the length of its projection on the plane $\phi = const.$ shown in Fig. 18.2. Magnetic field lines have mostly a toroidal component and the exact orientation of them as they touch the plate must be considered to determine the plasma wall interaction. We will return later in this chapter on this point. We anticipate that L_c will be the characteristic variation length along the magnetic field. The length along the magnetic field of the divertor region is typically much smaller than L_c but this depends on the magnetic geometry of the divertor.

In the direction perpendicular to the last closed magnetic surface the SOL width is typically of the order of a few millimetres and is determined by the characteristic decay length of the various quantities. A common approximation consists in assuming an exponential decay of density (with characteristic e-folding length λ_n), temperature (with characteristic e-folding length λ_T) and heat flux (with characteristic e-folding length λ_q). The e-folding lengths are determined by the balance between parallel and perpendicular transport.

As an example let us consider the main SOL region and the continuity equation in steady state conditions. In the absence of ionisation and recombination sources the continuity equation reduces to

$$\nabla \cdot (n_i \mathbf{v}) = \nabla \cdot (n_i \mathbf{v}_{||i}) + \nabla \cdot (n_i \mathbf{v}_{\perp i}) = 0 \tag{18.2}$$

therefore in each volume of the main SOL region the flux of particles entering via diffusion from the magnetic separatrix must be equal to the flux of particles moving to the divertor plates. A simple estimate for λ_n is therefore obtained by assuming

[1] It is important to stress that the temperature of the plasma *at the plate* is of the order of few eV and so much larger than the temperature *of the plate* that is at most of a few thousand degrees.

$\nabla \cdot (n_i \mathbf{v}_{||i}) \approx n_i C_s / L_c$ and $(n_i \mathbf{v}_{\perp i}) = -D \nabla n_i \approx D n_i / \lambda_n$, with D the perpendicular diffusion coefficient, yielding

$$\frac{Dn}{\lambda_n^2} = \frac{nC_s}{L_c} \tag{18.3}$$

or $\lambda_n = (DL_c/C_s)^{1/2}$. The perpendicular diffusion coefficient is usually approximated by the estimate given by Bohm $D \approx D_B = T/(eB)$ [5] that represents the upper bound for the turbulence diffusion in a plasma. Thus, the density e-folding length is expected to be of the order of $(\rho_i L_c)^{1/2}$.

The SOL dynamics can be conveniently described by the two-fluids equations. In the following the coordinate along the magnetic field will be denoted by s. The continuity equation for the ions can be written as

$$\frac{d(n_i v_{||i})}{ds} - \frac{D_B n_i}{\lambda_n^2} = 0 \tag{18.4}$$

where again ionisation and recombination in the SOL have been neglected. The parallel ion velocity is in turn determined by the momentum equation

$$n_i m_i v_{||i} \frac{dv_{||i}}{ds} = -\frac{dp_i}{ds} + en_i E_{||} - m_i v_{||i} \frac{D_B n_i}{\lambda_n^2} \tag{18.5}$$

with the last term in Eq. (18.5) being the parallel momentum sink due to perpendicular diffusion. The parallel momentum equation for electrons in the limit of negligible inertia reduces to the force balance along B

$$0 = -\frac{dp_e}{ds} - en_e E_{||} \tag{18.6}$$

Equation (18.6) can be used to eliminate the electric field in Eq. (18.5) that, combined with the continuity equation Eq. (18.4), yields

$$\frac{d}{ds}(p_e + p_i + n_i m_i v_{||i}^2) = 0 \tag{18.7}$$

that can be easily integrated. If we make, for simplicity, the assumption of constant electron and ion temperatures along B, Eq. (18.7) (together with the charge neutrality condition $n_e = n_i = n$) provides a link between the density and the parallel velocity in the SOL

$$n = \frac{n_o}{1 + M^2} \tag{18.8}$$

where n_o is the density at the stagnation point (where $v_{||i} = 0$) and the Mach number $M \equiv v_{||i}/C_s$ has been introduced. The dependence of M on s can be determined by introducing Eq. (18.8) into the ion momentum equation yielding

$$\frac{dM}{ds} = -\frac{D_B}{\lambda_n^2 C_s} \frac{1 + M^2}{1 - M^2} \tag{18.9}$$

that can be integrated by separation of variables. We will set the origin of the coordinate along B ($s = 0$) at the stagnation point. Symmetry considerations suggest that for the simplified geometry of Fig. 18.2 the stagnation point is equidistant from the two plates. As already anticipated, at the plate $s = L_c/2$ the Mach number must be one. Thus the function $M(s)$ can be obtained in implicit form as

$$\left(\frac{\pi}{2} - 1\right) \frac{2s}{L_c} = -M + 2\text{arctg}(M) \tag{18.10}$$

Note however that this expression is valid only if the temperature is constant along s. Finally, from Eq. (18.6) it is possible to determine also the electric field and the electrostatic potential $E_{||} = -d\Phi/ds$.

$$\Phi = -\frac{T_e}{e}\ln(1 + M^2) \tag{18.11}$$

As the plate is approached and $M \to 1$ the electrostatic potential takes the value $\Phi_{se} = -(T_e/e)\ln 2$, where the suffix se denotes the sheath edge. This is the value at the boundary between the main SOL and the sheath (the so-called *pre-sheath* region). To close the description we need to solve the equations inside the sheath region. Here the dynamics is entirely dominated by the gradients along B and the perpendicular dynamics can be neglected. The equations are the same as those discussed above (with the perpendicular dynamics, i.e. the term related to D_B neglected) but with a significant difference: in the sheath charge neutrality does not hold and the electron and ion densities must be separately determined and then combined to determine the charge distribution in the equation for the electric induction. For the sake of simplicity it is convenient to assume that the ion temperature is much lower than the electron temperature. The electron density is again determined by Eq. (18.6) yielding

$$n_e = n_{se} e^{\frac{e(\Phi - \Phi_{se})}{T_e}} \tag{18.12}$$

The continuity equation can be trivially integrated yielding

$$n_i v_{||i} = n_{se} v_{||se} \tag{18.13}$$

Note that the ion and electron density at the sheath edge are assumed identical since in that region charge neutrality must hold. The parallel ion momentum equation Eq. (18.7) (simplified by neglecting the ion pressure) yields a simple expression for the parallel ion velocity in terms of the electrostatic potential

$$m_i v_{||i}^2 + e\Phi = m_i v_{||se}^2 + e\Phi_{se} \tag{18.14}$$

that finally provides an expression for the ion density in the sheath

$$n_i = \frac{n_{se}}{[1 + \frac{2T_e}{m_i v_{||se}^2} \frac{e(\Phi_{se} - \Phi)}{T_\epsilon}]^{1/2}} \tag{18.15}$$

Upon substituting the expressions for n_i and n_e into the equation for electric induction

$$\frac{d^2\Phi}{ds^2} = e\frac{n_e - n_i}{\epsilon_o} \tag{18.16}$$

and defining the new dimensionless variable $\Xi \equiv e(\Phi_{se} - \Phi)/T_e$, Eq. (18.16) can be cast in te following form

$$\lambda_D^2 \frac{d^2\Xi}{ds^2} = \frac{1}{[1 + \frac{2}{M_{se}^2}\Xi]^{1/2}} - e^{-\Xi} \tag{18.17}$$

with $M_{se} \equiv v_{||se}/C_s$. It is possible now to show that $M_{se} = 1$ without explicitly solving Eq. (18.17). Upon multiplying by $d\Xi/ds$ and integrating the following invariant is obtained [4]

$$\frac{\lambda_D^2}{2}\left(\frac{d\Xi}{ds}\right)^2 = M_{se}^2\left[\left(1 + \frac{2\Xi}{M_{se}^2}\right)^{1/2} - 1\right] + e^{-\Xi} - 1 \tag{18.18}$$

where the condition $d\Xi_{se}/ds = 0$ has been assumed. The l.h.s. is always positive therefore also the r.h.s must be positive. Upon expanding the r.h.s. for small values of Ξ, at the lowest order we obtain the condition

$$M_{se}^2\left[1 + \frac{2\Xi}{M_{se}^2}\right]^{1/2} + e^{-\Xi} - 1 \approx \Xi^2(1 - \frac{1}{M_{se}^2}) \geq 0 \tag{18.19}$$

This conditions requires $M_{se} \geq 1$. However, Eq. (18.9) is free from singularities provided that $M \leq 1$. The two conditions are compatible provided

$$M_{se} = 1 \tag{18.20}$$

This is the *Bohm condition*. As a final point it is possible to evaluate the electrostatic potential of the divertor plate Φ_p. This is obtained by requiring that the parallel electron and ion currents are equal. The ion current is simply given by the ion density and velocity at the sheath entrance

$$j_{||i} = en_{se}C_s \tag{18.21}$$

whereas the electron current can be evaluated from the integral of the distribution function restricted to the domain of the electrons that have sufficient energy to overcome the electrostatic repulsion of the plate

$$j_{\|e} = en_{se} \int_{v_{min}}^{\infty} v_{\|} \frac{dv_{\|}}{\pi^{1/2} v_{te}} e^{-\frac{v_{\|}^2}{v_{te}^2}} = \frac{en_{se} v_{te}}{2\pi^{1/2}} e^{-\frac{v_{min}^2}{v_{te}^2}} \tag{18.22}$$

with $m_e v_{min}^2/2 = -e\Phi_p$. The condition $j_{\|e} = j_{\|i}$ yields

$$\Phi_p = \frac{T_e}{2e} \ln \left[\frac{4\pi m_e}{m_i} \left(1 + \frac{T_i}{T_e} \right) \right] \approx -\frac{3T_e}{e} \tag{18.23}$$

with the estimate $\Phi_p \approx -3T_e/e$ obtained for equal electron and ion temperatures at the plate and neglecting the secondary electron emission.

18.3 The Two-Point Model of SOL

To solve the equation in the SOL requires complex numerical codes. However, the main features of the various regimes can be captured by a simple model called the two point model of the SOL [1]. The two-point model reduces the set of partial differential equation to a set of algebraic equations involving four quantities: the density and the temperature at the stagnation point (n_u and T_u with the suffix u denoting the upstream plasma) and in front of the divertor plates (n_p and T_p). From the two fluids equations it is possible to determine the three conditions linking these four quantities. Thus, the upstream density n_u is usually considered as an independent variable. As a simplifying assumption it is usually assumed equal electron and ion temperatures.

The parallel momentum equation Eq. (18.7) together with the Bohm condition Eq. (18.20) provide a simple relation

$$n_u T_u = 2n_p T_p \tag{18.24}$$

Equation 18.1 provides the condition of heat transmission through the sheath. Denoting with P_{SOL} the power flowing through the separatrix in the SOL and A_q the area of the SOL perpendicular to B, this condition can be cast in the following form

$$\frac{P_{SOL} - P_{rad}}{A_q} = \gamma n_p T_p C_{sp} \tag{18.25}$$

where P_{rad} is the power lost via radiation in the SOL and we have neglected ϵ_{pot}. For the time being we will neglect P_{rad} (we will come back to this important contribution when we will consider divertor detachment). Equations (18.24)–(18.25) can be used to determine the plate density and temperature in terms of n_u and T_u. A straightforward calculation yields

$$T_p = \frac{m_i}{2}\left(\frac{2q_{||u}}{\gamma n_u T_u}\right)^2 \quad (18.26)$$

$$n_p = \frac{\gamma^2}{m_i}\frac{(n_u T_u)^3}{4q_{||u}^2} \quad (18.27)$$

with $q_{||u} = P_{SOL}/A_q$. The electron energy balance equation provides the last condition. Here, the heat transport is dominated by the parallel diffusion and, in the absence of sources or sinks, the equation can be simplified as follows

$$q_{||} \equiv n_e \chi_e \frac{dT_e}{ds} = q_{||u} \quad (18.28)$$

Note that with radiation neglected the parallel heat flux is constant ($q_{||p} = q_{||u}$). The electron thermal conductivity χ_e is proportional to $\lambda_{Coul}^2 \nu_{ee}$, with $\lambda_{Coul} = v_{te}/\nu_{ee}$ the mean free path for Coulomb collisions. Thus, we can define $n_e\chi_e = \kappa_o T_e^{5/2}$ with $\kappa_o \equiv (4\pi\epsilon_o)^2/(m_e^{1/2} \ln \Lambda e^4)$ independent of density and temperature.

Equation (18.28) can be integrated trivially yielding

$$T_u^{7/2} = T_p^{7/2} + \frac{7}{2}\frac{q_{||u} L_c}{2\kappa_o} \quad (18.29)$$

Equation 18.29 shows that two different regimes are possible (Fig. 18.3):

- The *conduction limited regime* characterised by $T_u \gg T_p$. This is the regime of interest since corresponds to low plasma temperatures at the plate. In this regime we can neglect T_p in the r.h.s. of Eq. (18.29) and determine the upstream temperature as

$$T_u \approx \left[\frac{7}{2}\frac{q_{||u} L_c}{2\kappa_o}\right]^{2/7} \quad (18.30)$$

substituting into the equations for n_p and T_p we obtain

$$T_p \propto q_{||u}^{10/7} L_c^{-4/7} n_u^{-2} \quad (18.31)$$

$$n_p \propto q_{||u}^{-8/7} L_c^{6/7} n_u^3 \quad (18.32)$$

This regime can be achieved if the upstream density is sufficiently large. Long connection lengths are also favourable.

- The *sheath limited regime* characterised by the negligible effect of the parallel electron thermal conduction yielding $T_u \approx T_p$. In this case the plate density is

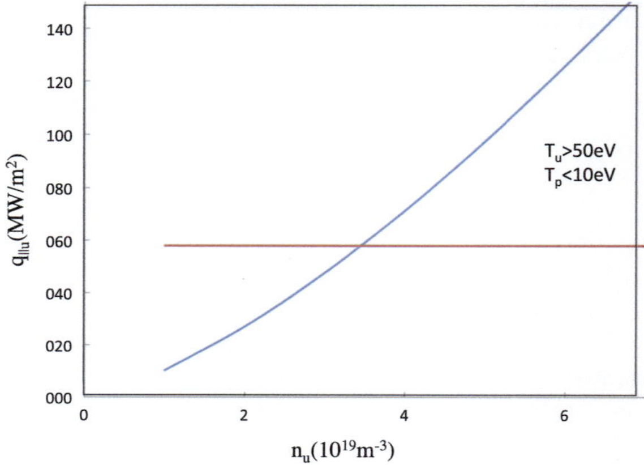

Fig. 18.3 The different SOL regimes corresponds to the regions in the figure. The condition $T = 10eV$ corresponds to the blue line. The horizontal red line corresponds to an upstream temperature of 50eV. A larger power in the SOL increases the upstream temperature above this limit. The interesting region corresponds to the high recycling regime with temperatures above 50eV because it allows to operate the SOL with high input powers and low temperatures at the plate

about half of the upstream density and the plate temperature is entirely determined by te balance between the power entering the SOL and the sheath transmission.

Note that the conduction limited regime is also called *high recycling regime* since, from Eq. (18.25), the particle flux has the following expression

$$\Gamma_p = \frac{q_{||u}}{\gamma T_p} \propto n_u^2 \tag{18.33}$$

Using Eqs. (18.30) and (18.31) it follows that the conduction limited regime requires

$$q_{||u} << (\frac{7}{4\kappa_o})^{1/4} L_c^{3/4} n_u^{7/4} \tag{18.34}$$

18.4 Divertor Detachment

Although the simple model shown in the previous section predicts the possibility of temperatures at the plate below 10 eV, these values would be still too high and would produce a rapid erosion of the divertor surface. Furthermore, the reduction of the heat load on the target would be achieved only through flux expansion and tile shaping but this would not be enough to achieve engineering manageable limits. The solution to this problem relies in the achievement of *detached divertor* regimes. These regimes consist in producing a region of high dissipation close to the divertor plates that leads to temperatures of the order of $1eV$ and a reduction of the plasma pressure. The high dissipation region extends along the magnetic field line up to a point, where a sharp increase in temperature takes place, called the *detachment front*. The detachment front can extend up to the $X - point$ (fully detached conditions) but as the high-radiation region gets closer to the separatrix a degradation in energy confinement takes place. Furthermore, detached conditions are also accompanied by a decrease of neutrals compression and reduced pumping efficiency. Therefore the position of the detachment front must be carefully controlled and maintained at a location that does not produce confinement degradation. The control is typically achieved by impurity seeding e.g. by monitoring the $X - point$ radiation. The achievements of detached conditions depend on the heat flux into the SOL and on the upstream density. At fixed heat flux a window in the upstream density values exists for the achievement of detachment. A review of the experimental findings on detachment can be found in [6–8]. A critical aspect for power plant applications is how the achievement of detachment depends upon the divertor magnetic configuration. We will come back to this point later in the chapter.

In order to understand how a dissipation region can lead to divertor detachment it is convenient to introduce correction factors in the two-points model shown above. In the presence of a dissipation region associated with hydrogen neutrals and impurities the power balance and momentum balance equations must be modified to account for radiation close to the plate and momentum losses due to friction between plasma particles and neutrals. Following [3] we replace these terms with two *ad hoc* coefficients f_{imp} and f_{mom} as follows

$$f_{mom} n_u T_u = 2 n_p T_p \tag{18.35}$$

$$q_{||u} = q_{||p} + f_{imp} q_{||u} \tag{18.36}$$

Solving for f_{imp} from Eq. (18.36), it is possible to obtain an equation for f_{imp} as a function of T_p

$$f_{imp} = 1 - \left(\frac{7L_c}{2\kappa_o}\right)^{2/7} \frac{f_{mom} n_u C_{sp}(\gamma T_p + \epsilon_{pot})}{2 T_p q_{||u}^{5/7}} \tag{18.37}$$

For $f_{mom} = 1$ f_{imp} has a maximum for $T_p = \epsilon_{pot}/\gamma$ with $f_{imp} \approx 0.7$ implying that if the power flux to the plate can be reduced at most by a factor three. However, for $f_{mom} = 0.1$ $f_{imp} \approx 0.95$ in a broad range of plate temperatures with a reduction by a factor twenty of the power to the plates, showing the importance of momentum losses.

The *ad hoc* factors f_{mom} f_{imp} must be related to actual physical mechanisms. It can be shown [3] that the quantity f_{mom} can be related to the ratio between the ionisation rate coefficient $< \sigma v >_{ion}$ and the momentum loss rate coefficient $< \sigma v >_{mom}$ and in a first approximation is a function of temperature only with values of f_{mom} below unity achievable for $T < 10eV$ as requested from the argument above. The quantity f_{imp} can be obtained by integrating the heat flux equation with a sink term given by the impurity radiation

$$\frac{dq_{||}}{ds} = -n_{imp} n L_{imp}(T) = -c_{imp} n^2 L_{imp}(T) \tag{18.38}$$

with c_{imp} the impurity concentration assumed constant along s and $L_{imp}(T)$ the radiated power coefficient [9] approximated as a constant for $T > 5eV$ and zero below. Upon multiplying by $q_{||}$ and integrating between the stagnation point and the boundary of the recycle region using Eqs. (18.28) and (18.7) the following results is obtained

$$q_{||u}^2 - q_{||r}^2 = \frac{1}{3}\kappa_o c_{imp} L_{imp} (n_u T_u)^2 (T_u^{3/2} - T_x^{3/2}) \tag{18.39}$$

with the suffix r indicating the quantities at the boundary of the recycle region and T_x the maximum between 5 eV and T_r. Neglecting T_x compared to T_u the following condition must be satisfied to obtain $f_{imp} = 1$

$$q_{||u} = \frac{14}{3} c_{imp} L_{imp} n_u^2 L_c \tag{18.40}$$

which shows that detached condition can be achieved via impurity injection, large upstream densities, long connection lengths or reduced power input to the SOL (e.g. via core radiation).

18.5 The Plate Wetted Surface

The challenge of heat exhaust in magnetically confined devices arises from the concentration of the large power conducted through the separatrix on a narrow area of the divertor plates. It is therefore mandatory to increase as much as possible the area wetted by the plasma in order to reduce the specific heat load. This result can be achieved by two methods: a) by relying on the flux expansion close to the X-point; and, b) by tilting the plate surface with respect to the magnetic field direction.

Making reference to Fig. 18.4, consider first the case of a target with a surface parallel to the *toroidal* magnetic field line. In this case the angle of incidence of the magnetic field line on the plate is given by $\theta_n = \arctan(B_{pn}/B_\phi)$ with B_{pn} being the component of the poloidal field normal to the surface. Since close to the X-point the poloidal field is small, the incidence angle is of the order of few degrees. Considering for a moment the case in which the poloidal field has no tangential component on the plate, the wetted area A_{wet} on the two divertor plates is given by $A_{wet} = 4\pi R_p \lambda_p$, with R_p the major radius of the divertor plates assumed for the sake of simplicity equal in this estimate. The wetted area should not be confused with the quantity A_q that is the area taken in the plane perpendicular to B and therefore is smaller than A_{wet} by a factor $\sin\theta_n$. The e-folding length λ_p evaluated at the plate is related to the value at the stagnation point λ_u by the flux expansion factor. As the X-point is approached the magnetic field line tend to thin out. After the X-point region the lines

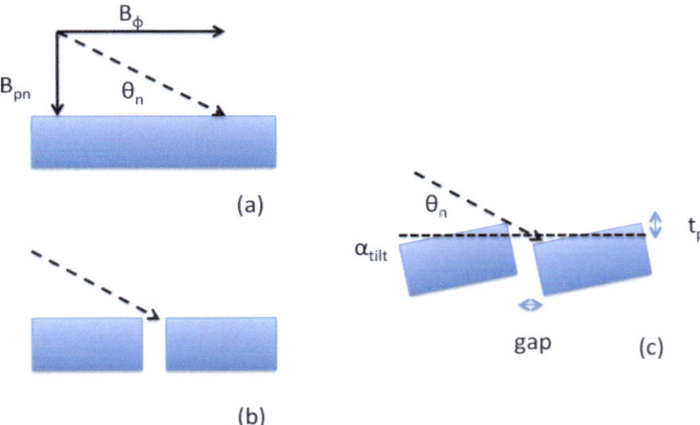

Fig. 18.4 The divertor wetted area depends can be determined from the geometrical parameter of the magnetic configuration. In the case of a continuous surface parallel to the toroidal field (**a**) the incidence angle is associated to the ratio between the normal component of the poloidal field and the toroidal field. For a segmentation in tiles this would produce exposed edges (**b**). To avoid the problem with exposed edges the tiles can be tilted by an angle α_{tilt} (**c**) but this reduces the wetted surface

thicken again but still there is a significant expansion with respect to the stagnation point. As a first approximation we can write

$$\lambda_p = \frac{(B_p/B)_u}{(B_p/B)_p}\lambda_u \equiv f_x\lambda_u \tag{18.41}$$

Typical values of the flux expansion are in the range 5–10.

One would be tempted to increase the wetted surface by making the plate almost parallel also to the poloidal field but as we will see in a moment this trick does not work in real stuations. Furthermore, even a small perturbation of the position of the X-point could lead to the loss of contact with the divertor plate.

In realistic situations, the divertor surface is made of tiles with finite size. This arrangements produces exposed edges that experience a very large localised heat flux. This is further exacerbated by the unavoidable misalignment between neighbouring tiles and requires tilting the surface by an angle α_{tilt} in the toroidal direction. The tilting angle is related to the gap between tiles, the tile width and the t_p by

$$\mathrm{tg}\alpha_{\mathrm{tilt}} = \frac{\mathrm{gap} \times \mathrm{tg}\theta_p + t_p}{\mathrm{width}} \tag{18.42}$$

However the toroidal tilting reduces the area wetted by the plasma with the reduction being larger at lower values of θ_p.

Finally it should be noted that the exponential decay that is observed at the stagnation point is somewhat modified as the plate is approached due to perpendicular diffusion in the private region [10]. At the plate a better fit is given by

$$q = \frac{q_o}{2} e^{(\frac{S}{2\lambda_q})^2 - \frac{y}{\lambda_q f_x}} erfc(\frac{S}{2\lambda_q} - \frac{y}{\lambda_q f_x}) + q_{BG} \tag{18.43}$$

with y the coordinate on the plate measured from the strike point position and S the *power spreading factor* that depends on the perpendicular diffusion coefficient. These two quantities ultimately matter to determine the heat load at the plate. Recent analyses [11] have pointed out that the quantity λ_q on ITER may be significantly smaller than previously expected, potentially raising the issue of high heat loads. The convolution with the terms dependent on S gives rise to an *integral power decay length* $\lambda_{int} \approx \lambda_q + 1.64S$, thus, even if $\lambda_q \to 0$ the heat load remains finite. The definition of a reliable scaling for the power spreading factor is presently a subject of active research [12].

18.6 Impurity Production at the Plates

The production of impurities at the divertor plates is described by the sputtering yield Y_{sp} defined as

$$\Gamma_z = Y_{sp} \Gamma_i \tag{18.44}$$

with Γ_z the flux at the plate of impurities of atomic number Z and Γ_i the flux at the plate of the ion species i. Impurities can be extracted as a result of the impact of main ions as well as of the impact of the same impurity species (*self-sputtering*). The sputtering yield is shown in Fig. 18.5 as a function of energy of the incoming particle

The sputtering yield is typically an increasing function of the energy. It exhibit a threshold in energy that increases with the mass number of the plate material. This can be understood by recalling that in an elastic collision between a projectile of mass m_1 and a particle at rest of mass m_2, the energy E_2 taken by the particle initially at rest is inversely proportional to the rest particle mass $E_2 \approx 4m_1m_2/(m_1 + m_2)^2 E_1$ with E_1 the initial energy of the projectile. Thus, the heavier is the plate material the lower will be the energy taken in a collision. Since the ejection of particles require to overcome the binding force, to kick out heavier particles require a larger initial energy of the projectile. Similarly, at fixed plate material heavier ions impinging on the plate have higher probability of producing sputtering. Note that the sputtering yield can exceed unity.

The energy E of the impinging particle is the sum of the thermal energy in front of the plate and of the energy absorbed during the acceleration in the sheath. A precise calculation would require a kinetic analysis. As a first approximation we can estimate the energy of the impinging particle as

$$E = 2T_i + 3ZT_e \tag{18.45}$$

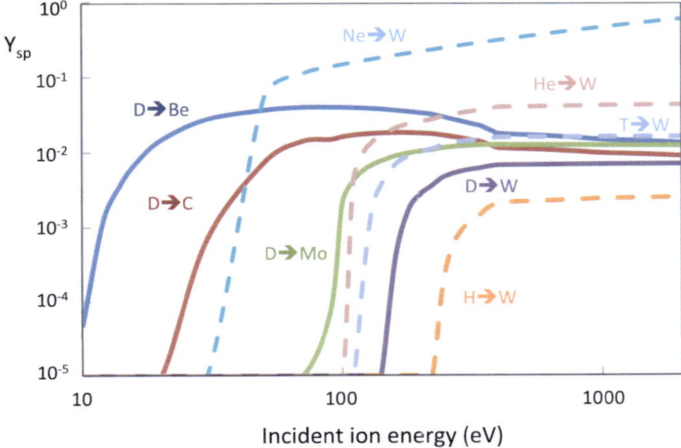

Fig. 18.5 Physical sputtering yield versus incident ion energy for different elements using the fits of Ref. [14]. Continuous lines correspond to deuterium impinging on different target elements. Dashed lines correspond to different elements impinging on tungsten

Numerical fits for the sputtering yield can be fund in Ref. [13]. For a review of the calculations of the sputtering yields we refer to Ref. [14].

18.7 The Engineering Challenges of the Divertor Plate

The limit to the power that can be safely exhausted is related to the limits of the materials of the plasma facing components. The solution used for the ITER divertor is the so called *W monoblock* schematically shown in Fig. 18.6. The power deposited on the exposed face of the mono block is transferred via thermal conduction to the cooling channel made of *CuCrZr* joined with a Cu interlayer to the mono block. A temperature gradient is established inside the monoblock. The gradient is a two-component vector but for the sake of simplicity we will assume that the dominant component is the one perpendicular to the plasma facing surface

$$\frac{dT}{dx} \approx \frac{T_{surf} - T_{wall}}{\Delta} = \frac{P}{kS} \tag{18.46}$$

with k the thermal conductivity of the monoblock.

Water at low temperature (inlet temperature T_{in} and outlet temperature T_{out}) flows in the cooling channel and the mass flow rate G determines the power P_{ex} that can be extracted

$$P_{ex}(W) = G(\text{kg/s})c_p(\text{J kg}^{-1}K^{-1})\Delta T(^\circ C) \tag{18.47}$$

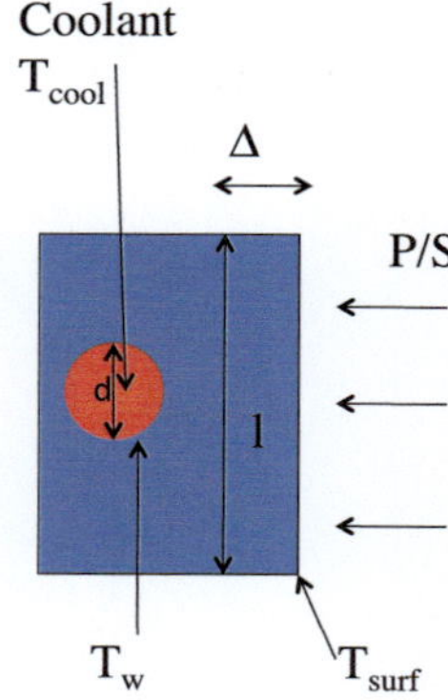

Fig. 18.6 Tungsten monoblock is the technology chosen for ITER and investigated for DEMO. The mock-up is shown in (**a**) (Reprinted from Ref. [18] with permission of Elsevier). The model used in the text is shown in (**b**)

with c_p the thermal capacity. The temperature at the surface of the cooling channel is determined by the heat transfer coefficient h

$$T_{wall} = T_{coolant} + \frac{q_{wall}}{h} \tag{18.48}$$

with $T_{coolant} \approx (T_{in} + T_{out})/2$ the average coolant temperature. The flux on the wall of the cooling channel must be equal to the power on the exposed surface corrected by the difference in the conduction surface

$$q_{wall} = \frac{P}{S}\frac{l}{d} \tag{18.49}$$

For the heat transfer coefficient the Colburn regression [15] will be assumed

$$h = 0.027\mathrm{Re}^{4/5}\mathrm{Pr}^{1/3}\frac{K}{d}S_w \tag{18.50}$$

with $\mathrm{Re} = \rho V d/\mu$ the Reynolds number, $\mathrm{Pr} = \mu c_p/K$ the Prandl number, K is the thermal conductivity of the coolant and the factor $S_w \approx 1.5$ to account for the enhancement due to the insertion of a swirl tape.

A maximum heat transfer coefficient $h = 2 \times 10^5 \mathrm{W/m^2}K$ can be achieved for wall temperatures above the saturation temperature at which nucleate boiling takes place [16] but below the temperatures for critical heat flux conditions [17].

As an example let us take $c_p = 4210\,\mathrm{J\,kg^{-1}}K^{-1}$ and $G = 10\,\mathrm{kg/s}$. To exhaust $1MW$ of power on each divertor sector the temperature difference from Eq. (18.47) is about 24 °C. With $T_{in} = 100\,°\mathrm{C}$, we obtain $T_{out} = 124\,°\mathrm{C}$. Let us assume an average heat flux on the divertor of $10MW/m^2$ and a peak flux of $20\,\mathrm{MW/m^2}$. For $l/d = 1.5$ the maximum heat flux on the cooling channel is $q_{wall} = 30\,\mathrm{MW/m^2}$. The resulting temperature at the channel wall from Eq. (18.48) is $T_w = 274\,°\mathrm{C}$. If the distance between the exposed surface and the divertor channel is 5 mm, taking the conductivity of tungsten $k = 100\,\mathrm{W/mK}$ the maximum temperature at the exposed surface is 1274 °C that is in the range of the recrystallisation temperature and below the melting temperature of tungsten 3422 °C.

This calculation shows that divertor conditions may become close to criticality An increase in the heat flux above the critical heat flux leads to a sudden reduction of the heat transfer coefficient and a failure of the system. Note also that the use of tungsten produce favourable conditions thanks to its large thermal conductivity and high melting temperature.

18.8 Plasma Facing Materials in a Power Plant

Plasma facing materials in a power plant must be chosen on the basis of a number of considerations (or a general review of the development of plasma facing materials we refer to [19, 20] and references therein):

- Capability of withstanding high heat loads. In steady state conditions the first wall is expected to withstand loads ≤ 1 MW/m^2 whereas the divertor should be capable of withstanding up to 20 MW/m^2. Furthermore, under transient conditions these loads can be considerably higher and appropriate strategies must be in place to deal with these events.
- High thermal conductivity, low sputtering yield and sufficient mechanical properties. These properties need to be maintained also in the presence of intense neutron irradiation that can produce a damage rate up to $7dpa/FPY$ in the divertor region [21]. Furthermore, radiation induced embrittlement (see Chap. 19) can affect the performance of plasma facing materials.
- Reduced tritium retention. Tritium tends to accumulate in the plasma facing materials. This accumulation is particularly large in the case of carbon since co-deposits can be formed that rapidly raise the tritium inventory in the vacuum vessel requiring frequent stops of operation for tritium removal. For this reason carbon has been excluded in ITER. Metals have a retention that is more than an order of magnitude lower than carbon and are the preferred solutions [22]. Due to the simultaneous request of low sputtering yield, high-Z materials are the best solution.
- Reduced activation. This specifically excludes or severely limits the use of Mo and Nb as alloying elements since they transmute to very long lived radioisotopes.
- Compatibility with plasma operation. This condition would favour low-Z materials and indeed most of the experiments that have so far operated have adopted low-Z materials. The amount of tungsten that can be allowed in a burning plasma corresponds to a concentration below 10^{-5}. Nevertheless substantial progress has been made on present generation devices to demonstrate that tungsten accumulation in the core can be avoided.

The considerations above lead to the choice of W as plasma facing material. Tungsten has a high melting temperature, high sputtering threshold, good thermal conductivity an tolerable nuclear trans mutants. To deal with the problem of maintaining good ductile properties three different lines are pursued: tungsten solid solutions/alloys using materials such as V, Ta or, possibly, Ti and Hf; materials with ultra fine grained structure; composite materials [19].

A different approach is pursuing the use of liquid metals such as lithium as plasma facing materials. We will come back to this possibility at the end of this chapter.

18.9 High Radiation Regimes

The considerations presented in this chapter show that the solution to the exhaust problem in a fusion power plant cannot be achieved simply through advances in material technology but rather require most of the exhaust to be redistributed on the main wall through radiation. DEMO will require a fraction well above 90% to be radiated. These highly radiating plasmas must be produced maintaining high energy confinement properties.

Regimes with localised highly radiating region have been observed in many devices since the very beginning of tokamak research. A phenomenon called *MARFE (Multifaceted Asymmetric Radiation From the Edge)* was first described in [23]. The interaction of plasma with neutral hydrogen and impurities produce a local cooling of the plasma and an increase in radiation. The phenomenon is linked to the tokamak density limit that may lead to disruption. More recently regimes with a *X-point radiator (XPR)* have been observed in devices with metallic walls with the radiating region located above the X-point and being produced by injected impurities such as N, Ne or Ar. The XPR appears to be a controllable phenomenon and an important tool to control plasma detachment [24].

18.10 Advanced Divertors

An area of active research involves the investigation of divertor configurations that can handle larger heat flux than those foreseen in ITER. Two lines of research are presently being pursued: a) Advanced magnetic configurations that allow to spread the heat on larger surfaces; b) advanced materials for the divertor plates such as liquid metals.

The first line of research has been the subject of intense research and a preliminary assessment of the suitability of the various configurations for power plant applications can be found Ref. [25].

- The X-divertor [26] corresponds to the formation of a second X-point below the divertor plates. The flux surfaces are flared outwards rather than contracted inward as the target is approached. This produces an increase of the connection length and a flux expansion that allows to spread power on a larger area. Furthermore, when the neutrals dynamics is considered, the flaring out geometry tends to decrease the area of plasma-neutral interaction towards the main plasma producing a stabilising effect on the detachment front location.

- In the case of the Snowflake configuration [27] a large flux expansion is produced through a second order null at the location of the main X-point (the name *snowflake* derives from the six-fold symmetry of the resulting magnetic configuration that resembles a snowflake). Two variants exist called *plus* and *minus* in which the second order null is replaced with two close X-points making the configuration stable to small perturbations. Note that although the recipe is somehow similar to the X-divertor the magnetic surfaces are convergent as the target is approached.
- In the case of the Super-X divertor [28] an additional poloidal field coil is placed at an external radius and a similar vertical location of the coil producing elongation. This makes the separatrix to turn outwards locating the strike point at a larger major radius R_{strike}. Since the plasma wetted area is proportional to R_{strike} the heat load can be substantially reduced.
- A Double Null configuration produces an up-down symmetric configuration with two divertors. As turbulence tends to be located in the outer part of the torus, this configuration reduces the load on the inner divertors. In addition, the inner wall tends also to be less loaded reducing the constraints on the protecting armour.

The feasibility of these configurations is often not straightforward. The coil current that are necessary to produce the configuration may go beyond standard operational limits if external coils are used. We refer to Ref. [29] for a detailed analysis. Figure 18.7 summarises the different variants for DEMO

The use of liquid metals has been also considered as they offer a possible solution to the problem of the damage of plasma facing components as they allow the surface to be continuously reformed. As discussed in [30] the main challenges arise from: (a) the liquid metal splashing under the $J \times B$ force from disruptions; (b) the large ion sputtering; (c) the heat removal and the prevention of strong evaporation; and (d) the tritium removal for elements such as lithium that have a strong affinity with hydrogen isotopes. In addition, surface corrosion may be a problem. Part of the

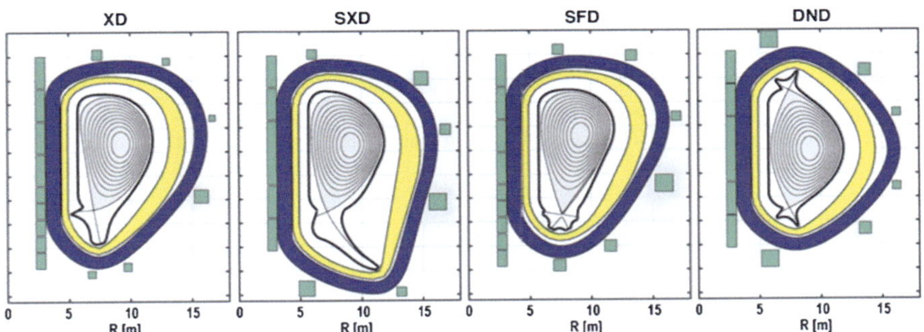

Fig. 18.7 Alternative divertor configurations for DEMO. Reprinted from [29] with permission from Elsevier

problem are solved using capillary porous systems in which the liquid metal surface is continuously reformed through a metal wire grid. Flowing liquid surfaces are also being proposed (see Ref. [31] for a critical review).

To test different solutions for advanced divertor configurations a dedicated Divertor Tokamak Test facility is presently under construction in Italy to start operation in the early 30's [32]

18.11 Suggestions for Further Readings

An extensive analysis of the subject presented in this chapter can be found in the monograph by Stangeby [1].

References

1. P.C. Stangeby, The plasma boundary of magnetic fusion devices. IOP (2000) ISBN 0 7503 0559 2
2. P.C. Stangeby, G.M. Mc Cracken, Plasma boundary phenomena in tokamaks. Nucl. Fusion **30**, 1225 (1990)
3. C.S. Pitcher, P.C. Stangeby, Experimental divertor physics. Plasma Phys. Control. Fusion **39**, 779 (1997)
4. K.-U. Riemann, The Bohm criterion and sheath formation. J. Phys. D: Appl. Phys. **24**, 493 (1991)
5. D. Bohm, *The Characteristics of Electrical Discharges in Magnetic Fields*, ch. 3, ed. by A. Guthry, R.K. Wakerling (MacGraw-Hill, New York, 1949), p. 77
6. A.W. Leonard, Plasma detachment in divertor tokamaks. Plasma Phys. Control. Fusion **60**, 044001 (2018)
7. P.C. Stangeby, Basic physical processes and reduced models for plasma detachment. Plasma Phys. Control. Fusion **60**, 044022 (2018)
8. B. Lipschultz et al., Sensitivity of detachment extent to magnetic configuration and external parameters. Nucl. Fusion **60**, 056007 (2018)
9. D.E. Post, A review of recent developments in atomic processes for divertors and edge plasmas. J. Nucl. Mater. **220–222**, 143–157 (1995)
10. F. Wagner, A study of the perpendicular transport properties in the scrape-off layer of ASDEX. Nucl. Fusion **25**, 525 (1985)
11. T. Eich et al., Scaling of the tokamak near the scrape-off layer H-mode power width and implications for ITER. Nucl. Fusion **53**, 093031 (2013)
12. M.A. Makowski et al., Analysis of a multi-machine database on divertor heat fluxes. Phys. Plasmas **19**, 056122 (2012)
13. W. Eckstein, C. Garcia-Rosales, J. Roth, W. Ottenberger, Sputtering data. IPP report IPP-9/82 Max Planck Institute for Plasma Physics (1993)
14. R. Beriech, W. Eckstein (eds.), *Sputtering by Particle Bombardment*. Springer (2007)
15. A.P. Colburn, A method of correlating forced convectin heat transfer data and a comparison with fluid friction. Trans. AIChE **29**, 174–210 (1933)
16. A.E. Bergles, W.M. Rosenhow, The determination of forced-convection surface-boiling heat transfer. J. Heat Transfer **86**, 365 (1964)
17. A.F. Raffray et al., Critical heat flux analysis and R&D for the design of the ITER divertor. Fusion Eng. Des. **45**, 377 (1999)

18. J.-H. You et al., High-heat-flux technologies for the European demo divertor targets: state-of-the-art and a review of the latest testing campaign. J. Nucl. Mater. **544**, 152670 (2021)
19. Ch. Linsmeier et al., Development of advanced high heat flux and plasma-facing materials. Nucl. Fusion **57**, 092007 (2017)
20. J. Roth et al., Recent analysis of key plasma wall interactions issues for ITER'. J. Nucl. Mater. **390–391**, 1 (2009)
21. M.R. Gilbert et al., An integrated model for materials in a fusion power plant: transmutation, gas production, and helium embrittlement under neutron irradiation. Nucl. Fusion **52**, 083019 (2012)
22. G.F. Matthews et al., Plasma operation with an all metal first wall: comparison of an ITER-like wall with carbon wall in JET. J. Nucl. Mater. **438**, S2-10 (2013)
23. B. Lipschultz et al., Marfe: an edge plasma phenomenon. Nucl. Fusion **24**, 977 (1984)
24. U. Stroth et al., Model for access and stability of the X-point radiator and the threshold for marfes in tokamak plasmas. Nucl. Fusion **62**, 076008 (2022)
25. H. Reimerdes et al., Assessment of alternative divertor configurations as an exhaust solution for DEMO. Nucl. Fusion **60**, 066030 (2020)
26. M. Kotschenreuther et al., On heat loading, novel divertors, and fusion reactors. Phys. Plasmas **14**, 072502 (2007)
27. D.D. Ryutov, Geometrical properties of a "snowflake" divertor. Phys. Plasmas **14**, 064502 (2007)
28. P.M. Valanju et al., Super X divertors for solving heat and neutron flux problems of fusion devices. Fusion Eng. Des. **85**, 46–52 (2010)
29. R. Ambrosino, et al., Evaluation of feasibility and costs of alternative magnetic divertor configurations for DEMO. Fusion Eng. Des. **146**, 2717–2720 (2019)
30. S.V. Mirnov, et al., Li experiments on T-11M and T-10 in support of a steady-state tokamak concept with Li closed loop circulation. Nucl. Fusion **51**, 073044 (2011)
31. Physics and PSI, R.E. Nygren, F.L. Tabarés, Liquid surfaces for fusion plasma facing components-a critical review. Part I. Nucl. Mater. Energy **9**, 6–21 (2016)
32. F. Romanelli, et al., Divertor Tokamak Test facility project: status of design and implementation. Nucl. Fusion **64**, 112015 (2024)

Chapter 19
The Nuclear Challenges of a Fusion Power Plant

Abstract *A short introduction to the problems related with tritium breeding and material damage by neutrons is presented.*

There are two main nuclear challenges in operating a fusion power plant. The first is the breeding of tritium in sufficient quantity. The second is the development of materials that can withstand the damage due to the interaction with the neutrons produced in the fusion reactions. Both aspects have an important consequence on the perspective of fusion as an energy source and will be examined in this chapter.

19.1 Tritium Balance

As described in Chap. 8 tritium is a radioactive isotope of hydrogen with a half-life of only 12.3y and on Earth it exists only in traces. The amount of tritium that is burned *daily* in a fusion power plant (and therefore the amount that must be produced on the same time scale) can be determined on the basis of the following estimate.

Taking a thermodynamic conversion efficiency of 33% the thermal fusion power in a $1GW_e$ fusion power plant is $3GW_{th}$. We assume that the thermal power is associated with the fusion-generated 14 MeV neutrons plus the 4.3 MeV energy released in the $n -^6 Li$ reaction (we neglect the power through the divertor as it will typically be extracted at low temperature). The corresponding neutron production is $\approx 10^{21} n/s$. An equivalent amount of 10^{21} tritium nuclei will be burnt each second in the fusion power plant. Integrating over an entire day with 100% plant availability the tritium consumption will be 0.440 kg/d.

For comparison, the *yearly* production of tritium in a 680MWe CANDU-6 reactor is of the order of 130 g [1].[1] In this case tritium, generated via the neutron capture by deuterium, must be extracted from the moderator. The CANDU reactors or possibly other sources [1–3] will have to produce the amount of tritium needed for initial

[1] At the moment there are 19 CANDU reactors operating in Canada, 3 in Korea and 2 in Romania [6]. Not all of them recover tritium. The amount recovered in one year is about 1.8kg.

F. Romanelli, *Physics of Nuclear Energy*, Springer Series in Plasma Science and Technology, https://doi.org/10.1007/978-981-97-9609-0_19

operation. At the moment an amount between 20 kg and 25 kg is globally available. The minimisation of the *start-up inventory*, i.e. the amount of tritium needed to start a new power plant, is therefore necessary to facilitate the deployment of fusion. The requirements on the tritium start-up inventory will be discussed in the next section.

As mentioned in Chap. 8, tritium is produced through the reaction between neutrons and lithium placed in a blanket around the plasma. Since the ^7Li reaction has a threshold at 2.8 MeV, the blanket must be enriched in ^6Li that has a large cross section at low energies. The key parameter in assessing the tritium self-sufficiency of a power plant is the *Tritium Breeding Ratio (TBR)* defined as the ratio between the rate of tritons produced in the blanket to the rate of tritons burned in the plasma. A TBR ≥ 1.1 must be obtained [4] to achieve tritium self sufficiency and produce in a reasonable amount of time the amount of tritium to start a new power plant.

Tritium balance is a critical aspect of a fusion system based on the DT reaction. Tritium must be continuously produced to keep the power plant going. In case of malfunctioning of the tritium reprocessing system a sufficiently large inventory must be available to ensure continued operation. Tritium must be also produced in excess to start a new power plant. These conditions set stringent limits to the start-up inventory and the TBR that must be ensured by the blanket system. In this section we determine both these quantities on the basis of a simplified model of the tritium processing system. A complete analysis can be found in Ref. [5] and references therein.

To fuel the plasma, tritium is injected in the vacuum vessel with a rate $\dot{T}_i$. However, only a small fraction of the injected flow is burned via fusion reactions. Such a fraction can be estimated as the product of the fraction η_f that reaches the plasma core and the fraction f_b of the tritium nuclei inside the plasma that undergo fusion reaction. The number of tritium nuclei burned in the plasma per unit time can be evaluated as $\dot{N}_T = \eta_f f_b \dot{T}_i$. The neutrons associated with these reactions are absorbed by the blanket and produce new tritium nuclei. To provide an estimate of the fuelling efficiency and of the burn fraction would require numerical codes that take into account realistic plasma conditions. In the following their product $\epsilon_{fb} = \eta_f f_b$ will be considered as an input parameter to the model. In the case of ITER $\epsilon_{fb} \approx 0.36\%$ and, as it will be shown later, a fusion power plant will require values in excess of 2%.

The tritium system can be modelled as shown in Fig. 19.1. The tritium storage (S) is typically made of uranium beds. The tritium inventory in the storage (I_S) is the result of a balance between the tritium losses due to the natural decay with half-life $\tau_{1/2}$ and the tritium input from the fuel cycles. It is convenient to distinguish between the *inner fuel cycle (IFC)* and the *outer fuel cycle (OFC)*. The IFC includes the pumping system of the divertor, the fuel cleanup and the isotope separation system and its role is to process the portion of the tritium fuels that does not undergo fusion reactions. The OFC includes the blanket and the tritium extraction system. The tritium production in the blanket depends on the TBR; as the achievable TBR is close to unity, it is convenient to define $TBR = 1 + \epsilon_{TBR}$ and treat ϵ_{TBR} as a small quantity. The IFC and the OFC differ in two respects. First, the rate of tritium recirculation in the OFC is much smaller than the rate in the IFC since $\epsilon_{fb} = \eta_f f_b << 1$. Second,

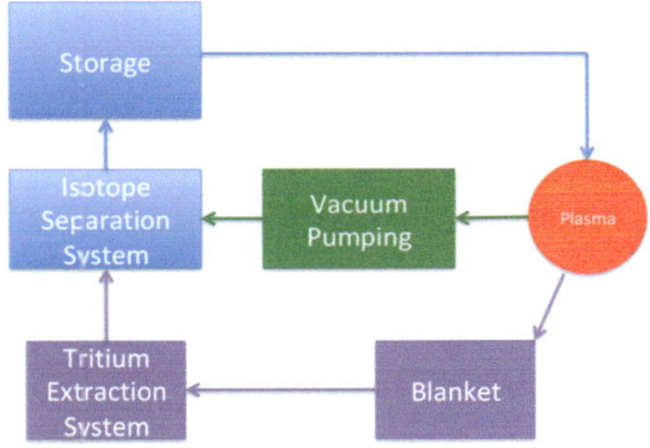

Fig. 19.1 Tritium cycle. The tritium management system can be divided into two circuits: the outer fuel cycle OFC (purple) and the inner fuel cycle IFC (green). They have common parts (blue) but are characterised by very different processing times (1 h for the IFC and 24 h for the OFC)

the time scale for processing tritium is longer in the OFC than in the IFC with the former being $\tau_{OFC} \approx 1d$ and the latter being $\tau_{IFC} \approx 1$ h.

The tritium inventories in the storage, the IFC and the OFC satisfy rate equations as follows

$$\frac{dI_S}{dt} = -\dot{T}_i + \frac{I_{IFC}}{\tau_{IFC}} + \frac{I_{OFC}}{\tau_{OFC}} - \frac{(\ln 2) I_S}{\tau_{1/2}} \tag{19.1}$$

$$\frac{dI_{IFC}}{dt} = -\frac{I_{IFC}}{\tau_{IFC}} + \dot{T}_i - \dot{N}_T \tag{19.2}$$

$$\frac{dI_{OFC}}{dt} = -\frac{I_{OFC}}{\tau_{OFC}} + (1 + \epsilon_{TBR})\dot{N}_T \tag{19.3}$$

For the sake of simplicity, losses of the various components are neglected.

In order to solve Eqs. (19.1)–(19.3) it is convenient to assume that the IFC reaches rapidly a secular equilibrium characterised by an inventory

$$I_{IFC} = \tau_{IFC}(1 - \epsilon_{fb})\dot{T}_i \tag{19.4}$$

The inventory in the OFC is obtained by direct integration of Eq. 19.3

$$I_{OFC} = \dot{N}_T \tau_{OFC}(1 + \epsilon_{TBR})(1 - e^{-t/\tau_{OFC}}) \tag{19.5}$$

Furthermore, the tritium decay inside the tritium storage can be neglected as it takes place on the time scale of the tritium half-life that is longer than the typical time scales of the fuel cycle system. With these assumptions the solution for the inventory in the storage is

$$I_S = I_S(0) + \epsilon_{TBR}\epsilon_{fb}\dot{T}_i t - (1 + \epsilon_{TBR})\epsilon_{fb}\dot{T}_i\tau_{OFC}(1 - e^{-t/\tau_{OFC}}) \tag{19.6}$$

The storage inventory Eq. (19.6) is shown in Fig. 19.2 versus time. The inventory initially decreases on the time scale of τ_{OFC} and achieves a minimum value before

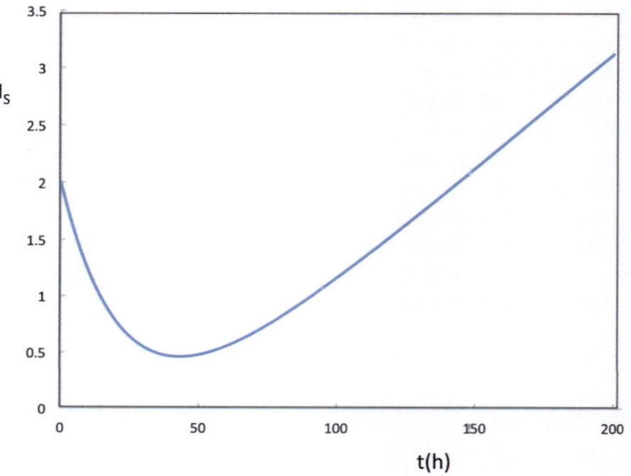

Fig. 19.2 Tritium storage inventory versus time. The initial decrease due to the finite processing time of the outer fuel cycle is balanced by the tritium production that eventually yield a linear increase of the inventory

the tritium production in the blanket takes over. The minimum is achieved at a time t_{min} and corresponds to a value I_{Smin} given by

$$t_{min} = \tau_{OFC} \ln(\frac{1}{\epsilon_{TBR}}) \tag{19.7}$$

$$I_{Smin} = I_S(0) - \dot{T}_i \tau_{OFC} \epsilon_{fb} (1 - \epsilon_{TBR} \ln(\frac{1}{\epsilon_{TBR}})) \tag{19.8}$$

The equations above assume that the availability factor A_F is 100% with A_F defined as the fraction of the time in which the power plant is producing energy. If the plant is available for a fraction $A_F < 1$, as long as the time scale is much smaller than $\tau_{1/2}$ the effect of a finite availability can be simply included by replacing the time t with $A_F t$.

We can now determine the start-up inventory and the TBR. The minimum inventory must be sufficient to ensure that in case of a malfunctioning of a fraction q of the tritium system there is enough fuel to keep the fusion reactions going for a time τ_R. The corresponding inventory is $q\dot{T}_i\tau_R$. The condition $I_{Smin} = q\dot{T}_i\tau_R$ yields

$$I_S(0) \approx \frac{\dot{N}_T \tau_{OFC}}{\epsilon_{fb}} \left(q \frac{\tau_R}{\tau_{OFC}} + \epsilon_{fb} \right) \tag{19.9}$$

The tritium inventory in the IFC given in Eq. (19.4) must be added to this number to have the total inventory. Since by assumption $\tau_{IFC} \ll \tau_{OFC}$ this provides a correction of the order 20–40% to Eq. (19.9). The use of techniques for direct recycling of the unreacted fuels [6] can further reduce this contribution. Furthermore, tritium production must be sufficient to provide the start-up inventory for a new power plant in a time τ_D (doubling time). Since the doubling time is of the order of a few years $\tau_{OFC} \ll \tau_D \ll \tau_{1/2}$, the condition $I_S(\tau_D) = 2I_S(0)$ yields

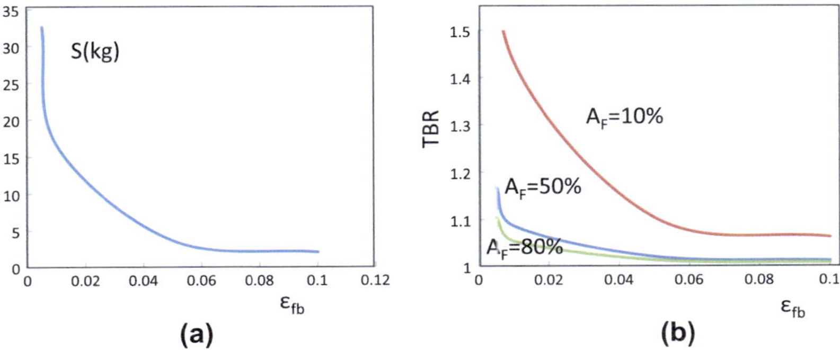

Fig. 19.3 The start-up inventory (**a**) and the required tritium breeding ratio (**b**) are shown as a function of the parameter ϵ_{fb}, the product between the fuelling efficiency and the burn-up efficiency, for a $3GW_{th}$ fusion power plant. The other parameters are: $q = 75\%$, $A_F = 50\%$, $\tau_{OFC} = 2\tau_R =$ 48 h, $\tau_D = 5y$. Note that for reasonable values of the availability factor for a DEMO power plant values of ϵ_{fb} in excess of 2% must be achieved

$$\epsilon_{TBR} = \frac{\tau_{OFC}}{A_F \tau_D \epsilon_{fb}} (q \frac{\tau_R}{\tau_{OFC}} + 2\epsilon_{fb}) \tag{19.10}$$

For illustration it is convenient to consider the case of a $3GW_{th}$ power plant introduced above for which $\dot{N}_T = \epsilon_{fb} \dot{T}_i = 0.440 kg/d$. Taking $q = 75\%$, $\tau_R = 24h$, $\tau_{OFC} = 48h$, $\tau_D = 5y$ and $A_F = 50\%$, Eqs. 19.9 and 19.10 yield start-up inventory of 17 kg and $TBR = 1.09$ for $\epsilon_{fb} = 1\%$ (3.6 kg and $TBR = 1.02$ for $\epsilon_{fb} = 5\%$). Figure 19.3 summarises the behaviour of these quantities.

The *required* TBR must be compared with the *achievable* TBR. The latter depends on the details of the blanket and can be determined only via sophisticated neutronic simulations. The results of several studies show that a TBR around 1.1 can be achievable by most of the blanket concepts [7] but these estimate bear a significant uncertainty and stringent conditions need to be imposed on the machine design. For example, the coverage of the vacuum vessel with the breeding blanket must be as close as possible to 100%, requiring to keep to minimum the port area.

19.2 Breeding Blanket Structure

The breeding blanket in a fusion power plant has three main functions:

- Breeding tritium to maintain tritium self sufficiency;
- Shielding the vacuum vessel and the magnet from the neutron radiation;
- Absorbing the neutron energy that will be extracted as thermal energy to power a turbine as in any thermal power plant.

All these functions are equally important. We have discussed in the previous section the challenge of tritium self sufficiency. Shielding is necessary to reduce to a minimum the amount of radioactive waste (that ideally would be limited to the blanket structure) and to minimise the damage to the tokamak components. To satisfy simultaneously all these objectives produces stringent requirements on the blanket components. Due to the damage related to neutron bombardment the blanket has to be periodically replaced. To maximise the plant availability the blanket replacement should be not too frequent and should take the lowest possible time. This sets strong requirements on the lifetime of the blanket materials and poses engineering challenges to the remote handling system. Furthermore, to maximise the thermodynamic efficiency the blanket temperature should be as high as possible posing again a strong requirement on the blanket materials and on the choice of the coolant.

There are several breeding blanket concepts but all of them have common features. Making reference to Fig. 19.5 the following main components can be identified

- The first wall protected by an armour (typically made of tungsten)
- The breeding zone that includes the breeder and the neutron multiplier
- The back shield
- The cooling systems for each zone
- The tritium recovery system

The first wall is the part of the blanket that has to withstand the largest power load. As this part does not breed tritium the amount of neutron absorption must be kept to a minimum, setting a limit to its thickness.

The breeding zone is the region where tritium is produced. The Li composition must be enriched in ^6Li with an enrichment varying between 40 and 90% to be compared with a natural ^6Li abundance of 7.3%. The enrichment process is still a subject of active R&D [8]. The column exchange *COLEX* process, originally employed for military applications, was abandoned due to its environmental impact as it involved the use of mercury with the risk of contaminating the air exhaust. Several other processes are under analysis. We refer to Ref. [8] for a discussion.

Since part of the neutrons are lost due to the absorption by other materials or through the ports, to maintain the needed TBR suitable neutron multipliers are inserted in the breeding zone. Two choices for the neutron multiplier are usually considered: beryllium and lead (Fig. 19.4). The $(n, 2n)$ reactions that these two materials undergo in the neutron multiplication process are endoergic and the incident neutron must have an energy above a certain threshold for the reaction to occur.

- ^9Be has a lower threshold energy of 1.7 MeV and therefore makes use of a larger part of the neutron spectrum but it is a scarce material and requires a special manipulation;
- Pb has a higher threshold energy of 7.4 MeV (thus a narrow window of operation in the neutron energy spectrum) but it has other advantages that will become apparent later in this chapter.

Also the reaction between neutrons and ^7Li produces an extra neutron with a low threshold energy and it takes part in the neutron multiplication process.

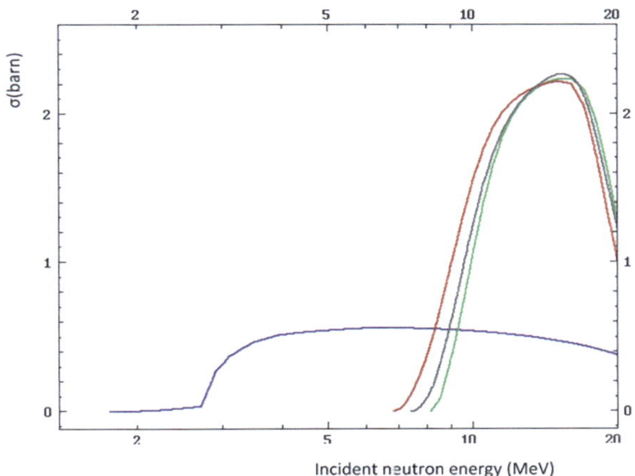

Fig. 19.4 The $(n, 2n)$ cross section from the ENDF library [9] for ^{9}Be (blue), ^{206}Pb (green), ^{207}Pb (red) and ^{208}Pb (grey) are shown versus the neutron energy

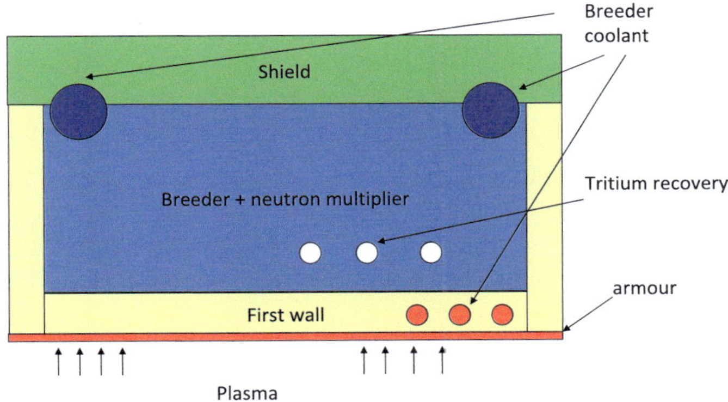

Fig. 19.5 All the breeding blanket concepts are made of the following elements: the primary first wall facing the plasma and protected by an armour, the breeding zone that contains the breeder and the neutron multiplier, and the back shield to protect the wall from the residual radiation. All these components must be actively cooled. In addition the tritium recovery system collects the tritium generated in the breeding zone and delivers it to the fuel reprocessing system

The breeder compound is characterised by a number of parameters

- The chemical and physical stability
- The thermal conductivity. This is important for the solid breeder (see below) as it allows to minimise the heat exchange surface
- The Li density
- The tritium retention
- The compatibility with the containment structure and the coolant
- The working temperature

Two classes of breeder are usually considered

- Liquid breeder. The breeder incorporates the neutron multiplier. The coolant can be either He, H_2O or the liquid breeder itself (self cooling). Liquid breeders are characterised by a high TBR. The flowing liquid can produce MHD effects with the generation of electric fields (orthogonal to both the flow and the magnetic field directions) that oppose the fluid motion. Furthermore, there are issues with corrosion and reactivity with water if the latter is used as a coolant. Examples of liquid breeders are:
  - The eutectic compound $Li_{17}Pb_{83}$ with Li atoms being 17% of the total number of atoms. The melting temperature is 235 °C. The compound is characterised by a large thermal conductivity (20 W/mK) [10] and a low reactivity with water. Furthermore, the tritium recovery can be performed more easily outside the vacuum vessel. On the negative side, the ^6Li enrichment required for the liquid breeder is high (90%), there are potential corrosion problems of the wall for temperatures above 400 °C and MHD effects must be dealt with.
  - Molten salts such as FLiBe $(LiF)_2(BeF_2)$ or FLiNaBe $(LIF - BeF_2 - NaF)$ that are characterised by a complex chemistry and have a low thermal conductivity ($\approx$1 W/mK) [11]. FLiBe has a melting temperature of 460 °C and FLiNaBe of 315 °C. They have been used in molten salt fission reactors.
- Solid breeder. The breeder and the multiplier are separated. The breeder is in the form of ceramic compounds Li_4SiO_4, Li_2TiO_3, Li_2O, Li_2ZrO_3 and $LiAlO_2$ fabricated in the form of spherical pebbles. Their advantage that with a lower Li enrichment 30–60% it is possible to obtain the required TBR and tritium can be extracted with a low pressure He flow circulating around the pebbles. On the negative side their thermal and structural performance are poor.

The coolant must be chosen according to the following characteristics

- High heat transfer coefficient
- Low pressure drop
- High thermal capacity
- Compatibility with the blanket materials
- Good nuclear properties
- Low working pressure

At the moment, excluding the case of the self cooled blanket, only two options exist: He and H_2O. The use of water is well established: the heat transfer coefficient is high and the pressure drop is low but the compatibility with the breeder and the multiplier must be considered. Furthermore a water leak in the vacuum vessel requires to stop operation even though the recovery of good discharge conditions is usually fast. The use of He on the other hand has the advantage of being compatible with any material but the heat transfer is low and the pressure drop high. It will be of interest for high temperature applications when materials that can work at high temperature will be available.

19.3 Interaction of Neutrons with Structural Materials

The fusion produced neutrons may collide with the nuclei of the structural materials of the reaction chamber producing either a displacement of the atom inside the crystal lattice or its transmutation. The displaced atom is a *primary knock-on atom (PKA)* that starts a cascade of displacements by interacting with the neighbouring atoms and produces self-interstitial atom defects and vacancies *(Frenkel pairs)* that, in turn, produce significant changes in the mechanical properties. The amount of damage depends on the number of displacements occurred in the material and is quantified in *displacements per atoms (dpa)*.

In addition to the damage due to the atom displacements, transmutations produce a change in the chemical composition of materials that has an impact on their properties. Specifically, DT neutrons are so energetic that hydrogen and helium can be produced via (n, p) and (n, α) reactions and accumulate in the lattice with a further degradation of the structural properties. Transmutations have also an impact on the waste management. In this respect fusion power plants have a clear advantage over fission power plants since the primary reaction does not produce radioactive materials and the neutron activation of the in-vessel component, with a proper choice of materials, decays to manageable levels within a time scale of the order of 100 years. However, although fusion does not produce high-level waste, it is still under investigation the production of intermediate level waste. A comparison between the neutron flux spectrum in thermal fission and fusion power plant is presented in Fig. 19.6 reproduced from Ref. [12]

As a rule of thumb, the amount of damage in a fusion power plant for steel under a neutron wall load of 1 MW/m^2 after one year is $10dpa$ and the He production

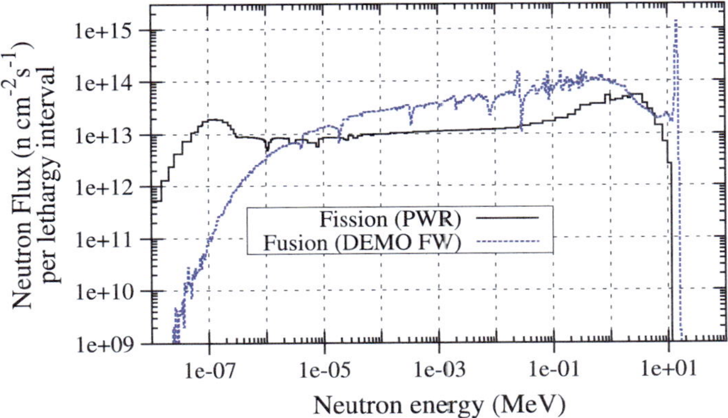

Fig. 19.6 Fission versus fusion neutron spectrum. Comparison between the neutron flux spectrum in a thermal fission power plant and in a DT fusion power plant at the first wall location. Reprinted from Ref. [12] with permission from IAEA. The peak in the case of fusion at neutron energies around 14 MeV makes it possible the (n, p) and (n, α) reactions

is $100appm$. Therefore in a power plant operating at full power for five years at a neutron load of 2 MW/m^2 the amount of damage is $100dpa$. For comparison, the amount of damage foreseen in ITER at the end of life is between 1 and $2dpa$.

A detailed treatment of the behaviour of materials under nuclear irradiation is beyond the scope of this book and we will give here only an outline (see [13] for a simple tutorial).

It is worthwhile recalling first the basic concepts of material structure. Materials for fusion have a crystalline structure. Out of the 14 Bravais lattices that describe the possible periodic structures, most of the materials of interest corresponds to three structures: body centred cubic (bcc), face centred cubic (fcc) and hexagonal close packed (hcp). The crystalline structure is not perfect. Defects exist either as point defects (vacancy and interstitials) or line defects (dislocations), see Fig. 19.7. Vacancies are vacant lattice sites whereas interstitials are atoms located at the interstices between two lattice atoms. Dislocations are half-planes that are terminated at some locations inside the lattice. When a sufficiently intense load is applied to the lattice, dislocations start to move by breaking and reattaching the atomic bonds over one atomic half plane. A plastic deformation of the material is produced that, in turn, increases the number of dislocations. This phenomenon leads to a strengthening of the material through a process known as *strain hardening*.

Structural materials are chosen on the basis of their capability of withstanding the large loads typical of fusion systems. Two kinds of behaviour can be distinguished

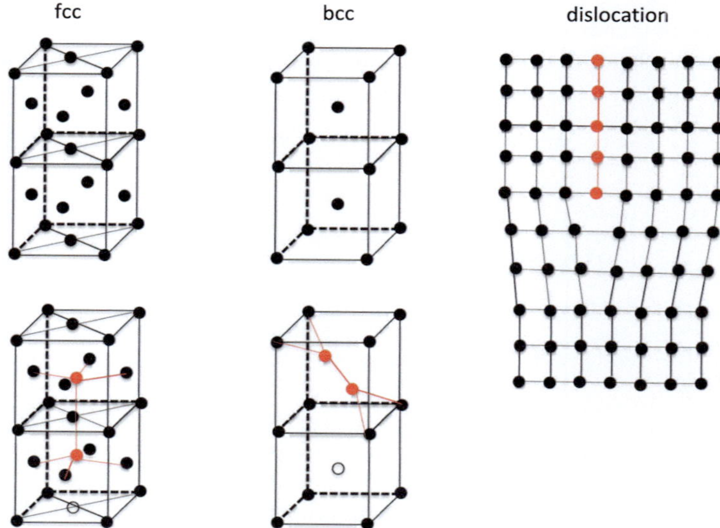

Fig. 19.7 Examples of lattice defects. The figures at the left and at the centre show point defects in the case of fcc and bcc materials respectively. Black circles correspond to occupied lattice sites, open circles to vacancies and red circles to interstitials. The figure at the right shows an example of dislocation

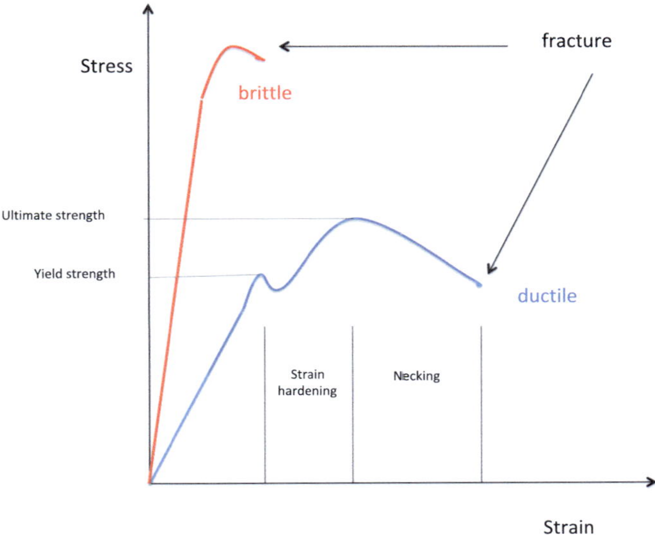

Fig. 19.8 The behaviour of brittle (red) and ductile (blue) materials is shown in the stress-strain diagram. For low values of strain the stress-strain relation is linear (elastic behaviour). Above a critical value plastic deformation takes place. For brittle materials the range of plastic deformation is minimal whereas for ductile materials the plastic region is extended. The amount of energy involved in the deformation is equal to the area below the curve

as exemplified in the stress-strain diagram (Fig. 19.8): materials that exhibit a brittle fracture and those that exhibit a ductile fracture. The stress-strain diagrams displays a linear part corresponding to the material elastic behaviour associated to the *Young modulus*. This region extends up to a maximum stress value corresponding to the *yield strength* and it is followed by the region of plastic deformation in which the stress first increases *(strain hardening)* up to the *ultimate strength* and then decreases *(necking)* up to a critical value of the strain at which the fracture takes place. In the case of ductile materials the crack tip is characterised by a large plastic deformation and associated energy, leading to a predictable failure mode. For this reason ductile materials are preferred. Conversely, in the case of brittle materials the maximum strength is large but the extent of the plastic deformation region is limited. This behaviour can lead to unpredictable catastrophic failure and it is clearly undesirable. While fcc materials exhibit ductile behaviour at all temperatures, in the case of bcc metals a characteristic temperature exists that separates the low-temperature brittle behaviour from the high-temperature ductile behaviour. Therefore the design must ensure that the material operates above such a *ductile-to-brittle transition temperature (DBTT)*. Unfortunately, neutron irradiation causes a shift in the $DBTT$ with the consequence of reducing the window of operations.

At high temperature material performance is limited by *creep* i.e. the tendency to undergo a slow deformation under the effect of continuous mechanical stresses.

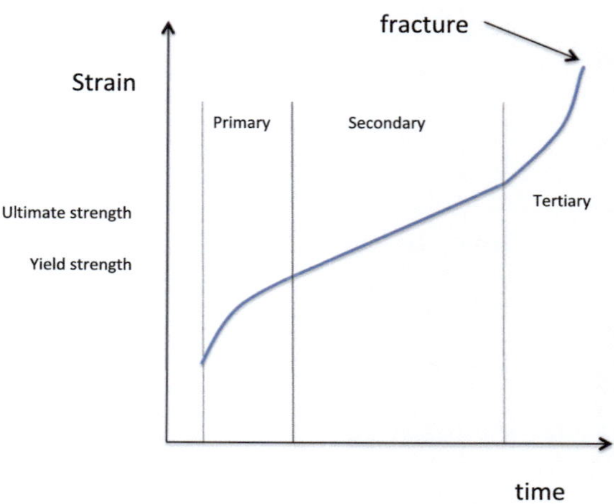

Fig. 19.9 The effect of creep can be shown in the plot of strain versus time

It is possible to distinguish three main stages of the creep (Fig. 19.9). In the first stage (primary creep) the strain rate is a function of time. This phase is followed by a constant strain rate (secondary creep). The third phase corresponds to an exponential increase of the strain rate as a result of necking, void formation or internal cracks, eventually ending in the rupture. The resilience to creep can be measured by the *creep rupture strength*, i.e. the stress that leads to rupture at a given temperature in a given amount of time (e.g. 10^5 h). At high temperature materials are limited by *thermal creep* that for fusion relevant structural materials occurs in the range 550–650 °C. Neutron bombardment produces *irradiation creep* that occurs at lower temperatures than thermal creep [14].

The dynamics of the neutron damage can be described as follows. A neutron of energy E_n can transfer sufficient energy to a nucleus to produce a *primary knock-on atom (PKA)* of energy E that interacts with the surrounding lattice atoms to displace them from their equilibrium position creating defects. The PKA energy depends on the neutron energy and the atomic mass number A_s of the structural material and can be as high as 2–7% of the impinging neutron energy. In the case of 10 MeV neutrons and a target of ^{56}Fe the maximum transferred energy in an elastic collision is $E = 4A/(A+1)^2 \approx 6.9\%$ corresponding to PKA energies in the range 700 keV. Part of the PKA energy is dissipated via electron excitation. The remaining part (the *damage energy* $\hat{E}$) can be evaluated using a modified Lindhard partition function (see [15] for details)

$$\hat{E} = \frac{E}{1 + kg(\epsilon)} \tag{19.11}$$

$$g(\epsilon) = 3.4008\epsilon^{1/6} + 0.40244\epsilon^{3/4} + \epsilon \tag{19.12}$$

$$k = 0.1337Z_1^{1/6}(\frac{Z_1}{A_1})^{1/2} \tag{19.13}$$

$$\epsilon = \frac{A_2 E}{A_1 + A_2} \frac{a}{Z_1 Z_2 m_e c^2} \tag{19.14}$$

$$a = (\frac{9\pi^2}{128})^{1/3} \frac{1}{\alpha^2} (Z_1^{2/3} + Z_2^{2/3})^{-1/2} \tag{19.15}$$

with α the fine structure constant and the suffix 1 and 2 referring to the projectile and the target, respectively.

The mean free path of neutrons in solids is of the order of few centimetres whereas the range of a PKA is of the order of $100nm$. Therefore the interaction produces spatially isolated displacement cascade regions.

The displacement damage is quantified in dpa, with a dose of $1dpa$ corresponding to stable displacement of <u>all</u> lattice atoms during the irradiation without accounting for the recovery due to the recombination of point defects due to their thermally activated diffusion. For this reason the amount of dpa corresponds more to a measure of the *exposure* of the material rather than of its effective damage and, indeed, the neutron irradiation effects discussed below can be cured by performing irradiation at sufficiently high temperature to facilitate the defect recombination.

The number of atoms instantaneous knocked-off by a PKA is about 100. During this *thermal spike* phase that typically lasts a few picoseconds the cascade region can achieve temperatures above the melting temperature but heat transfer rapidly cool down the region. During the recovery phase, lasting about $10ps$, most of the displaced atoms find an empty lattice site but a small fraction is unable to do so and a certain number of stable Frenkel pairs are produced with the vacancies typically located in the centre of the displacement cascade region. On a longer time scale the defects can migrate and annihilate at the sinks.

In addition to the damage due to the atom displacements, DT neutrons are so energetic that hydrogen and helium can be produced via (n, p) and (n, α) reactions. The cross section for these processes is typically smaller than the cross section for radiative capture (see Fig. 19.10) but their impact is substantial since the H/He atoms accumulate in the lattice producing swelling and embrittlement of the material with a further degradation of the structural properties.

The interaction of neutrons with materials results in the modification of materials properties as summarised in the Table 19.1 (with T_M the melting temperature)

- At low temperature irradiation produces a high concentration of sessile defects that reduce the dislocation motion and produce significant hardening. Already at low damage level the effect of neutron irradiation produces low-temperature embrittlement of bcc metals with a shift of the *DBTT* from values below 0 to 200 °C or above. This effect however depends on the temperature at which irradiation is performed as the recombination of defects inside the lattice is a temperature dependent phenomenon and the irradiation effect can be annealed out at sufficiently large temperature.

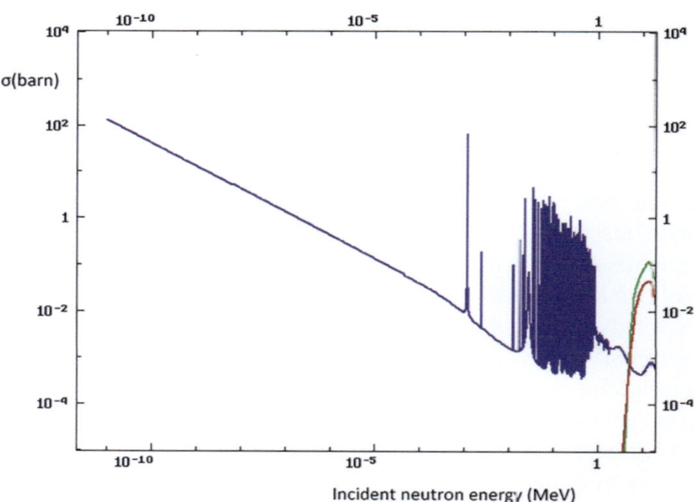

Fig. 19.10 The cross sections from the ENDF library [9] for the (n, p) (green), (n, α) (red) and (n, γ) (blue) of ^{56}Fe are shown as a function of the neutron energy. Note the threshold for the (n, p) and (n, α) reactions at high neutron energy

Table 19.1 Damage dose and temperature range for the onset of the various radiation damage effects [13] with T_M the melting temperature

Effect	Radiation damage threshold (*dpa*)	Temperature
Low-temperature radiation hardening and embrittlement	0.1	$T < 0.35T_M$
Phase instabilities from radiation induced segregation and radiation induced precipitation	10	$0.3T_M < T < 0.6T_M$
Radiation induced thermal creep	10	$T > 0.45T_M$
Volumetric swelling from void formation	10	$0.3T_M < T < 0.6T_M$
High-temperature He embrittlement of grain boundaries	10	$T > 0.5T_M$

- Radiation induced segregation and precipitation occur at intermediate temperatures and produce a local change in the elemental composition of the alloy near grain boundaries leading to grain boundary embrittlement.
- As to volumetric swelling, for austenitic steels (fcc) this phenomenon can lead to a severe loss of dimensional stability ($\approx$10% at 70dpa) whereas ferritic steels (bcc) the effect is substantially reduced.
- Irradiation creep occurs at temperatures at which thermally activated creep is small [16, 17] an can produce expansion along the direction of high stress.

- At very high temperature and for concentrations above $100appm$ (corresponding to $10dpa$) the helium produced by the (n, α) reactions can migrate to the grain boundaries forming cavities and causing inter granular fracture. Together with the thermal creep strength, the high-temperature He embrittlement (HTHE) of grain boundaries defines the high temperature operational limit of the material.

As a result, materials exhibit a window of operating temperatures between the $DBTT$ and the temperature at which creep/HTHE becomes significant. In the next section a short overview of the structural material development is provided.

Having defined the temperature/dpa domain in which significant degradation can occur it is necessary to quantify the dpa production rate. To this aim the approach of Ref. [15] will be used (see also Ref. [12]).

Various reactions can take place between a neutron of energy E_n and the target nucleus. Each reaction type will be labelled with the index j. Each reactions produces a PKA of energy $E_j(E_n)$ and, in turn, a number of Frenkel pairs N_{dj} given by [15]

$$N_{dj}(E_n) = \frac{0.8\hat{E}_j(E_n)}{2E_d} \tag{19.16}$$

with $\hat{E}_j$ given in terms of E_j by Eq. (19.11). The displacement threshold energy E_d is assumed $31eV$ for Be, $40eV$ for Fe, Cr, V, Nb, Zr, $60eV$ for Mo and $90eV$ for W and Ta. It is important to note that the displacement dose rate depends not only on the PKA energy but also on the material.

To obtain the total dpa production rate the above expression must be summed over all the neutron energies [12]

$$dpa\ per\ second = \int_0^\infty dE_n \phi(E_n) \sum_j \sigma_j(E_n) N_{dj}(E_n) \tag{19.17}$$

with σ_j the cross section for reaction j and $\phi(E_n)$ the neutron flux per unit energy. Taking as an example a neutron load of 1 MW/m^2 corresponding to $4.4 \times 10^{17} n/(m^2 s)$, and a PKA energy of 486 keV (half of the maximum energy that a 14 MeV neutron can transfer to a ^{56}Fe nucleus at rest) the annual displacement dose in iron is $12dpa$ if the average cross section is $4barn$.

In the calculations outlined in this chapter it is necessary to make use of averages over the neutron spectrum. This in turn requires specific neutronic analysis to determine the neutron flux per unit energy. The results of MCNP calculations for a fusion power plant show that typically the neutron flux per unit energy at the first wall has a peak at 14 MeV that includes about 25–30% of the total flux. Below 5 MeV the spectrum is similar to that of a thermal fission reactor with the characteristic $1/E$ behaviour (i.e. approximately constant for lethargy interval). The low energy tail is less pronounced for blanket concepts that use Pb as multiplier instead of Be that has a better neutron moderation effect. As we move inside the blanket the 14 MeV peak

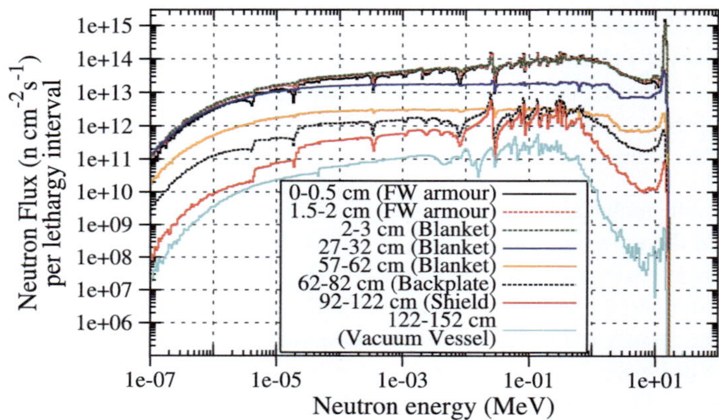

Fig. 19.11 Neutron flux spectrum at different locations inside the blanket. Reprinted from Ref. [12] with permission from IAEA. As we move from the first wall to the vacuum vessel the high energy component decreases and the spectrum resembles that of a thermal fission power plant

becomes less and less pronounced. The flux is reduced by four orders of magnitude at the shield location with a constant flux for unit lethargy almost constant for energies below 1 MeV.

In Fig. 19.11 from Ref. [12] the neutron flux spectrum at different locations inside the blanket is shown.

Using the calculated fusion spectrum and model of the blanket in terms of a set of homogenous layers from the first wall to the vacuum vessel it is possible to determine the amount of damage in the various regions. An example of this calculation is reported in Fig. 19.12 from Ref. [18] for the DEMO water-cooled lithium-lead blanket design model described in [19]. The maximum damage rate is localised at the first wall and corresponds to $10dpa$ for full power year (FPY) which means that after five years at full power the accumulated damage is about $50dpa$. However, the damage decreases by one order of magnitude in the middle of the breeding zone (about one third of the blanket depth) and at the vacuum vessel is in the range $10^{-2}dpa/FPY$ leaving a substantial margin from the maximum allowed damage of 1dpa at the end of life (10appm of He) set by the ITER design guidelines.

19.4 Candidate Structural Materials for Fusion Reactors

The development of candidate materials for fusion application is one of the most active research fields [14, 20]. The challenge posed by fusion are more difficult than those of fission [13] and require a dedicated approach. Materials in the first commercial fission power plant were exposed to $1dpa$ at a maximum temperature of 300 °C. The second generation fission plants were characterised by exposure doses

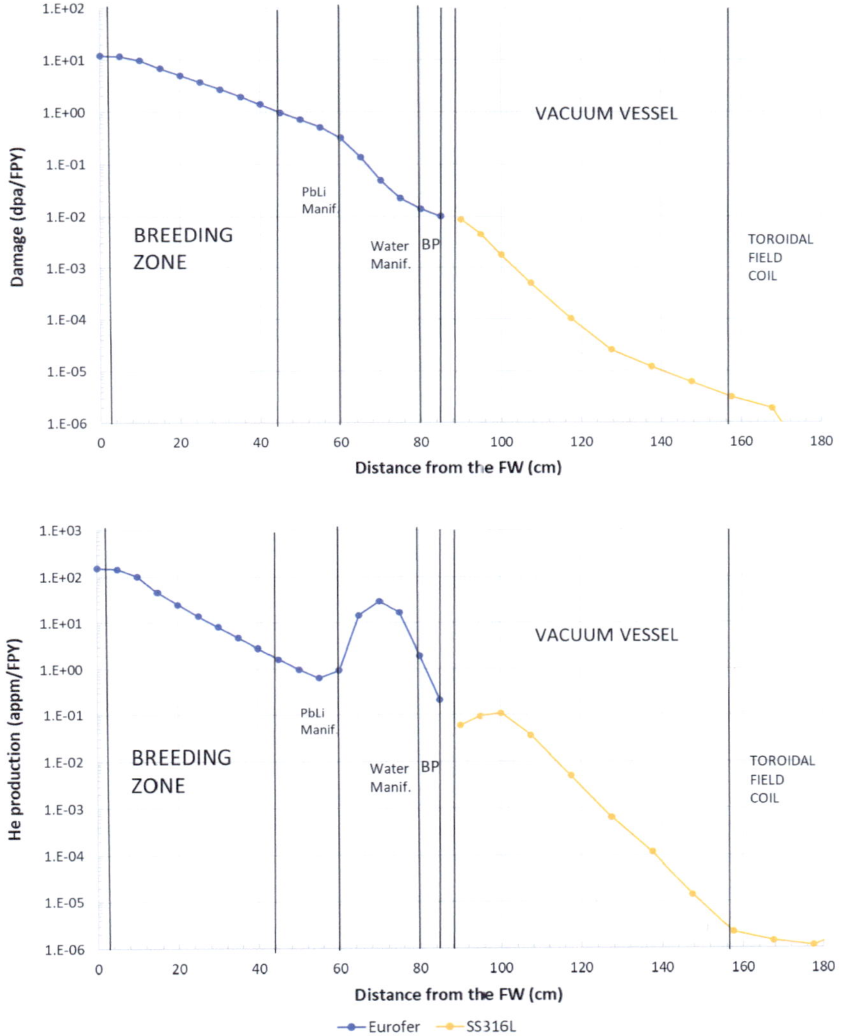

Fig. 19.12 Radial profile of the damage rate (top) and He production rate (bottom) in the inboard side of DEMO. The distances are measured from the first wall. The blue lines correspond to the blanket where EUROFER is used as structural material whereas the yellow lines correspond to the vacuum vessel where 316LN is used. Reprinted from Ref. [18] with permission from Elsevier

up to 30*dpa* and temperatures below 350 °C. Fast breeder reactors are expected to have displacement doses up to 100*dpa* and temperature of the order of 600 °C. The first demonstration fusion power plant is expected to have displacement doses in the range 50–150 *dpa* and maximum temperature above 550 °C. These conditions are similar to those expected in Generation IV fission power plants but, in addition, the high-energy fusion neutrons produce He and H that act against the recombination of point defects and further contribute to material embrittlement.

The modification of material properties under neutron irradiation depends on the neutron energy spectrum. Thus, in order to investigate and qualify fusion relevant materials using material testing facilities the relevant neutron energy spectrum must be used. In a fusion power plant neutrons are generated at 14 MeV and then slow down due to the interaction with the materials of the reaction chamber. A spallation source produces a spectrum with a substantial amount of neutrons with energy well above 14 MeV, whereas in fission reactors the amount of neutrons with a energy above 2 MeV is negligible (see e.g. Eq. (9.17). For this reason a dedicated facility, the *International Fusion Material Irradiation Facility (IFMIF)*, has been proposed to produce a realistic spectrum where materials can be tested in fusion relevant conditions [21]. IFMIF is made of two 125 mA, 40 MeV deuteron beams impinging on a lithium target. The stripping reaction between deuterium and lithium produces neutron in the relevant energy range. Material specimens in a volume of $0.5l$ will receive a damage dose up to $20dpa$ per year. In the absence of a dedicated facility the investigation of the effects of neutron irradiation is made in fission reactors where some materials have been irradiated already to a damage dose of a few tens of dpa. The He production is simulated either via the (n, α) reaction by ^{10}B doping of the material or via ion-beam implantation. These methods are not fully representative as boron doping can overestimate the effect of helium embrittlement and ion implantation is limited to the material surface.

The development of materials for fusion has been the subject of various reviews. The most recent assessment has been conducted in Europe in connection with the elaboration of the fusion roadmap [22] and is summarised in Refs. [23] and [24]. The material assessment has used project-based and systems-engineering approaches, the concept of Technology Readiness Levels, and considered lessons learned from fission reactor material development. A set of possible baseline materials have been identified. Since these materials bear significant risks, a list of risk mitigation materials has also been recommended.

It is important to emphasise that the development of fusion relevant materials has a crucial impact on the economy of a fusion power plant but it has limited impact on the plant licensing as the primary safety boundary of a fusion power plant is the vacuum vessel that is shielded by the blanket and has to withstand a neutron flux orders of magnitude lower than the first wall and characterised by a softer neutron spectrum. Therefore the vacuum vessel can be made of conventional austenitic steels already available on the market. The development of fusion relevant materials is on the contrary necessary for the blanket in order to avoid its too frequent replacement due to the loss of structural properties.

The baseline materials for the blanket steel are the reduced activation ferritic martensitic steels (RAFM). A version of RAFM has been developed by the European industry under the name EUROFER. Similar materials have been developed in Japan (F82H) and in US (1537-1539 and Mod-NF616) [23]. The composition of EUROFER and F82H is shown in Table 19.2 [20].

Table 19.2 Chemical composition of EUROFER. Reprinted from Ref. [20] with permission of IAEA. In the first column the limits imposed by radiological considerations are shown. The second column list the ideal composition and the third column what has been achieved so far. The last column shows the composition of the Japanese RAFM F82H

	Radiologically desired mass (%)	EUROFER-97 specified mass (%)	EUROFER-97 achieved mass (%)	F82Hmod.Heat 9741 mass (%)
C		0.11	0.11–0.12	0.09
Cr		9.0	8.82–8.96	7.7
W		1.1	1.07–1.15	1.94
Mn		0.4	0.38–0.49	0.16
V		0.15–0.25	0.18–0.20	0.16
Ta		0.12	0.13–0.15	0.02
N_2		0.03	0.018–0.034	0.006
P		<0.005	0.004–0.005	0.002
S		<0.005	0.003–0.004	0.002
B		<0.001	0.0005–0.0009	0.0002
O_2		<0.01	0.0013–0.0018	(0.01)
Nb	<0.000001	<0.001	0.0002–0.0007	0.0001
Mo	<0.0001	<0.005	0.001–0.0032	0.003
Ni	<0.001	<0.005	0.007–0.0028	0.02
Cu	<0.001	<0.005	0.0015–0.022	0.01
Al	<0.0001	<0.01	0.006–0.009	0.003
Ti	<0.02	<0.01	0.005–0.009	0.01
Si	<0.04	<0.05	0.04–0.07	0.11
Co	<0.001	<0.005	0.003–0.007	0.05

EUROFER has a lattice bcc structure that provides good stability under fission neutron irradiation. It can be manufactured on an industrial scale and has shown good mechanical properties and sufficient corrosion resistance to liquid metals as required for the lithium-lead blanket concept. It is prone to low-temperature embrittlement and this requires operation above 350 °C which reduces the available temperature window to 350–550 °C. Above 550 °C the creep rupture strength falls below $10^4 h$ for stresses in the range of $100 MPa$ that are of the same order of those proposed as maximum primary stress for fusion application of fission codes such as RCC-MRx [23]. Data on neutron irradiation have been obtained up to $80 dpa$ in the fast fission reactor $BOR-60$ between 300 and 330 °C [25] and there is some evidence that the damage can be annealed by post-irradiation operation at high temperature [26] but the effect on mobility of the He from the (n, α) reactions is unknown.

Among the risk mitigation materials it is worth mentioning the *Oxide Dispersion Strengthened (ODS)* steels that have shown improved irradiation resistance and a better performance at high temperature. Vanadium alloys [27] and silicon carbide composites [28] may also offer interesting alternatives.

One of the main advantages of fusion is the absence of high-level waste that require a geological disposal. As shown in [29], the existing version of EUROFER 97 shows good radiological properties with the surface gamma dose decreasing down to the intermediate-level waste in about 100 years (the ideal EUROFER composition would achieve the hands-on level in about 300 years). The main elements of EUROFER are Fe and Cr. The contact dose rate of these elements as defined in Eq. (7.42) decays in a few tens of years. After five years of irradiation at $\approx$2.4 MW/m^2 Fe reaches the hands-on limit ($10\mu Sv/h$) in 100 years and Cr in 30 years. Alloying elements such as Ni, Nb, Mo and Al may cause long term activation and have been replaced with elements such as Ta and W that reach acceptable levels on time scales similar to that of Fe and Cr.

Fusion waste, although produced in a quantity per unit energy that is larger than that of fission plants, have significantly lower radioactivity content. Most of the waste arise from the structural materials of the vacuum vessel and of the blanket. The most important radioisotopes (besides tritium that must be recovered as much as possible to fuel new power plants) are the long-lived ^{14}C, ^{94}Nb, ^{59}Ni, ^{99}Mo radionuclides [30, 31].

Even though the radiological behaviour of EUROFER97 is already acceptable when only the gamma dose is considered, it should be noted that more recent calculations [30] that have included also beta emission show that in some cases the waste of a fusion power plant could exceed the low-level waste classification required in some European country unless mitigation techniques such as isotope tailoring or decarburisation are employed.

19.5 Suggestions for Further Readings

A recent overview of the tritium breeding challenges can be found in [5]. For material development the results of a dedicated study has been published in 2017 (see [24] for the introduction to the special Nuclear Fusion issue and [23] for a summary of the findings). An excellent introduction to materials for fusion can be found in [13].

References

1. M. Ni, Y. Wang, B. Yuan, J. Jiang, Y. Wu, Tritium supply assessment for iter and demonstration power plant. Fusion Eng. Des. **88**(9), 2422–2426 (2013)
2. R.J. Pearson, A.B. Antoniazzi, W.J. Nuttall, Tritium supply and use: a key issue for the development of nuclear fusion energy. Fusion Eng. Des. **136**, 1140–1148 (2018)
3. M. Kovari, M. Coleman, I. Cristescu, R. Smith, Tritium resources available for fusion reactors. Nucl. Fusion **58**, 026010 (2017)

4. S. Zheng, T.N. Todd, Study of impacts on tritium breeding ratio of a fusion DEMO reactor. Fusion Eng. Des. **98–99**, 1915–1918 (2015)
5. M. Abdou et al., Physics and technology considerations for the deuterium-tritium fuelcycle and conditions for tritium fuel self sufficiency. Nucl. Fusion **61**, 013001 (2021)
6. C. Day, T. Giegerich, The direct internal recycling concept to simplify the fuel cycle of a fusion power plant. Fusion Eng. Des. **88**, 616–20 (2013)
7. M.E. Sawan, M.E. Abdou, Physics and technology conditions for attaining tritium self-sufficiency for the DT fuel cycle. Fusion Eng. Des. **81**, 1131–1144 (2006)
8. T. Giegerich, K. Battes, J.C. Schwenzer, C. Day, Development of a viable route for lithium-6 supply of DEMO and futurefusion power plants. Fusion Eng. Des. **149**, 111339 (2019)
9. Evaluated Nuclear Data Files (ENDF) https://www-nds.iaea.org/exfor/endf.htm
10. G. Kuhlbörsch, F. Reiter, Physical properties and chemical reaction behaviour of $Li_{17}Pb_{83}$ related to its use as fusion reactor blanket material. Nucl. Eng. Des./Fusion **1**, 195–203 (1984)
11. D. Williams, L. Toth, K. Clarno, *Assessment of Candidate Molten Salt Coolants for the Advanced High Temperature Reactor (AHTR)* (Department of Energy, United States, 2006)
12. M.R. Gilbert et al., An integrated model for materials in a fusion power plant: transmutation, gas production, and helium embrittlement under neutron irradiation. Nucl. Fusion **52**, 083019 (2012)
13. S.J. Zinkle, Fusion materials science: overview of challenges and recent progress. Phys. Plasmas **12**, 058101 (2005)
14. S.J. Zinkle, et al., Development of next generation tempered and ODS reduced activation ferritic/martensitic steels for fusion energy applications. Nucl. Fusion **57**, 092005 (2017)
15. M.J. Norgett, M.T. Robinson, I.M. Torrens, A proposed method of calculating displacement dose rates. Nucl. Eng. Des. **33**, 50–54 (1975)
16. J.R. Matthews, M.W. Finnis, Irradiation creep models—an overview. J. Nucl. Mater. **159**, 257 (1988)
17. H. Tanigawa et al., Development of benchmark reduced activation ferritic/martensitic steels for fusion energy applications. Nucl. Fusion **57**, 092004 (2017)
18. S. Noce et al., Nuclear analysis of the single module segment WCLL DEMO. Fusion Eng. Des. **147**, 111207 (2019)
19. A. Del Nevo et al., WCLL breeding blanket design and integration for DEMO 2015: status and perspectives. Fus. Eng. Des **124**, 682–686 (2017)
20. A. Möslang et al., Towards reduced activation structural materials data for fusion DEMO reactors. Nucl. Fusion **45**, 649–655 (2005)
21. Current status and future options, J. Knaster, et al., IFMIF, the European-Japanese efforts under the Broader Approach agreement towards a Li(d, xn) neutron source. Nucl. Mater. Energy **9**, 46–54 (2016)
22. F. Romanelli, et al., Fusion electricity. A roadmap to the realization of fusion energy. European Fusion Development Agreement, EFDA, ISBN 978-3-00-040720-8 (2012)
23. D. Stork, et al., Materials R&D for a timely DEMO: Key findings and recommendations of the EU Roadmap Materials Assessment Group. Fusion Eng. Des. **89**, 1586–1594 (2014)
24. D. Stork, S.J. Zinkle, Introduction to the special issue on the technical status of materials for a fusion reactor. Nucl. Fusion **57**, 092001 (2017)
25. B. van der Schaaf et al., High dose, up to 80 dpa, mechanical properties of Eurofer 97. J. Nucl. Mater. **386–388**, 236–240 (2009)
26. E. Gaganidze, J. Arktaa, Assessment of neutron irradiation effects on RAFM steels. Fusion Eng. Des. **88**, 118–128 (2013)
27. R.J. Kurtz et al., J. Nucl. Mater. **283–287**, 70 (2000)
28. B. Riccardi et al., J. Nucl. Mater. **329–333**, 56 (2004)
29. R. Lindau et al., Present development status of EUROFER and ODS-EUROFER for application in blanket concepts. Fusion Eng. Des. **75–79**, 989–996 (2005)
30. M.R. Gilbert et al., Waste implications from minor impurities in European DEMO materials. Nucl. Fusion **59**, 076015 (2019)
31. S.M. Gonzalez de Vicente, et al., Overview on the management of radioactive waste from fusion facilities: ITER, demonstration machines and power plants. Nucl. Fusion **62**, 085001 (2022)

Chapter 20
Solutions to Selected Problems

Abstract *The solutions to selected problems are presented in this chapter.*

Problems of Chap. 1

1.1 Hydrogen has two stable isotopes: protium ($A = 1$) and deuterium ($A = 2$). The hydrogen atomic weight is $1.00794AMU$. Determine the amount of deuterium in a litre of water.

If f is the fraction of deuterium we have

$$1.00794 \approx (1 - f) \times (1.00728 + 0.00055) + f \times 2.014 \tag{20.1}$$

which yields $f \approx 1.09 \times 10^{-4}$

Taking into account that the fraction of hydrogen is water is about $2/18$, the amount of deuterium in one litre of water is

$$1\,\text{kg} \times 2/18 \times 1.09 \times 10^{-4} \times 2 \approx 24\,\text{mg} \tag{20.2}$$

1.2 Given a cathode ray tube $1m$ long and an electric field of $10kV/m$ the deflection of an electron along y can be compensated by adding a magnetic field along z of intensity $1G(=10^{-4}T)$. Determine the deflection in the absence of the magnetic field. Determine the electron velocity along x.

$$v = \frac{E}{B} = \frac{10\,\text{kV/m}}{10^{-4}T} = 10^{8}\,m/s \tag{20.3}$$

$$T = \frac{L}{v} = \frac{1\,m}{10^{8}\,m/s} = 10^{-8}s \tag{20.4}$$

$$a = \frac{qE}{m} = \frac{1.6 \times 10^{-19}C10kV/m}{9 \times 10^{-31}kg} = 1.76 \times 10^{15}m/s^{2} \tag{20.5}$$

F. Romanelli, *Physics of Nuclear Energy*, Springer Series in Plasma Science and Technology, https://doi.org/10.1007/978-981-97-9609-0_20

$$\Delta y = \frac{aT^2}{2} = 8.8\,\text{cm} \tag{20.6}$$

Problems of Chap. 2

2.1 Evaluate the average velocity in the three directions and the average of the kinetic energy for a distribution function given by a shifted Maxwellian $f(\mathbf{v}) = n_o/(v_t\pi^{1/2})^3\exp(-(\mathbf{v}-V\mathbf{x})^2/v_t^2)$

$$\int_{-\infty}^{+\infty} dv_x \int_{-\infty}^{+\infty} dv_y \int_{-\infty}^{+\infty} dv_z (\frac{m}{2\pi T})^{3/2} e^{-\frac{m((v_x-V)^2+v_y^2+v_z^2)}{2T}} g(\mathbf{v}) = \tag{20.7}$$

$$= \int_{-\infty}^{+\infty} du_x \frac{e^{-(u_x-U)^2}}{\pi^{1/2}} \int_{-\infty}^{+\infty} du_y \frac{e^{-u_y^2}}{\pi^{1/2}} \int_{-\infty}^{+\infty} du_z \frac{e^{-u_z^2}}{\pi^{1/2}} g(\mathbf{u}v_t) \tag{20.8}$$

$$< \mathbf{v} >= \int_{-\infty}^{+\infty} du_x \frac{e^{-(u_x-U)^2}}{\pi^{1/2}} \int_{-\infty}^{+\infty} du_y \frac{e^{-u_y^2}}{\pi^{1/2}} \int_{-\infty}^{+\infty} du_z \frac{e^{-u_z^2}}{\pi^{1/2}} \mathbf{u}v_t \tag{20.9}$$

$w_x = u_x - U$

$$= \int_{-\infty}^{+\infty} dw_x \frac{e^{-w_x^2}}{\pi^{1/2}} \int_{-\infty}^{+\infty} du_y \frac{e^{-u_y^2}}{\pi^{1/2}} \int_{-\infty}^{+\infty} du_z \frac{e^{-u_z^2}}{\pi^{1/2}} (w_x + U,\ u_y,\ u_z) v_t \tag{20.10}$$

$$\int_{-\infty}^{+\infty} dt \frac{e^{-t^2}}{\pi^{1/2}} = 1 \quad \int_{-\infty}^{+\infty} dt \frac{e^{-t^2}}{\pi^{1/2}} t = 0 \tag{20.11}$$

$$< \mathbf{v} >= (0 + Uv_t,\ 0,\ 0) = (V,\ 0,\ 0) \tag{20.12}$$

$$< v^2 >= \int_{-\infty}^{+\infty} du_x \frac{e^{-(u_x-U)^2}}{\pi^{1/2}} \int_{-\infty}^{+\infty} du_y \frac{e^{-u_y^2}}{\pi^{1/2}} \int_{-\infty}^{+\infty} du_z \frac{e^{-u_z^2}}{\pi^{1/2}} (u_x^2 + u_y^2 + u_z^2) v_t^2 \tag{20.13}$$

$w_x = u_x - U$

$$= \int_{-\infty}^{+\infty} dw_x \frac{e^{-w_x^2}}{\pi^{1/2}} \int_{-\infty}^{+\infty} du_y \frac{e^{-u_y^2}}{\pi^{1/2}} \int_{-\infty}^{+\infty} du_z \frac{e^{-u_z^2}}{\pi^{1/2}} [(w_x^2 + U^2 + 2w_x U) + u_y^2 + u_z^2] v_t^2 \tag{20.14}$$

$$\int_{-\infty}^{+\infty} dt \frac{e^{-t^2}}{\pi^{1/2}} = 1 \quad \int_{-\infty}^{+\infty} dt \frac{e^{-t^2}}{\pi^{1/2}} t = 0 \quad \int_{-\infty}^{+\infty} dt \frac{e^{-t^2}}{\pi^{1/2}} t^2 = \frac{1}{2} \tag{20.15}$$

$$< v^2 >= v_t^2 \, [(\frac{1}{2} + U^2 + 0) + \frac{1}{2} + \frac{1}{2}] = \frac{3}{2} v_t^2 + V^2 \tag{20.16}$$

2.2 Evaluate the thermal velocity of ^4He ions at a temperature $T = 10\,\text{keV}$. The ^4He atomic mass is

$$m_{He} = 4.0026AMU = 4.0026 \times 1.660539 \times 10^{-27} kg = 6.6465 \times 10^{-27} kg \tag{20.17}$$

The mass of ^4He ion is

$$m'_{He} = m_{He} - 2m_e = 6.6465 \times 10^{-27} kg - 2 \times 9.1094 \times 10^{-31} kg = 6.6446 \times 10^{-27} kg \tag{20.18}$$

$$v_{tHe} = \left(\frac{2T}{m'_{He}}\right)^{1/2} = \left(\frac{2 \times 10^4 \times 1.6022 \times 10^{-19} J}{6.6446 \times 10^{-27} kg}\right)^{1/2} = 6.9444 \times 10^5 \, m/s \tag{20.19}$$

2.3 Evaluate the average kinetic energy (in eV) of neutrons at $T = 300\,°\text{C}$ and their thermal velocity.

$$m_n = 1.00866AMU = 1.6749 \times 10^{-27} kg \tag{20.20}$$

$$300\,°C \rightarrow \frac{(300 + 273.15)K}{11604\, K/eV} = 0.0494eV \tag{20.21}$$

$$v_{tn} = \left(\frac{2T}{m_n}\right)^{1/2} = \left(\frac{2 \times 0.0494 \times 1.6022 \times 10^{-19} J}{1.6749 \times 10^{-27} kg}\right)^{1/2} = 3073.9\, m/s \tag{20.22}$$

2.4 Evaluate the most probable velocity and energy of a Maxwell-Boltzmann distribution at given T.

$$F(E) = 2\left(\frac{E}{\pi T^3}\right)^{1/2} e^{-\frac{E}{T}} \tag{20.23}$$

$$\frac{\partial F}{\partial E} = 0 \quad E = T/2 \tag{20.24}$$

Problems of Chap. 3

3.1 A gas of ionised nuclei at $T = 20\,\text{keV}$ is made of 50% deuterium and 50% tritium. The gas density is $n = 10^{20}m^{-3}$. Determine the number of fusion events per seconds.

$$\nu = n < \sigma v >= (0.5)^2 \times 10^{20}m^{-3} \times 4 \times 10^{-22}m^3/s = 10^{-2}s^{-1} \tag{20.25}$$

3.2 The particles of species A can be absorbed by particles of species B with a cross section $\sigma = 100b$. A beam of particles A impinges on a target of particles B. The beam has a density $n_A = 10^{20}m^{-3}$ and intensity $10^{10}cm^{-2}s^{-1}$. The target has a density $10^{23}m^{-3}$. Evaluate the beam attenuation after crossing a width of $10cm$, the attenuation length λ and the velocity v_A of the beam particles.

$$\lambda = (n_b\sigma)^{-1} = (10^{23}m^{-3} \times 100 \times 10^{-28}m^2)^{-1} = 10^3\,m \tag{20.26}$$

$$\frac{\Delta\Phi}{\Phi} = 1 - e^{-\frac{\Delta x}{\lambda}} \approx \frac{\Delta x}{\lambda} = \frac{0.1\,m}{10^3\,m} = 10^{-4} \tag{20.27}$$

$$v_A = \frac{\Phi}{n_A} = 10^{-4}cm/s \tag{20.28}$$

3.3 Species A and species B interact through elastic collisions with cross section $\sigma_e = 10b$ and react (forming species C) with cross section $\sigma_r = 0.1b$. Evaluate how many elastic collisions are needed to get a reaction $a + b \to c$.

$$\nu_{ab,e} = n_b\sigma_e u \tag{20.29}$$

$$\nu_{ab,r} = n_b\sigma_r u \tag{20.30}$$

$$\frac{\nu_{ab,e}}{\nu_{ab,r}} = \frac{\sigma_e}{\sigma_r} = 100 \tag{20.31}$$

3.4 A nucleus of ^{238}U decays in a ^{234}Th nucleus emitting an α particle. The energy released in the decay is 4.3 MeV. Determine the energy of the α particle.

Assume $m_1 = 234$ and $m_2 = 4$

$$m_1\mathbf{v}_1 + m_2\mathbf{v}_2 = 0 \qquad |v_1|/|v_2| = m_2/m_1 \tag{20.32}$$

$$\frac{E_1}{E_2} = \frac{m_1 v_1^2/2}{m_2 v_2^2/2} = m_2/m_1 = 4/234 \tag{20.33}$$

$$E_1 + E_2 = 4.3\,\text{MeV} = (4/234) \qquad E_2 + E_2 = (238/234)E_2 \tag{20.34}$$

$$E_2 = \frac{234 \times 4.3\,\text{MeV}}{238} = 4.23\,\text{MeV} \tag{20.35}$$

3.5 A ^5He nucleus decays in a neutron and an α particle releasing 17.5 MeV. Evaluate the energy of the two fragments.

Assume $m_1 = 1$ and $m_2 = 4$

$$m_1\mathbf{v}_1 + m_2\mathbf{v}_2 = 0 \qquad |v_1|/|v_2| = m_2/m_1 \tag{20.36}$$

$$\frac{E_1}{E_2} = \frac{m_1 v_1^2/2}{m_2 v_2^2/2} = m_2/m_1 = 4 \tag{20.37}$$

$$E_1 + E_2 = 17.5\,\text{MeV} = 4E_2 + E_2 = 5E_2 \tag{20.38}$$

$$E_2 = \frac{17.5\,\text{MeV}}{5} = 3.5\,\text{MeV} E_1 = 4 \times 3.5\,\text{MeV} = 14\,\text{MeV} \tag{20.39}$$

3.6 How many collisions are needed to slow a 1 MeV neutron down to thermal energy ($T = 300\,°\text{C}$) using hydrogen ($A = 1$) or carbon ($A = 12$) as moderator?

The lethargy variation in the two cases is $\xi_H = 1$ and $\xi_C = 0.15777$. Thus

$$< lnE_n >=< lnE_o > -n\xi \tag{20.40}$$

with $E_o = 1\,\text{MeV}$ and $E_n = (3/2)T = (3/2)0.0494eV = 0.0741eV$, yielding $n_H = 16$ and $n_C = 104$.

Problems of Chap. 4

4.1 The Ge crystal lattice has a constant $d = 5.658\text{Å}$. The electromagnetic radiation is diffracted at an angle $\theta = 30°$ at the second order. Evaluate the wavelength and the frequency of the electromagnetic radiation.

From the Bragg relation

$$2d \sin\theta = n\lambda \tag{20.41}$$

we have

$$2 \times 5.658\text{Å} \sin(30°) = 2\lambda \tag{20.42}$$

Thus $\lambda = 2.829$Å and $\nu = c/\lambda = 1.06 \times 10^{18}$ Hz.

4.2 Consider the dispersion relation given in Eq. (4.47). Evaluate the phase and the group velocity for $\omega = 2 \times \omega_p$.

The dispersion relation Eq. (4.47) is

$$\omega = (\omega_p^2 + k^2c^2)^{1/2} \tag{20.43}$$

The wave with $\omega = 2 \times \omega_p$ corresponds to $k = 3^{1/2}\omega_p/c$. The phase velocity is

$$\frac{\omega}{k} = \frac{(\omega_p^2 + k^2c^2)^{1/2}}{k} = c\left(1 + \frac{\omega_p^2}{k^2c^2}\right)^{1/2} = \left(\frac{4}{3}\right)^{1/2} c > c \tag{20.44}$$

whereas the group velocity is

$$\frac{\partial\omega}{\partial k} = 2kc^2/[2(\omega_p^2 + k^2c^2)^{1/2}] = \frac{c}{(1 + \frac{\omega_p^2}{k^2c^2})^{1/2}} = \left(\frac{3}{4}\right)^{1/2} c < c \tag{20.45}$$

4.3 Evaluate the rest energy of the deuterium nucleus and compare it to the sum of the rest energy of the proton and the neutron.

The atomic mass of deuterium is $m_D(at.) = 2.014102AMU$. Taking into account that the electron mass is $m_e = 5.5 \times 10^{-4}AMU$ the mass of the deuterium nucleus is $m_D(nuc.) = 2.0135532AMU$ and its rest energy is $m_Dc^2 = 1875.58$ MeV.

The proton and neutron rest energies are $m_pc^2 = 938.26$ MeV and $m_nc^2 = 939.55$ MeV. Thus the difference is

$$1875.58\,\text{MeV} - (938.26\,\text{MeV} + 939.55) = -2.22\,\text{MeV} \tag{20.46}$$

the difference is negative since the deuterium is a bound system. The binding energy per nucleon is 1.11 MeV.

4.4 Evaluate the velocity (as a fraction of the velocity of light of an electron and a proton both with energy 1 GeV.

The electron rest energy is $m_ec^2 = 0.511$ MeV whereas the proton rest energy is $m_pc^2 = 938.23$ MeV. From the Einstein relation

$$E = \frac{mc^2}{(1 - \frac{v^2}{c^2})}^{1/2} \tag{20.47}$$

we have

$$\frac{v}{c} = \left[1 - \left(\frac{mc^2}{E}\right)^2\right]^{1/2} \tag{20.48}$$

Therefore, $v_e/c \approx 0.99999987$ and $v_p/c \approx 0.346$. Taking the non relativistic limit for the proton, the result is $v_p/c = [2(1000 - 938,23)/938,23]^{1/2} \approx 0.363$ not too far from the correct result.

Problems of Chap. 5

5.1 Rutherford had made the hypothesis that the neutron was a proton-electron pair. What would be the electron energy if it had to be confined within a nuclear radius ($R \approx 10^{-15}m$)?

If the electron is confined within a distance $\Delta x = 10^{-15}m$ on the basis of the uncertainty principle his momentum has an uncertainty Δp given by

$$\Delta p = \frac{h}{4\pi \Delta x} = 5.3 \times 10^{-20} kgm/s \tag{20.49}$$

For such a large value of momentum we can use the ultra relativistic limit to determine the energy $E = \Delta pc = 99\,\text{MeV}$. Note that this is much larger than the Coulomb attraction energy $e^2/(4\pi\epsilon_o \Delta x) = 1.4\,\text{MeV}$

5.2 Evaluate the energy variation of a photon scattered at 30° by an electron initially at rest. Assume a photon in the UV rage and in the gamma range.

For $\nu = 10^{16}$ Hz ($h\nu = 41$ eV) we have $\nu\prime = 0.99999 \times 10^{16}$ Hz, i.e. the frequency of the diffused photon is almost the same of the incident photon. For $\nu = 10^{22}$ Hz ($h\nu = 41$ MeV) the frequency is significantly changed $\nu\prime = 8.4 \times 10^{20}$ Hz.

5.3 Evaluate the de Broglie wavelength of an electron of energy 100 keV and 1 MeV.

For $E = 100\,\text{keV}$ we can use the non relativistic limit yielding $p = (2mE)^{1/2} = 1.7 \times 10^{-22} kgm/s$ and the de Broglie wavelength is $\lambda = h/p = 3.9 \times 10^{-12} m$. For $E = 1\,\text{MeV}$ we have to use the relativistic expressions yielding $p = (E^2 - m_e^2 c^4)^{1/2}/c = 4.6 \times 10^{-22} kgm/s$. In this case $\lambda = h/p = 1.4 \times 10^{-12} m$.

Problems of Chap. 6

6.1 Evaluate the energy of the bound states of an electron in an infinite potential well of width $2a_o$ with a_o the Bohr radius.

Using Eqs. (6.23–6.24), the energy of the first three states is $E_1 = 33.47\,\text{eV}$, $E_2 = 133.87\,\text{eV}$ and $E_3 = 301.2\,\text{eV}$

6.2 Evaluate the transmission coefficient of an electron of energy equal to the thermal energy at 0 °C through a barrier of width a_o and height 10 eV.

Since the problem is $1D$ the thermal energy is $E = T/2 = 0.0118 eV$ and the corresponding free-particle wave vector is $k = 2\pi(2mE)^{1/2}/h = 5.6 \times 10^8 m^{-1}$. Inside the barrier the modulus of the wave vector given by Eq. (6.40) is $k_o = 2\pi(2m(V_o - E))^{1/2}/h = 1.6 \times 10^{10} m^{-1}$. Note that in this case $k_o a_o = 0.86$ is not

$\gg 1$ and we need to use the complete expression of the transmission coefficient Eq. (6.42), yielding

$$T = \frac{16(k/k_o)^2}{(1-k^2/k_o^2)^2[e^{2k_oa}+e^{-2k_oa}-2]+4k^2/k_o^2[e^{2k_oa}+e^{-2k_oa}+2]} \quad (20.50)$$

Upon substituting the values we obtain $T = 5 \times 10^{-3}$. Using the asymptotic expression Eq. (6.43) would have given $T = 3.4 \times 10^{-3}$.
6.3 On the basis of the general expression for the alpha decay constant evaluate the value for the decay of a ^{238}U nucleus with the emission of a 4.2 MeV alpha particle. Assume that the well width is the sum of the radius of the alpha particle and that of the nucleus after the decay using the estimate $R = 1.2 \times 10^{-15} m A^{1/3}$ and that the well corresponds to a potential of -15 MeV.

The alpha particle mass is $m_\alpha = 6.64466 \times 10^{-27} kg$. The reduced mass of the α-^{234}Th system is $m_r = m_\alpha 234/238 = 6.53298 \times 10^{-27} kg$. The velocity of the alpha particle in the potential well (taking into account the depth of the well) is

$$v_\alpha = (2(4.2+15) \times 10^6 \times 1.6 \times 10^{-19}/m_r)^{1/2} = 3 \times 10^7\, m/s \quad (20.51)$$

i.e. $1/10$ of the speed of light. The sum of the two radii is $R = 1.2 \times 10^{-15}(234^{1/3} + 4^{1/3})m = 9.29957 \times 10^{-15} m$. The classical reflection point x_m is located at

$$x_m = \frac{2 \times 90 e^2}{4\pi\epsilon_o E} = 6.17136 \times 10^{-14} m \quad (20.52)$$

The other quantities appearing in the Gamow factor Eq. (6.53) are $y = R/x_m = 0.388187$ yielding $G = 44.68762$ and the decay constant $\lambda = 5.00781 \times 10^{-18} s^{-1}$. The half-life estimated with this simple model is $T_{1/2} = 4.39 \times 10^9 y$ whereas the experimental value is $T_{1/2\,\exp} = 4.47 \times 10^9 y$ in good agreement.

Problems of Chap. 7

7.1 Using the Weizsäcker formula evaluate the binding energy of the following nuclei indicating the most stable nucleus and discuss which of the contributions in the mass formula make the other nuclei less stable
$^{27}Mg(Z = 12)$ $^{27}Al(Z = 13)$ $^{27}Si(Z = 14)$

The various terms of the mass formula are summarised in the following table

	^{27}Mg	^{27}Al	^{27}Si
A	27	27	27
Z	12	13	14
N	15	14	13
E_{volume}(MeV)	421.2	421.2	421.2
$E_{surface}$(MeV)	−154.8	−154.8	−154.8
$E_{Coulomb}$(MeV)	−33.6	−39.43	−45.73
$E_{symmetry}$(MeV)	−7.76	−0.86	−0.86
$E_{pairing}$(MeV)	0	0	0
$E_{binding}$(MeV)	225.03	226.1	219.8

Therefore the most bound state corresponds to Aluminium.

7.3 If $1g$ of ^{226}Ra has an activity of $1Ci(=3.7 \times 10^{10}Bq)$ evaluate the half life of ^{226}Ra.

The ^{226}Ra atomic weight is 226, $0254AMU$ corresponding to $3.7532 \times 10^{-25}kg$. Thus, the number of atoms in $1g$ is $N_{Ra} = \frac{10^{-3}}{3.75\times10^{-25}} = 2.6644 \times 10^{21}$. Taking into account that $1Ci = 3.7 \times 10^{10}Bq = \lambda N_{Ra}$, it is possible to determine the decay constant λ

$$\lambda = 1.39 \times 10^{-11}s^{-1} \tag{20.53}$$

and in turn the half-life $T_{1/2} = 1583y$.

7.5 Plot in the (Z, N) plane the curves corresponding to the stable nuclei and the proton/neutron drip lines. Compare the result with the nuclide chart.

The proton drip line can be obtained from the condition $Q > 0$ with

$$\frac{Q}{c^2} = m(Z, A) - m(Z-1, A-1) - m_p \tag{20.54}$$

Neglecting for the sake of simplicity the pairing term, we have

$$Q = -b_1 + \frac{2}{3}b_2\left(A^{-1/3} - \frac{1}{3}\right)b_3\frac{Z^2}{A^{4/3}} + b_4\left(1 - \frac{4Z^2}{A^2}\right) + (m_p - m_n)c^2$$
$$+ 2b_3\frac{Z}{A^{1/3}} - 4b_4\left(1 - \frac{2Z}{A}\right) \tag{20.55}$$

Problems of Chap. 8

8.1 Using the Weizsäcker formula evaluate the energy produced in the fission of ^{235}U in two nuclei with $A = 90$ and $A = 142$. Evaluate the energy released in the beta decays of the fission fragments.

The neutrons emitted in the fission process are $4 = 235 - 90 - 142$. The fission product can be:

$({}^{90}_{35}Br, {}^{142}_{57}La)({}^{90}_{36}Kr, {}^{142}_{56}Ba)({}^{90}_{37}Rb, {}^{145}_{52}Cs)({}^{90}_{38}Sr, {}^{142}_{54}Xe)({}^{90}_{39}Y, {}^{142}_{53}I)({}^{90}_{40}Zr, {}^{142}_{52}Te)$.

The stable nuclei with $A = 90$ and $A = 142$ have $Z = 40$ and $Z = 58$ (using the Weizsäcker formula we would obtain $Z = 39$ e $Z = 59$).

For all these channels the energy released is 181.5 MeV.

8.2 Evaluate the number of DT fusion reactions per unit volume and unit time in a plasma with $n_D = n_T = 0.5 \times 10^{20} m^{-3}$ and temperature $T = 20$ keV.

The Maxwellian DT reactivity at 20 keV is $< \sigma v >= 4 \times 10^{-22} m^3/s$. The number of DT fusion reactions per unit volume and unit time is therefore 10^{18}.

8.3 Using a file excel plot the Maxwellian reactivity for the DT reaction using the Bosch-Hale fit in the range 1–30 keV.

Check the numbers with the table given in the Bosch-Hale paper

8.4 A target made of $1g$ of ${}^{59}Co$ is irradiated with a flux of thermal neutrons with intensity $10^{12} cm^{-2} s^{-1}$. In the reaction the radioactive isotope ${}^{60}Co$ is produced which has a half life of $5.27y$. Evaluate the activity of ${}^{60}Co$ after one year of irradiation assuming a cross section for the transmutation process $\sigma = 40b$.

The evolution equation for the two radionuclides are

$$\frac{dN_{59}}{dt} = -\sigma \Phi N_{59} \tag{20.56}$$

with $\sigma \Phi = 4 \times 10^{-11} s^{-1}$, and

$$\frac{dN_{60}}{dt} = -\lambda_{60} N_{60} + \sigma \Phi N_{59} \approx \sigma \Phi N_{59} \tag{20.57}$$

with $\lambda_{60} = 4.17 \times 10^{-9} s^{-1}$, to be integrated with the initial conditions $N_{59}(0) = 1.02 \times 10^{22}$ and $N_{60}(0) = 0$. Note that since the decay time of ${}^{60}Co$ is much longer than the irradiation time the number of ${}^{59}Co$ nuclei transmuted is roughly equal to the number of ${}^{60}Co$ nuclei present after one year

$$N_{59}(1y) - N_{59}(0) = 1.29 \times 10^{19} \ N_{60}(1y) \tag{20.58}$$

Thus $A_{60}(1y) = 5.37 \times 10^{10} Bq$. The result without approximations is $A_{60}(1y) = 5.03 \times 10^{10} Bq$.

8.5 In a fusion reactor tritium is produced via the ${}^{6}Li(n, \alpha)T$ reaction. Evaluate the amount of tritium produced in one year from 10^{21} ${}^{6}Li$ atoms assuming a neutron flux of $5 \times 10^{14} n/(cm^2)s$ and a cross section for the reaction (averaged over the neutron spectrum) of $10b$. How much tritium will be available at the end of the irradiation?

The number N_{6Li} of ${}^{6}Li$ nuclei evolves according to

$$\frac{dN_{6Li}}{dt} = -\sigma \Phi N_{6Li} \tag{20.59}$$

with $\sigma\Phi = 5 \times 10^{-9} s^{-1}$ and $1y = 3.15 \times 10^7 s$. The evolution of the tritium nuclei is given by

$$\frac{dN_T}{dt} = -\lambda_T N_T + \sigma\Phi N_{6Li} \approx \sigma\Phi N_{6Li} \tag{20.60}$$

with $\lambda_T = 1.8 \times 10^{-9} s^{-1}$. The amount of tritium produced in $1y$ is

$$N_{6Li}(1y) - N_{6Li}(0) = 1.46 \times 10^{20} \tag{20.61}$$

If we take into account the decay of tritium, the amount of tritium present after one year is $N_T(1y) = 1.42 \times 10^{20}$.

Problems of Chap. 10

10.1 Show that the using the Rutherford expression for the deflection angle the logarithmic divergence at small impact parameters disappear.

From $tg(\chi/2) = b_o/(2b)$ and $1 - \cos\chi = 2(1 - \cos^2(\chi/2))$ it follows (with $t = (2b/b_o)^2$)

$$2\pi \int_0^{b_{max}} bdb(1 - \cos\chi) = 2\pi \frac{b_o^2}{4} \int_0^{(2b_{max}/b_o)^2} \frac{dt}{1+t} = \pi b_o^2 ln(\frac{2b_{max}}{b_o}) \tag{20.62}$$

that coincides with the expression obtained in the small deflection angle limit Eq. (10.11) provided $b_{min} = b_o/2$.

10.3 Starting from the expression of the range of protons in air $R_p(m) = (E_p(\text{MeV})/9.3)^{1.8}$ determine the range of ^3He particles with energy 1 and 10 MeV in Al.

First we need to determine the equivalent energy of a proton $E_{equiv} = E/A$ corresponding to 0.333 MeV and 3.33 MeV, respectively. The range in air must be multiplied by the factor $A/Z^2 = 0.75$. The resulting range in air is $1.87mm$ and $0.118m$ for the two energies. To determine the range in aluminium we have to multiply by the density of air ($1.22mg/cm^3$) and divide by the density of aluminium ($2.7g/cm^3$, to obtain $0.844\mu m$ and $0.053mm$, respectively.

Problems of Chap. 11

11.3 Order the relevant plasma space and time scales for parameters typical of a magnetically confined fusion plasma $n = 10^{20}m^{-3}$, $T = 10\,\text{keV}$, $B = 5T$, $L = 1m$.

$n^{-1/3} = 2.15 \times 10^{-7}m$, $\lambda_D = 7.43 \times 10^{-5}m$, $\rho_e = 6.73 \times 10^{-5}m$, $\rho_i = 2.88 \times 10^{-3}m$, $\lambda_c = 10^4 m$

$\omega_{pe} = 5.64 \times 10^{11} s^{-1}$, $\Omega_e = 8.79 \times 10^{11} rad/s$, $\Omega_p = 4.79 \times 10^8 rad/s$, $v_{te}/L = 5.93 \times 10^7 s^{-1}$, $v_{tp}/L = 1.38 \times 10^6 s^{-1}$

For the electron-electron collision frequency we have the simple estimate in Eq. 10.8 (with $b_{max} = \lambda_D$ and $b_{min} = b_o$, see Chap. 12) $\nu_{ee} = 6.28 \times 10^3 s^{-1}$ and the more precise estimate given e.g. in [1] $\nu_{ee} = 5.84 \times 10^3 s^{-1}$

Problems of Chap. 12

12.1 Determine the slowing down time of an alpha particle for parameters typical of a magnetically confined fusion plasma $n = 10^{20} m^{-3}$, $T = 10\,\text{keV}$.

Using Eq. (12.13) with $A_1 = 4$ and $Z_1 = 2$ we have $\tau_{SD} = 0.38 s$.

Problems of Chap. 13

13.1 Determine the explicit expression of the magnetic field for the current distribution shown in Fig.13.1.

Taking $J = J_o$ for $R \le a$ and $J = 0$ otherwise the application of Stokes theorem on a circular contour $R < a$ yields

$$2\pi R B_\phi(R) = \mu_o 2\pi \int_0^R R' dR' J_o = \mu_o J_o \pi R^2 \tag{20.63}$$

therefore

$$B_\phi(R) = \frac{\mu_o J_o R}{2} \tag{20.64}$$

For $R > a$ we have

$$2\pi R B_\phi(R) = \mu_o J_o \pi a^2 \tag{20.65}$$

or

$$B_\phi(R) = \frac{\mu_o J_o a^2}{2R} \tag{20.66}$$

13.2 Determine the explicit expression of the magnetic field for a finite cylindrical solenoid of length L.

The expression can be obtained from Eq. (13.9)

$$\int d^3r\prime \frac{(\mathbf{r} - \mathbf{r}\prime) \times \mathbf{j}(\mathbf{r}\prime)}{|\mathbf{r} - \mathbf{r}\prime|^3} = \int d^2 r_\perp\prime dz\prime \frac{(\mathbf{r}_\perp - \mathbf{r}_\perp\prime) \times \mathbf{j}(\mathbf{r}\prime) + (Z - Z\prime)\hat{Z} \times \mathbf{j}(\mathbf{r}\prime)}{[|\mathbf{r}_\perp - \mathbf{r}_\perp\prime|^2 + (Z - Z\prime)^2]^{3/2}} \tag{20.67}$$

However, in the case of a finite solenoid the integral in Z' has to be made between the limits $-L/2$ and $L/2$ and except for $Z = 0$ the second term in the numerator does not vanish.

The integral of the first term can be expressed in terms of trigonometric functions and the second term can be easily integrated yielding (with $B_{Zo} = \mu_o J_o (R_{out} - R_{in})$ see Eq. (13.8))

$$B_Z(R, Z) = \frac{B_{Zo}}{4\pi} \int_0^{2\pi} \frac{d\phi R_o (R\cos\phi - R_o)}{R^2 + R_o^2 - 2RR_o\cos\phi} \times$$
$$\left[\frac{\frac{L}{2} - Z}{(R^2 + R_o^2 + (\frac{L}{2} - Z)^2 - 2RR_o\cos\phi)^{1/2}} + \frac{\frac{L}{2} + Z}{(R^2 + R_o^2 + (\frac{L}{2} + Z)^2 - 2RR_o\cos\phi)^{1/2}} \right] \tag{20.68}$$

$$B_R(R, Z) = \frac{B_{Zo} R_o}{2\pi} \int_0^{2\pi} d\phi \times$$
$$\left[\frac{1}{(R^2 + R_o^2 + (\frac{L}{2} - Z)^2 - 2RR_o\cos\phi)^{1/2}} - \frac{1}{(R^2 + R_o^2 + (\frac{L}{2} + Z)^2 - 2RR_o\cos\phi)^{1/2}} \right] \tag{20.69}$$

13.3 Assuming that the poloidal field is generated by a current density along z $j_z = j_o$ for $r \leq a$ and $j_z = 0$ for $r \geq a$, evaluate the radial profile of the rotational transform and of the safety factor.

Using the result of the problem above we can explicitly write the expression of the safety factor as follows (we use r as radial variable instead of R). For $r \leq a$ we have

$$q(r) = \frac{B_z r}{B_p R_o} = \frac{2B_z}{\mu_o J_o R_o} \tag{20.70}$$

independent of r. For $r > a$ we have

$$q(r) = \frac{B_z r}{B_p R_o} = \frac{2B_z r^2}{\mu_o J_o a^2 R_o} = q(a)\frac{r^2}{a^2} \tag{20.71}$$

Problems of Chap. 14

14.4 Determine the ideal ignition temperature of a plasma made mostly of deuterium with a concentration $c_T = 10^{-2}$ of tritium. Quantify the relative importance of DD and DT reactions.

The charge neutrality condition requires $c_D = 99\%$. The power in the form of charged particles is associated with the sum of the 1.01 MeV tritons and the 3.02

MeV protons generated in the $D(d,p)T$ branch and the 0.82 MeV ^3He generated in the $D(d,n)^3He$ branch with each branch having 50% probability. The $T(d,n)^4He$ reaction produces power in the form of 3.5 MeV α particles. Therefore the charged particle power for homogeneous profiles is

$$P_{charge} = n_e^2(\frac{c_D^2}{2}(50\% \times 4.03\,\text{MeV} + 50\% \times 0.82\,\text{MeV}) < \sigma v >_{DD} + \\ + c_D c_T 3.5\,\text{MeV} < \sigma v >_{DT})V = 1.6 n_{e20}^2 \times 10^{27}(c_D^2 1.21 < \sigma v(m^3/s) >_{DD} + \\ + c_D c_T 3.5 < \sigma v(m^3/s) >_{DT})V(m^3) \tag{20.72}$$

Bremmstrahlung must be calculated taking $Z_{eff} = 1$ since we are not including impurities

$$P_{Brem}(MW) = 1.69 \times n_{e20}^2\, T(eV)^{1/2} V(m^3) \tag{20.73}$$

Taking $< \sigma v >_{DD} = 10^{-26} T(\text{keV})^2 m^3/s$ and $< \sigma v >_{DT} = 10^{-24} T(\text{keV})^2 m^3/s$ the ideal ignition temperature is 17 keV with the largest contribution coming from the DT reactions. If the DT contribution is neglected the ideal ignition temperature rises to about 42 keV.

14.5 Determine the triple product of a DT plasma and compare with the value expected in ITER.

From Eq. (14.46) with $c_1 = c_D = 50\%$, $c_2 = c_T = 50\%$ and $< \sigma v > \approx 10^{-22}(T/10\,\text{keV})^2 m^3/s$ valid in the range 10–20 keV, we have

$$n_e \tau_E T = \frac{3T}{2} \frac{1 + 0.5 + 0.5}{0.5^2 \frac{3.5\,\text{MeV}}{T} 10^{-22}(T/10\,\text{keV})^2\, m^3/s} = 3.43 \times 10^{21} m^{-3} s\,\text{keV} \tag{20.74}$$

This has to be compared with the expected ITER parameters $n_o = 1.2 \times 10^{20} m^{-3}$, $\tau_E = 3.5s$ and $T_o = 20\,\text{keV}$ corresponding to $n_e \tau_E T = 8.4 \times 10^{21} m^{-3} s\,\text{keV}$

Problems of Chap. 15

15.1 Estimate the electric field generated by the charge separation in a purely toroidal magnetic field and the time to push the plasma out of the confinement region (hint: describe the charge separation as the charging of a capacitor with planar plates at the top and the bottom of the plasma)

The charge imbalance at the location of the two plates (top and bottom of the toroidal plasma) is given by the charge density times the volumes where charge neutrality no longer applies. The volume on each side is given by $V(t) = 2\pi R \pi a \Delta(t)$ where $\Delta(t) = v_{drift} t$ is the position of the vertically drifting centroid of the positive and negative charge distribution.

Assuming that this equivalent to a capacitor with distance $2a$ between the plates and plate surface $2\pi R 2a$ the electric potential Φ is given by $\Phi = Q/C$ with the charge $Q = neV(t)$ and the capacitance $C = \epsilon_o 2\pi R2a/(2a)$. The electric field is given by $E = \Phi/(2a)$.

Upon combining the expressions above we obtain

$$E = \frac{\Phi}{2a} = \frac{ne2\pi R\pi a v_{drift} t}{\epsilon_o 2\pi R} \frac{1}{2a} = \frac{ne}{\epsilon_o} \frac{\pi}{2} v_{drift} t \tag{20.75}$$

The electric field produces an $E \times B$ drift that increases linearly with time. The time t_d needed to achieve a horizontal displacement equal to a is given by

$$\frac{ne}{\epsilon_o} \frac{\pi}{4B} v_{drift} t_d^2 = a \tag{20.76}$$

For typical parameters of a fusion device t_d is in the range of fractions of μs

15.2 Determine the shift of the drift orbits of an electron and an ion with energy equal to T and a ratio $v_{||}/v_{\perp}$ equal to 0.5 and 2 times the value corresponding to the boundary of the trapping cone for ITER parameters $T = 10\,\text{keV}$, $B = 5T$ and $R = 6.3m$.

The boundary of the trapping come corresponds to the condition

$$\frac{v_{||}}{v_{\perp}} = \left(\frac{B_{max}}{B_{min}} - 1\right)^{1/2} = \alpha_c \tag{20.77}$$

For illustration we take $\epsilon = 1/5$ that is a typical value around half radius. Therefore we take the case of a circulating particles with $v_{||}/v_{\perp} = 2\alpha_c$ and of a trapped particle with $v_{||}/v_{\perp} = 0.5\alpha_c$.

For the circulating particle we have

$$v_{||} = 2\left(\frac{B_{max}}{B_{min}} - 1\right)^{1/2} v_{\perp} = 2\left(\frac{2\epsilon}{1-\epsilon}\right)^{1/2} v_{\perp} \tag{20.78}$$

If the particle energy is equal to T the velocity is equal to $v = v_t$, thus

$$v_{\perp}^2 + v_{||}^2 = v_{\perp}^2 + 4\frac{2\epsilon}{1-\epsilon} v_{\perp}^2 = \frac{1+7\epsilon}{1-\epsilon} v_{\perp}^2 = v_t^2 \tag{20.79}$$

from the equation above we can determine $v_{\perp}$

$$v_{\perp} = \left(\frac{1-\epsilon}{1+7\epsilon}\right)^{1/2} v_t \quad v_{||} = 2\left(\frac{2\epsilon}{1+7\epsilon}\right)^{1/2} v_t \tag{20.80}$$

We now can determine v_D, ω_t and the drift surface shift v_D/ω_t

$$v_D = \left(\frac{v_{\perp}^2}{2} + v_{||}^2\right) \frac{1}{\Omega R_o} = \frac{1+15\epsilon}{2(1+7\epsilon)} \frac{v_t^2}{\Omega R_o} \tag{20.81}$$

$$\omega_t = \frac{v_{||}}{qR_o} = 2\left(\frac{2\epsilon}{1+7\epsilon}\right)^{1/2}\frac{v_t}{qR_o} \tag{20.82}$$

$$\frac{v_D}{\omega_t} = \frac{qv_t}{4\Omega}\frac{1+15\epsilon}{(2\epsilon(1+7\epsilon))^{1/2}} = \left(\frac{25}{24}\right)^{1/2}\frac{qv_t}{\Omega} \approx 0.6cm \tag{20.83}$$

The above estimate is for a deuterium ion. In the case of electrons the result is a factor $(m_e/m_D)^{1/2} \approx 1/60$ smaller. For the trapped particle we have

$$v_{||} = 0.5\left(\frac{B_{max}}{B_{min}} - 1\right)^{1/2} v_\perp = 0.5\left(\frac{2\epsilon}{1-\epsilon}\right)^{1/2} v_\perp \tag{20.84}$$

$$v_\perp^2 + v_{||}^2 = v_\perp^2 + \frac{\epsilon}{2(1-\epsilon)}v_\perp^2 = \frac{1-\epsilon/2}{1-\epsilon}v_\perp^2 = v_t^2 \tag{20.85}$$

$$v_\perp = \left(\frac{1-\epsilon}{1-\epsilon/2}\right)^{1/2} v_t \quad v_{||} = 0.5(\frac{2\epsilon}{1-\epsilon/2})^{1/2} v_t \tag{20.86}$$

$$\Delta r = \frac{v_{||}}{eB_p/m} = 0.5\left(\frac{2\epsilon}{1-\epsilon/2}\right)^{1/2}\frac{v_t}{\Omega}\frac{q}{\epsilon} = \frac{5}{3}\frac{qv_t}{\Omega} \approx 1cm \tag{20.87}$$

$$\frac{1}{eB_p/m} = \frac{B}{B_p}\frac{1}{\Omega} = \frac{q}{\epsilon}\frac{1}{\Omega} \tag{20.88}$$

Problems of Chap. 16

16.1 Derive Eq. (16.47) using the properties of the modified Bessel functions of integer order.

Integrating by part Eq. (16.47) we obtain

$$U = \frac{\mu_o}{2\pi}\int Z_M(R)\frac{dR}{R}I_M^2 = \frac{\mu_o I_M^2}{2\pi}\left[Z_M \ln\left(\frac{R}{R_M}\right)|_{R_{in}}^{R_{out}} - \int \frac{dZ_M}{dR}\ln\left(\frac{R}{R_M}\right)dR\right] \tag{20.89}$$

taking into account that $R_{in} = R_M e^{-\kappa}$ and that $Z_M(R_{in}) = \pi\kappa R_M I_1(\kappa)$ and changing variable to $R = R_M e^{\kappa\cos\phi}$ we obtain

$$U = \frac{\mu_o I_M^2}{2\pi}\kappa^2 R_M\left[\pi I_1(\kappa) + \int_0^\pi d\phi \cos^2\phi e^{\kappa\cos\phi}\right] \tag{20.90}$$

Upon using the identity

$$e^{\kappa \cos\phi} = I_o(\kappa) + 2\sum_1^{\infty} I_n(\kappa)\cos(n\phi) \tag{20.91}$$

the integral over ϕ can be performed explicitly yielding

$$U = \frac{\mu_o I_M^2}{4}\kappa^2 R_M[2I_1(\kappa) - I_o(\kappa) + I_2(\kappa)] \tag{20.92}$$

16.2 Using Eq. 16.15 determine the forces between two coils with coil currents $1MA$ with radius $R = 1m$ and an axial distance of $1m$.

Equation 16.15 for two coils of identical radius is a function of the parameter $\xi = (Z_i - Z_k)/(2R)$ and takes the form $F_{ik} = \mu_o I_i I_k f(\xi)$ with $f(\xi) < 0$ to be evaluated in terms of complete elliptic integrals and $|f(\xi)|$ a monotonically decreasing function of ξ. Therefore the force is attractive for currents of the same sign and repulsive for currents of opposite signs. For ξ such that $f(\xi) \approx -1$ the force is of the order of $4\pi \times 10^5 N$.

16.3 Evaluate the inductance of a coil with dimensions equal to the ITER plasma and an internal inductance $l_i = 0.5$.

$$L = \mu_o R_o\left(\ln\frac{8R_o}{a} - 2 + \frac{l_i}{2}\right) = 11.3 \times 10^{-6} H \tag{20.93}$$

Problems of Chap. 17

17.1 Using Eq. 17.35 determine the dependence of the safety factor at 95% of the separatrix flux on elongation and aspect ratio and compare with the fit shown in Eq. (13.31)

The line element can be written as

$$dl = dR\left[1 + \left(\frac{d\hat{Z}}{dR}\right)^2\right]^{1/2} \tag{20.94}$$

with $\hat{Z}(R, \psi)$ given by Eq. (17.36). Upon evaluating B_p from its definition

$$RB_p = [(\frac{\partial\psi}{\partial R})^2 + (\frac{\partial\psi}{\partial Z})^2]^{1/2} \tag{20.95}$$

it can be easily shown that the expression for the safety factor reduces to

$$q = \frac{F(\psi)}{2\pi} \int_{R_-}^{R_+} \frac{dl}{R\zeta_2(R)\hat{Z}(R,\psi)} \tag{20.96}$$

with $R_\pm$ the solution of the equation $\hat{Z}(R,\psi) = 0$. This expression must be in general integrated numerically. For the special case $B_{Sol} = a_2$ the expression for $R_\pm$ can be easily obtained

$$R_\pm = R_X(1 + y_\pm)^{1/2} \tag{20.97}$$

with

$$y_\pm = \frac{-a_2 Z_X^2 \pm [a_2^2 Z_X^4 + 4(\psi - \psi_b)(A_{Sol} R_X^2 + 2a_2)\frac{R_X^2}{8}]^{1/2}}{2(A_{Sol} R_X^2 + 2a_2)\frac{R_X^2}{8}} \tag{20.98}$$

The expression for $\hat{Z}(R,\psi)$ reduces to a simple algebraic function

$$\hat{Z}(R,\psi) = \left(\frac{(A_{Sol} R_X^2 + 2a_2)\frac{R_X^2}{8}}{-a_2 y}\right)^{1/2} (y_+ - y)^{1/2}(y - y_-)^{1/2} \tag{20.99}$$

and the expression for the safety factor reduces to

$$q = \frac{F(\psi)}{4\pi[-a_2(A_{Sol} R_X^2 + 2a_2)\frac{R_X^2}{8}]^{1/2}} \int_{y_-}^{y_+} \frac{dy}{(1+y)[y(y_+ - y)(y - y_-)]^{1/2}} \tag{20.100}$$

We are interested in the limit $\psi \to \psi_b$ in which $y_- \to 0$ and $y_+ \to y_S \equiv (R_S^2/R_X^2 - 1)$. The integral over y can be divided in two parts: an integral between y_- and δ and an integral between δ and y_+ with $y_- \ll \delta \ll y_+$. Then we can approximate the integrand as follows

$$\int_{y_-}^{y_+} \frac{dy}{(1+y)[y(y_+ - y)(y - y_-)]^{1/2}} = \int_{y_-}^{\delta} \frac{dy}{[y(y_+)(y - y_-)]^{1/2}} + \\ + \int_{\delta}^{y_+} \frac{dy}{(1+y)y(y_+ - y)^{1/2}} \tag{20.101}$$

The two integrals can be made analytically. The results, in the limit $y_- \ll \delta \ll y_+$, is

$$\int_{y_-}^{y_+} \frac{dy}{(1+y)[y(y_+ - y)(y - y_-)]^{1/2}} = \frac{2\ln 2 + \ln\delta - \ln y_-}{y_+^{1/2}} +$$

$$-\frac{\ln\delta - 2\ln 2 - \ln y_+}{y_+^{1/2}} + \frac{1}{(1+y_+)^{1/2}}\ln\left|\frac{(y_+ + 1)^{1/2} - y_+^{1/2}}{(y_+ + 1)^{1/2} + y_+^{1/2}}\right| \tag{20.102}$$

As expected, the result does not depend on the choice of the parameter δ. Note the logarithmic divergence as $\psi \rightarrow \psi_b$ ($y_- \rightarrow 0$) is due to the fact that the poloidal field vanishes at the $X - point$. For this reason it is customary to express the edge safety factor as the value at $\psi = 95\%\psi_b$.

The expression of q depends on the value of the quantity $A_{Sol}R_X^2$. This can be expressed in terms of the plasma current as follows

$$\mu_o I_p \equiv \int_{S_{plasma}} dRdZj_\phi = 2\int_{R_X}^{R_S} \frac{dR}{R} Z(\hat{R})(A_{Sol}R^2 + B_{Sol}) \tag{20.103}$$

Again, in the limit $B_{Sol} = 2a_2$ the integral can be performed analytically yielding

$$\mu_o I_p = \frac{A_{Sol}R_X^3}{2}(\epsilon\kappa^2)^{1/2}\times$$
$$\times\left[\frac{2}{3}y_S^{3/2} - \frac{2(1+y_S)^{1/2}}{1+\epsilon\kappa^2}\left(\ln((1+y_s)^{1/2} + y_S^{1/2}) - \frac{y_S^{1/2}}{(1+y_S)^{1/2}}\right)\right] \tag{20.104}$$

where use has been made of the definition of Z_X and R_S in terms of ϵ and κ.

At this point it is convenient to take the large aspect ratio limit. The expression of q_{95} reduces to

$$q_{95} = \frac{2\pi F(\psi_b)\epsilon^2}{\mu_o I_p} C\kappa^2(1+\epsilon\kappa^2) \tag{20.105}$$

with C a constant of order unity.

17.2 Using Eq. 17.34 determine the dependence of the Shafranov shift on the plasma beta.

It is convenient to define the poloidal beta as follows

$$\beta_{po} = \frac{2\mu_o p(\psi_o)}{B_Z^2(R = R_X, Z = 0)} \tag{20.106}$$

with ψ_o the flux at the magnetic axis ($R = R_o, Z = 0$). From Eq. (17.34) it is possible to find the relation between B_{Sol} and A_{Sol} corresponding to a given R_o. The relation is a liner relation of the form $B_{Sol} = -\lambda A_{Sol}$. Upon substituting in the definition of the poloidal beta the results is independent of A_{Sol} and provides a function $\beta_{po}(R_o, \kappa, \epsilon)$ that can be graphically inverted to provide $R_o(\beta_{po})$.

Reference

1. Naval Research Laboratory 2019 Plasma formulary. https://library.psfc.mit.edu/catalog/online_pubs/NRL_FORMULARY_19.pdf

Appendix A
Binding Energy for Deformed Nuclei

The modification of the binding energy if the nucleus is deformed from a spherical to an ellipsoidal shape at constant nuclear volume can be evaluated by assuming a deformation from a sphere of radius R to an ellipsoid with major radius $a = R(1+\epsilon)$ and minor radius $b = R/(1+\epsilon)^{1/2}$ (Fig. 8.2). We focus on the change in the surface and Coulomb terms.

The surface of an ellipsoid with section given by the following equation

$$\left(\frac{z}{a}\right)^2 + \left(\frac{r}{b}\right)^2 = 1 \tag{A.1}$$

can be evaluated as

$$S = 2\pi \int r(z) dl \tag{A.2}$$

with

$$dl = [dz^2 + dr^2]^{1/2} = \left[1 + \left(\frac{dz}{dr}\right)^2\right]^{1/2} dr \tag{A.3}$$

Upon making the change of variable $t = (r/b)^2$ we obtain

$$S = 2\pi b^2 \int dt \frac{(1 + (3\epsilon + 3\epsilon^2 + \epsilon^3)t)^{1/2}}{(1-t)^{1/2}} \tag{A.4}$$

For small deformations ($\epsilon << 1$) the quantity S can be expanded as

$$S = 4\pi R^2 \left(1 + \frac{2\epsilon^2}{5} - \frac{52\epsilon^3}{105} + O(\epsilon^4)\right) \tag{A.5}$$

F. Romanelli, *Physics of Nuclear Energy*, Springer Series in Plasma Science and Technology, https://doi.org/10.1007/978-981-97-9609-0

The surface of the ellipsoid is larger than the surface of the sphere and this reduces the binding energy.

The Coulomb term can be evaluated assuming a uniform charge density inside the ellipsoid

$$W = \frac{\epsilon_o}{2} \int d\mathbf{r} E^2 = \int dr \rho(\mathbf{r}) \Phi(\mathbf{r}) \tag{A.6}$$

with the electrostatic potential given by

$$\Phi(\mathbf{r}) = \pi a b^2 \rho_o \int_0^\infty \frac{dv}{(b^2+v)(a^2+v)^{1/2}} [1 - (\frac{r^2}{b^2+v} + \frac{z^2}{a^2+v})] \tag{A.7}$$

Upon inserting Eq. (A.7) into Eq. (A.6) the expression of the Coulomb energy becomes

$$W = \frac{3}{5} \frac{Z^2 e^2}{4\pi\epsilon_o R} (1+\epsilon)^2 \int_0^\infty \frac{du}{(1+\nu v)(1+u)^{1/2}} \left[\frac{u}{1+u} + \frac{4}{5} \left(\frac{1}{1+u} - \frac{1}{2(1+\nu u)} \right) \right] \tag{A.8}$$

with $\nu \equiv (1+\epsilon)^3$. Equation (A.8) can now be expanded in powers of ϵ. After a tedious but straightforward algebra

$$W = \frac{3}{5} \frac{Z^2 e^2}{4\pi\epsilon_o R} \left(1 - \frac{\epsilon^2}{5} - \frac{4}{7}\epsilon^3 + O(\epsilon^4) \right) \tag{A.9}$$

The variation of the mass of the nucleus due to the deformation up to order ϵ^3 is therefore given by

$$\delta M(Z, A)c^2 = -\delta E_{binding} = b_2 A^{2/3} \frac{2\epsilon^2}{5} - \frac{b_3 Z^2}{A^{1/3}} \frac{\epsilon^2}{5} - \\ - \left(\frac{52}{105} b_2 A^{2/3} + \frac{4}{7} \frac{b_3 Z^2}{A^{1/3}} \right) \epsilon^3 \tag{A.10}$$

Appendix B
Collision Integrals

The expression fo the stopping power is

$$< \frac{dE_1}{dt} >\approx -\frac{q_1^2 q_2^2}{4\pi\epsilon_o^2} \ln \Lambda n_2 \left(\frac{\mathbf{v}_1 \cdot \mathbf{e}_v}{m_r} - \frac{\psi_v}{m_1} \right) \tag{B.1}$$

with

$$\mathbf{e}_v \equiv \int f(v_2) d^3 v_2 \frac{(\mathbf{v}_1 - \mathbf{v}_2)}{|\mathbf{v}_1 - \mathbf{v}_2|^3} \tag{B.2}$$

$$\psi_v \equiv \int f(v_2) d^3 v_2 \frac{1}{|\mathbf{v}_1 - \mathbf{v}_2|} \tag{B.3}$$

Let us consider first the quantity $\mathbf{e}_v$.The integral can be easily evaluated by noting that the contribution of the volume elements with $v_2 \geq v_1$ vanishes. Indeed by considering the spherical surface $v_2 = const$ and taking two elements on that surface corresponding to the same spherical angle $d\Omega$ (see Fig. B.1) it is easily seen that the two contributions are equal in magnitude and opposite in sign

$$u^2 d\Omega \frac{\mathbf{u}}{u^3} + u\prime^2 d\Omega \frac{\mathbf{u}\prime}{u\prime^3} = 0 \tag{B.4}$$

Thus, only the volume corresponding to $v_2 \leq v_1$ contributes to the integral. Again, it is convenient to consider two elements on the spherical surface $v_2 = const.$ and corresponding to the same spherical angle. Now the two contributions are equal and of the same sign. The surface integral can be made easily using spherical coordinates with v_1 along the z axis. In this case the interaction along ϕ is between 0 and 2π whereas the integration along θ is between 0 and θ_M with $\sin\theta_M = v_2/v_1$. From

F. Romanelli, *Physics of Nuclear Energy*, Springer Series in Plasma Science and Technology, https://doi.org/10.1007/978-981-97-9609-0

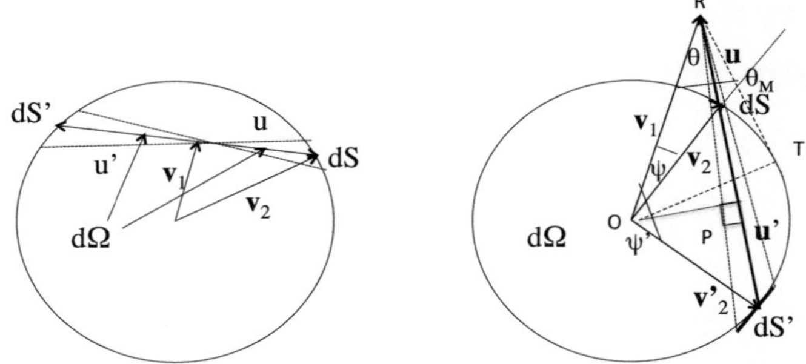

Fig. B.1 Evaluation of $\mathbf{e}_v$ and ψ_v. On the left the integration for $v_2 \geq v_1$. The contribution to the integral of the two elements dS and $dS\prime$ of the spherical surface $v_2 = const$ is equal and opposite. This implies that the contribution to $\mathbf{e}_v$ vanishes while the contribution to ψ_v is due to the sum of the two roots Eq. (B.9). On the right the integration for $v_2 \leq v_1$. In this case the contributions from dS and $dS\prime$ have the same sign. Now the contribution to ψ_v is due to the difference of the two roots Eq. (B.9).

Fig. B.1 looking at the triangles ORT and ORP the following relations can be derived

$$v_2 \sin\psi = u \sin\theta \quad v_2 \cos\psi + u\cos\theta = v_1 \quad v_2 \sin(\theta + \psi) = v_1 \sin\theta \tag{B.5}$$

with $\mathbf{u} = \mathbf{v}_1 - \mathbf{v}_2$. Furthermore, the surface elements can be written as $dS = u^2 \sin\theta d\theta d\phi / \cos(\theta + \psi)$. The $\mathbf{e}_v$ integral can then be cast in the following form

$$\mathbf{e}_v = \int_0^{v_1} f(v_2) dv_2 \int_0^{\theta_M} \frac{d\theta u^2 \sin\theta}{\cos(\theta+\psi)} \int_0^{2\pi} d\phi \frac{2\mathbf{u}}{u^3} \tag{B.6}$$

The integral over ϕ averages out the component of $\mathbf{u}$ orthogonal to $\mathbf{v}_1$ leaving only the component parallel to $\mathbf{v}_1$

$$\int_0^{2\pi} d\phi \frac{2\mathbf{u}}{u} = 2\pi \frac{2\mathbf{v}_1}{v_1} \frac{v_1 - v_2\cos\psi}{u} = 2\pi \frac{2\mathbf{v}_1}{v_1} \cos\theta \tag{B.7}$$

Using Eq. (B.5) the integral over θ can be done analytically

$$\mathbf{e}_v = 4\pi \frac{\mathbf{v}_1}{v_1} \int_0^{v_1} f(v_2) dv_2 \int_0^{\theta_M} \frac{d\theta \sin\theta}{\cos(\theta+\psi)} \cos\theta = \frac{\mathbf{v}_1}{v_1^3} \int_0^{v_1} f(v_2) 4\pi v_2^2 dv_2 \tag{B.8}$$

Let us consider now the quantity ψ_v. In this case both the domain $v_2 \leq v_1$ and $v_2 \geq v_1$ provide a contribution but it is convenient to consider these two contributions separately. We will call them ψ_v^+ for $v_2 \geq v_1$ and ψ_v^- for $v_2 \leq v_1$. From the definition of the relative velocity $\mathbf{u}$ it follows that $v_2^2 = v_1^2 + u^2 - 2uv_1 \cos\theta$ which yields $u = u_\pm$ with

$$u_\pm = v_1 \cos\theta \pm (v_2^2 - v_1^2 \sin^2\theta)^{1/2} \tag{B.9}$$

The two solutions for u correspond to the the two intersections of the line at fixed θ with the circle $v_2 = const$ and will be associated with the contributions of the element dS and $dS\prime$. Considering first the domain $v_1 \geq v_2$ from the equations above we have

$$\psi_v^- = \int f(v_2) dv_2 dS \frac{1}{u} = \int f(v_2) dv_2 \int_0^{\theta_M} \frac{d\theta \sin\theta}{\cos(\theta + \psi)} \int_0^{2\pi} d\phi (u_+ + u_-) =$$

$$= \int f(v_2) dv_2 \int_0^{\theta_M} d\theta 4\pi v_1 \frac{\cos\theta \sin\theta}{\cos(\theta + \psi)} \tag{B.10}$$

where in performing the integral on the spherical surface we have summed the contributions of dS and $dS\prime$. Upon using the definition of $\cos(\theta + \psi)$ the integral over θ can be easily performed yielding

$$\psi_v^- = \frac{1}{v_1} \int f(v_2) 4\pi v_2^2 dv_2 = \Phi\left(\frac{v_1}{v_t}\right) - \frac{2v_1}{\pi^{1/2} v_t} e^{-\frac{v_1^2}{v_t^2}} \tag{B.11}$$

with $\Phi(x)$ the error function. The contribution from the domain $v_1 \leq v_2$ can be evaluated in a similar way. However, making reference again to Fig. B.1, the contribution from dS and $dS\prime$ have opposite sign since the integral over $d\phi$ is made in opposite directions, yielding

$$\psi_v^+ = \int f(v_2) dv_2 dS \frac{1}{u} = \int f(v_2) dv_2 \int_0^{\theta_M} \frac{d\theta \sin\theta}{\cos(\theta + \psi)} \int_0^{2\pi} d\phi (u_+ - u_-) =$$

$$= \int f(v_2) dv_2 \int_0^{\theta_M} d\theta 4\pi (v_2^2 - v_1^2 \sin^2\theta)^{1/2} \frac{\sin\theta}{\cos(\theta + \psi)} \tag{B.12}$$

Again, using the definition of $\cos(\theta + \psi)$ the integral over θ can be easily performed

$$\psi_v^+ = \int f(v_2) 4\pi v_2^2 dv_2 \frac{1}{v_2} = \frac{2}{\pi^{1/2} v_t} (1 - e^{-\frac{v_1^2}{v_t^2}}) \tag{B.13}$$

Upon combining the expressions above we obtain

$$< \frac{dE_1}{dt} >= -\frac{q_1^2 q_2^2}{4\pi\epsilon_o^2 v_1 m_2} \ln \Lambda n_2 \left[\Phi \left(\frac{v_1}{v_{t2}} \right) - \left(1 + \frac{m_2}{m_1} \right) \frac{2v_1}{\pi^{1/2} v_{t2}} e^{-\frac{v_1^2}{v_{t2}^2}} \right] \quad \text{(B.14)}$$

The energy transfer between thermal species can be obtained by averaging Eq. (B.14) over the distribution function of species 1 assumed to be a Maxwell distribution

$$P_{12} \equiv \int d^3 v_1 f(v_1) < \frac{dE_1}{dt} > \quad \text{(B.15)}$$

The two integrals can be done as follows

$$\int_0^\infty v_1^2 dv_1 \frac{e^{-v_1^2/v_{t1}^2}}{v_1} \frac{2v_1}{\pi^{1/2} v_{t2}} e^{-\frac{v_1^2}{v_{t2}^2}} = \frac{v_{t1}^3 v_{t2}^2}{2(v_{t1}^2 + v_{t2}^2)^{3/2}} \quad \text{(B.16)}$$

and

$$\int_0^\infty v_1^2 dv_1 \frac{e^{-v_1^2/v_{t1}^2}}{v_1} \frac{2}{\pi^{1/2}} \int_0^{v_1/v_{t2}} dx e^{-x^2} = \frac{v_{t1}^2}{\pi^{1/2} v_{t2}} \int_0^\infty v_1 dv_1 e^{-v_1^2/v_{t1}^2} e^{-v_1^2/v_{t2}^2}$$
$$= \frac{v_{t1}^2}{2v_{t2}} \frac{v_{t1} v_{t2}}{(v_{t1}^2 + v_{t2}^2)^{1/2}} \quad \text{(B.17)}$$

where in Eq. (B.17) an integration by parts has been performed. By combining these results we obtain

$$P_{12} = -\frac{q_1^2 q_2^2}{4\pi\epsilon_o^2} \ln \Lambda \frac{n_1 n_2}{m_1 m_2} \frac{4}{\pi^{1/2} (v_{t1}^2 + v_{t2}^2)^{3/2}} (T_1 - T_2) \quad \text{(B.18)}$$

The integral related with resistivity is given by Eq. (12.20)

$$\mathbf{R}_{12} = n_1 \int d^3 v_1 \frac{e^{-v_1^2/v_{t1}^2}}{\pi^{3/2} v_{t1}^3} \frac{2\mathbf{v}_1 \cdot \mathbf{U}_1}{v_{t1}^2} < \frac{d\mathbf{p}_1}{dt} > \quad \text{(B.19)}$$

without loss of generality we can take $\mathbf{U}_1$ along the z axis. Using spherical coordinates the above integral becomes

$$\mathbf{R}_{12} = -n_1 \frac{q_1^2 q_2^2}{4\pi\epsilon_o^2 m_r} \ln \Lambda n_2 \int_0^\infty v_1^2 dv_1 \int_{-1}^{+1} d(\cos\theta)$$

$$\times \int_{-\pi}^{\pi} d\phi \frac{e^{-v_1^2/v_{t1}^2}}{\pi^{3/2} v_{t1}^3} \frac{2v_1 \cos\theta U_1}{v_{t1}^2} \frac{\mathbf{v}_1}{v_1^3} \int_0^{v_1} \frac{e^{-v_2^2/v_{t2}^2}}{\pi^{3/2} v_{t2}^3} 4\pi v_2^2 dv_2 \tag{B.20}$$

where the cartesian component of $\mathbf{v}_1$ are $v_1(\sin\theta\cos\phi, \sin\theta\sin\phi, \cos\theta)$. The integral in $d\phi$ averages out the first two component leaving only the component along $\mathbf{U}_1$ times a factor 2π. A further numerical factor $2/3$ arises from the $d(\cos\theta)$ integral, yielding

$$\mathbf{R}_{12} = -n_1 \frac{q_1^2 q_2^2}{4\pi\epsilon_o^2 m_r} \ln \Lambda n_2 \mathbf{U}_1 \int_0^\infty v_1 dv_1 \frac{8\pi}{3} \frac{e^{-v_1^2/v_{t1}^2}}{\pi^{3/2} v_{t1}^3} \frac{1}{v_{t1}^2} \int_0^{v_1} \frac{e^{-v_2^2/v_{t2}^2}}{\pi^{3/2} v_{t2}^3} 4\pi v_2^2 dv_2 \tag{B.21}$$

Upon performing an integration by part Eq. (B.21) reduces to

$$\mathbf{R}_{12} = -n_1 \frac{q_1^2 q_2^2}{4\pi\epsilon_o^2 m_r} \ln \Lambda n_2 \mathbf{U}_1 \frac{16}{3\pi v_{t1}^3 v_{t2}^3} \int_0^\infty v_1^2 dv_1 e^{-v_1^2/v_{t1}^2} e^{-v_1^2/v_{t2}^2} =$$

$$= -n_1 \frac{q_1^2 q_2^2}{(2\pi\epsilon_o)^2 m_1} \ln \Lambda n_2 \mathbf{U}_1 \frac{4\pi^{1/2}}{3v_{t1}^3} (\frac{m_r}{m_1})^{1/2} = n_1 m_1 < \nu_{12} > \mathbf{U}_1 \tag{B.22}$$

with

$$< \nu_{12} >= \frac{q_1^2 q_2^2}{(2\pi\epsilon_o)^2 m_1^2} \ln \Lambda n_2 \frac{4\pi^{1/2}}{3v_{t1}^3} \left(\frac{m_r}{m_1}\right)^{1/2} = \frac{Z_2^2 n_2}{n_e \tau_{ee}} \left(\frac{m_r m_e}{m_1^2}\right)^{1/2} \tag{B.23}$$

Appendix C
Elliptic Integrals

Complete elliptic integrals are defined as [1]

$$K(k^2) = \int_0^{\pi/2} d\theta (1 - k^2 \sin^2\theta)^{-1/2} \tag{C.1}$$

$$E(k^2) = \int_0^{\pi/2} d\theta (1 - k^2 \sin^2\theta)^{1/2} \tag{C.2}$$

Expansion for small arguments

$$K(m) \approx \frac{\pi}{2}\left[1 + \left(\frac{1}{2}\right)^2 m + \left(\frac{1}{2}\right)^2 \left(\frac{3}{4}\right)^2 m^2 + ...\right] \tag{C.3}$$

$$E(m) \approx \frac{\pi}{2}\left[1 - \left(\frac{1}{2}\right)^2 \frac{m}{1} - \left(\frac{1}{2}\right)^2 \left(\frac{3}{4}\right)^2 \frac{m^2}{3} + ...\right] \tag{C.4}$$

The following polynomial approximation can be used for both $K(m)$ and $E(m)$

$$\begin{aligned}&[a_o + a_1(1-m) + a_2(1-m)^2 + a_3(1-m)^3 + a_4(1-m)^4]+\\&+ [b_o + b_1(1-m) + b_2(1-m)^2 + b_3(1-m)^3 + b_4(1-m)^4] \ln\left(\frac{1}{1-m}\right)\end{aligned} \tag{C.5}$$

F. Romanelli, *Physics of Nuclear Energy*, Springer Series in Plasma Science and Technology, https://doi.org/10.1007/978-981-97-9609-0

with

$$a_o = 1.38629; \quad a_1 = 0.09666; \quad a_2 = 0.0359; \quad a_3 = 0.03742; \quad a_4 = 0.01451$$
$$b_o = 0.5; \quad b_1 = 0.12498; \quad b_2 = 0.0688; \quad b_3 = 0.03328 \quad b_4 = 0.00441 \quad (C.6)$$

for $K(m)$ and

$$a_o = 1; \quad a_1 = 0.44325; \quad a_2 = 0.0626; \quad a_3 = 0.04757; \quad a_4 = 0.01736$$
$$b_o = 0; \quad b_1 = 0.24998; \quad b_2 = 0.092; \quad b_3 = 0.04069 \quad b_4 = 0.00526 \quad (C.7)$$

for $E(m)$.

Reference

1. M. Abramowitz, I.A. Stegun, (eds.), *Handbook of Mathematical Functions with Formulas, Graphs, and Mathematical Tables. Applied Mathematics Series*, vol. 55 (Ninth reprint with additional corrections of tenth original printing with corrections (December 1972); first ed.) (United States Department of Commerce, National Bureau of Standards; Dover Publications, Washington D.C., New York, 1983)

Index

F. Romanelli, *Physics of Nuclear Energy*, Springer Series in Plasma Science and Technology, https://doi.org/10.1007/978-981-97-9609-0

MIX
Papier aus verantwortungsvollen Quellen
Paper from responsible sources
FSC® C105338

If you have any concerns about our products,
you can contact us on
ProductSafety@springernature.com

In case Publisher is established outside the EU,
the EU authorized representative is:
Springer Nature Customer Service Center GmbH
Europaplatz 3, 69115 Heidelberg, Germany

Printed by Libri Plureos GmbH
in Hamburg, Germany